Chemistry Essentials: Organic and Inorganic

Chemistry Essentials: Organic and Inorganic

Edited by
Rhett McAllister

Larsen & Keller
www.larsen-keller.com

Chemistry Essentials: Organic and Inorganic
Edited by Rhett McAllister
ISBN: 978-1-63549-209-5 (Hardback)

© 2017 Larsen & Keller

Larsen & Keller

Published by Larsen and Keller Education,
5 Penn Plaza,
19th Floor,
New York, NY 10001, USA

Cataloging-in-Publication Data

Chemistry essentials : organic and inorganic / edited by Rhett McAllister.
    p. cm.
Includes bibliographical references and index.
ISBN 978-1-63549-209-5
1. Chemistry, Organic. 2. Chemistry, Inorganic. 3. Chemistry. I. McAllister, Rhett.
QD251.3 .C44 2017
547--dc23

For more information regarding Larsen and Keller Education and its products, please visit the publisher's website www.larsen-keller.com

Table of Contents

Preface

This book attempts to understand the multiple branches that fall under the discipline of chemistry and how such concepts have practical applications. It describes in a detailed manner the theories and concepts of various different branches associated with chemistry. It specifically talks about the fundamental principles of organic and inorganic chemistry. The topics included in the text are designed to provide students with a clear understanding of this complex subject. This textbook on the essentials of chemistry is of utmost significance and is bound to provide incredible insights to readers. Coherent flow of topics, student-friendly language and extensive use of examples make this textbook an invaluable source of knowledge.

A foreword of all Chapters of the book is provided below:

Chapter 1 - Organic chemistry as explained in the chapter is a sub-discipline of chemistry, which involves the scientific study of the structure, properties and reactions of organic compounds. The chapter on organic chemistry offers an insightful focus, keeping in mind the complex subject matter; **Chapter 2** - The key concepts explained in this chapter are organic compounds, organic reactions, organic matter and the structural formula of a compound. This chapter explains to the reader the basic characteristics of organic chemistry which is the science concerned with all aspects of organic compounds; **Chapter 3** - Organic chemistry is best understood in confluence with the major topics listed in the following text. Some of the topics discussed are function groups, aliphatic compounds, aromatic hydrocarbons, fullerene, organosilicon etc. Functional groups are particular groups of atoms that are responsible for the characteristics of these molecules. The major categories of organic chemistry are elucidated in this section; **Chapter 4** - Compounds that are not organic are known as inorganic compounds. The study of these inorganic compounds is termed as inorganic chemistry. Organic compounds can be made from inorganic ones. This chapter educates the reader with the basics of inorganic chemistry and its progress in contemporary times; **Chapter 5** - The nomenclature of organic chemistry is commonly referred to as the blue book and is explained in this chapter along with IUPAC nomenclature of organic chemistry, international chemical identifier and simplified molecular-input line- entry system. The topics discussed in the chapter are of great importance to broaden the existing knowledge on organic chemistry and inorganic chemistry; **Chapter 6** - Inorganic chemistry has a number of applications. Some of these are catalysis, materials science and pigment. Chemical reactions increase with the participation of added substances. This process is known as the catalysis. The reader is provided with an in-depth understanding of the applications of inorganic chemistry; **Chapter 7** - One of the chemical elements is carbon. It is denoted by the symbol C and atomic number 6. Some of the allotropes of carbon are diamond and graphite. The other major components discussed are isotopes of carbon, the carbon cycle and the alpha and beta carbon. This chapter will not only provide an overview, it will also delve deep into the topics related to it; **Chapter 8** - Chemical compounds contain within themselves at least one bond. The bond is shared between a carbon atom and a metal. This is known as organometallic chemistry. Some of the allied fields of organic and inorganic chemistry explained in this chapter are bioorganometallic chemistry, medicinal chemistry, biochemistry and bioinorganic chemistry. This text provides a plethora of interdisciplinary topics for better comprehension of organic and inorganic chemistry; **Chapter 9 -** Chemical synthesis has a branch of study known as organic synthesis. It deals with organic compounds which are more complex than inorganic compounds. The techniques discussed within this text are retrosynthetic analysis, enantioselective synthesis, electro synthesis and drug design. This chapter will provide an integrated understanding of organic synthesis.

I would like to thank the entire editorial team who made sincere efforts for this book and my family who supported me in my efforts of working on this book. I take this opportunity to thank all those who have been a guiding force throughout my life.

Editor

Introduction to Organic Chemistry

Organic chemistry as explained in the chapter is a sub-discipline of chemistry, which involves the scientific study of the structure, properties and reactions of organic compounds. The chapter on organic chemistry offers an insightful focus, keeping in mind the complex subject matter.

Organic Chemistry

Organic chemistry is a chemistry subdiscipline involving the scientific study of the structure, properties, and reactions of organic compounds and organic materials, i.e., matter in its various forms that contain carbon atoms. Study of structure includes many physical and chemical methods to determine the chemical composition and the chemical constitution of organic compounds and materials. Study of properties includes both physical properties and chemical properties, and uses similar methods as well as methods to evaluate chemical reactivity, with the aim to understand the behavior of the organic matter in its pure form (when possible), but also in solutions, mixtures, and fabricated forms. The study of organic reactions includes probing their scope through use in preparation of target compounds (e.g., natural products, drugs, polymers, etc.) by chemical synthesis, as well as the focused study of the reactivities of individual organic molecules, both in the laboratory and via theoretical (in silico) study.

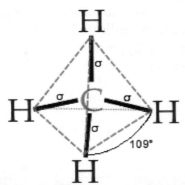

Methane, CH_4, in line-angle representation showing four carbon-hydrogen single (σ) bonds in black, and the 3D shape of such tetrahedral molecules, with ~109° interior bond angles, in green.

The range of chemicals studied in organic chemistry include hydrocarbons (compounds containing only carbon and hydrogen), as well as myriad compositions based always on carbon, but also containing other elements, especially oxygen, nitrogen, sulfur, phosphorus (these, included in many organic chemicals in biology) and the radiostable elements of the halogens.

In the modern era, the range extends further into the periodic table, with main group elements, including:

- Group 1 and 2 organometallic compounds, i.e., involving alkali (e.g., lithium, sodium, and potassium) or alkaline earth metals (e.g., magnesium)

- Metalloids (e.g., boron and silicon) or other metals (e.g., aluminium and tin)

In addition, much modern research focuses on organic chemistry involving further organometallics, including the lanthanides, but especially the transition metals; (e.g., zinc, copper, palladium, nickel, cobalt, titanium and chromium)

 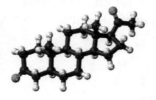 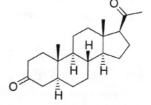

Space-filling representation   Ball-and-stick representation   Line-angle representation

Three representations of an organic compound, 5α-Dihydroprogesterone (5α-DHP), a steroid hormone. For molecules showing color, the carbon atoms are in black, hydrogens in gray, and oxygens in red. In the line angle representation, carbon atoms are implied at every terminus of a line and vertex of multiple lines, and hydrogen atoms are implied to fill the remaining needed valences (up to 4).

Finally, organic compounds form the basis of all earthly life and constitute a significant part of human endeavors in chemistry. The bonding patterns open to carbon, with its valence of four—formal single, double, and triple bonds, as well as various structures with delocalized electrons—make the array of organic compounds structurally diverse, and their range of applications enormous. They either form the basis of, or are important constituents of, many commercial products including pharmaceuticals; petrochemicals and products made from them (including lubricants, solvents, etc.); plastics; fuels and explosives; etc. As indicated, the study of organic chemistry overlaps with organometallic chemistry and biochemistry, but also with medicinal chemistry, polymer chemistry, as well as many aspects of materials science.

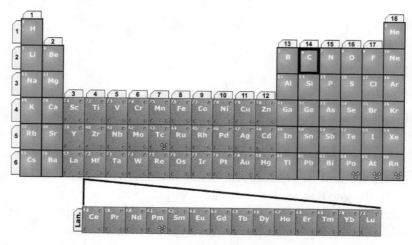

Periodic table of elements of interest in organic chemistry. The table illustrates all elements of current interest in modern organic and organometallic chemistry, indicating main group elements in orange, and transition metals and lanthanides (Lan) in grey.

History

Before the nineteenth century, chemists generally believed that compounds obtained from living organisms were endowed with a vital force that distinguished them from inorganic compounds. According to the concept of vitalism (vital force theory), organic matter was endowed with a "vital force". During the first half of the nineteenth century, some of the first systematic studies of organic compounds were reported. Around 1816 Michel Chevreul started a study of soaps made from various fats and alkalis. He separated the different acids that, in combination with the alkali, produced the soap. Since these were all individual compounds, he demonstrated that it was possible to make a chemical change in various fats (which traditionally come from organic sources), producing new compounds, without "vital force". In 1828 Friedrich Wöhler produced the organic chemical urea (carbamide), a constituent of urine, from the inorganic ammonium cyanate NH_4CNO, in what is now called the Wöhler synthesis. Although Wöhler was always cautious about claiming that he had disproved the theory of vital force, this event has often been thought of as a turning point.

Friedrich Wöhler

In 1856 William Henry Perkin, while trying to manufacture quinine, accidentally manufactured the organic dye now known as Perkin's mauve. Through its great financial success, this discovery greatly increased interest in organic chemistry.

The crucial breakthrough for organic chemistry was the concept of chemical structure, developed independently and simultaneously by Friedrich August Kekulé and Archibald Scott Couper in 1858. Both men suggested that tetravalent carbon atoms could link to each other to form a carbon lattice, and that the detailed patterns of atomic bonding could be discerned by skillful interpretations of appropriate chemical reactions.

The pharmaceutical industry began in the last decade of the 19th century when the manufacturing of acetylsalicylic acid—more commonly referred to as aspirin—in Germany was started by Bayer. The first time a drug was systematically improved was with arsphenamine (Salvarsan). Though numerous derivatives of the dangerous toxic atoxyl were examined by Paul Ehrlich and his group, the compound with best effectiveness and toxicity characteristics was selected for production.

An example of an organometallic molecule, a catalyst called Grubbs' catalyst, as a ball-and-stick model based on an X-ray crystal structure. The formula of the catalyst is often given as $RuCl_2(PCy_3)_2(=CHPh)$, where the ruthenium metal atom, Ru, is at very center in turquoise, carbons are in black, hydrogens in gray-white, chlorine in green, and phosphorus in orange. The metal ligand at the bottom is a tricyclohexyl phosphine, abbreviated PCy, and another of these appears at the top of the image (where its rings are obscuring one another). The group projecting out to the right has a metal-carbon double bond, as is known as an alkylidene. Robert Grubbs shared the 2005 Nobel prize in chemistry with Richard R. Schrock and Yves Chauvin for their work on the reactions such catalysts mediate, called olefin metathesis.

Early examples of organic reactions and applications were often found because of a combination of luck and preparation for unexpected observations. The latter half of the 19th century however witnessed systematic studies of organic compounds. The development of synthetic indigo is illustrative. The production of indigo from plant sources dropped from 19,000 tons in 1897 to 1,000 tons by 1914 thanks to the synthetic methods developed by Adolf von Baeyer. In 2002, 17,000 tons of synthetic indigo were produced from petrochemicals.

In the early part of the 20th Century, polymers and enzymes were shown to be large organic molecules, and petroleum was shown to be of biological origin.

The multiple-step synthesis of complex organic compounds is called total synthesis. Total synthesis of complex natural compounds increased in complexity to glucose and terpineol. For example, cholesterol-related compounds have opened ways to synthesize complex human hormones and their modified derivatives. Since the start of the 20th century, complexity of total syntheses has been increased to include molecules of high complexity such as lysergic acid and vitamin B_{12}.

R = 5'-deoxyadenosyl, Me, OH, CN

The total synthesis of vitamin B_{12} marked a major achievement in organic chemistry.

The development of organic chemistry benefited from the discovery of petroleum and the development of the petrochemical industry. The conversion of individual compounds obtained from petroleum into different compound types by various chemical processes led to the birth of the petrochemical industry, which successfully manufactured artificial rubbers, various organic adhesives, property-modifying petroleum additives, and plastics.

The majority of chemical compounds occurring in biological organisms are in fact carbon compounds, so the association between organic chemistry and biochemistry is so close that biochemistry might be regarded as in essence a branch of organic chemistry. Although the history of biochemistry might be taken to span some four centuries, fundamental understanding of the field only began to develop in the late 19th century and the actual term *biochemistry* was coined around the start of 20th century. Research in the field increased throughout the twentieth century, without any indication of slackening in the rate of increase, as may be verified by inspection of abstraction and indexing services such as BIOSIS Previews and Biological Abstracts, which began in the 1920s as a single annual volume, but has grown so drastically that by the end of the 20th century it was only available to the everyday user as an online electronic database.

Characterization

Since organic compounds often exist as mixtures, a variety of techniques have also been developed to assess purity, especially important being chromatography techniques such as HPLC and gas chromatography. Traditional methods of separation include distillation, crystallization, and solvent extraction.

Organic compounds were traditionally characterized by a variety of chemical tests, called "wet methods", but such tests have been largely displaced by spectroscopic or other computer-intensive methods of analysis. Listed in approximate order of utility, the chief analytical methods are:

- Nuclear magnetic resonance (NMR) spectroscopy is the most commonly used technique, often permitting complete assignment of atom connectivity and even stereochemistry using correlation spectroscopy. The principal constituent atoms of organic chemistry – hydrogen and carbon – exist naturally with NMR-responsive isotopes, respectively ^{1}H and ^{13}C.

- Elemental analysis: A destructive method used to determine the elemental composition of a molecule.

- Mass spectrometry indicates the molecular weight of a compound and, from the fragmentation patterns, its structure. High resolution mass spectrometry can usually identify the exact formula of a compound and is used in lieu of elemental analysis. In former times, mass spectrometry was restricted to neutral molecules exhibiting some volatility, but advanced ionization techniques allow one to obtain the "mass spec" of virtually any organic compound.

- Crystallography can be useful for determining molecular geometry when a single crystal of the material is available and the crystal is representative of the sample. Highly automated software allows a structure to be determined within hours of obtaining a suitable crystal.

Traditional spectroscopic methods such as infrared spectroscopy, optical rotation, UV/VIS spectroscopy provide relatively nonspecific structural information but remain in use for specific classes of compounds. Traditionally refractive index and density were also important for substance identification.

Properties

Physical properties of organic compounds typically of interest include both quantitative and qualitative features. Quantitative information includes melting point, boiling point, and index of refraction. Qualitative properties include odor, consistency, solubility, and color.

Melting and Boiling Properties

Organic compounds typically melt and many boil. In contrast, while inorganic materials generally can be melted, many do not boil, tending instead to degrade. In earlier times, the melting point (m.p.) and boiling point (b.p.) provided crucial information on the purity and identity of organic compounds. The melting and boiling points correlate with the polarity of the molecules and their molecular weight. Some organic compounds, especially symmetrical ones, sublime, that is they evaporate without melting. A well-known example of a sublimable organic compound is para-dichlorobenzene, the odiferous constituent of modern mothballs. Organic compounds are usually not very stable at temperatures above 300 °C, although some exceptions exist.

Solubility

Neutral organic compounds tend to be hydrophobic; that is, they are less soluble in water than in organic solvents. Exceptions include organic compounds that contain ionizable (which can be converted in ions) groups as well as low molecular weight alcohols, amines, and carboxylic acids where hydrogen bonding occurs. Organic compounds tend to dissolve in organic solvents. Solvents can be either pure substances like ether or ethyl alcohol, or mixtures, such as the paraffinic solvents such as the various petroleum ethers and white spirits, or the range of pure or mixed aromatic solvents obtained from petroleum or tar fractions by physical separation or by chemical conversion. Solubility in the different solvents depends upon the solvent type and on the functional groups if present.

Solid State Properties

Various specialized properties of molecular crystals and organic polymers with conjugated systems are of interest depending on applications, e.g. thermo-mechanical and electro-mechanical such as piezoelectricity, electrical conductivity, and electro-optical (e.g. non-linear optics) properties. For historical reasons, such properties are mainly the subjects of the areas of polymer science and materials science.

Nomenclature

$= C_5H_4N\text{-}3\text{-}CO_2H =$ pyridine-3-carboxylic acid $=$ niacin $=$ vitamin B_3

Various names and depictions for one organic compound.

The names of organic compounds are either systematic, following logically from a set of rules, or nonsystematic, following various traditions. Systematic nomenclature is stipulated by specifications from IUPAC. Systematic nomenclature starts with the name for a parent structure within the molecule of interest. This parent name is then modified by prefixes, suffixes, and numbers to unambiguously convey the structure. Given that millions of organic compounds are known, rigorous use of systematic names can be cumbersome. Thus, IUPAC recommendations are more closely followed for simple compounds, but not complex molecules. To use the systematic naming, one must know the structures and names of the parent structures. Parent structures include unsubstituted hydrocarbons, heterocycles, and monofunctionalized derivatives thereof.

Nonsystematic nomenclature is simpler and unambiguous, at least to organic chemists. Nonsystematic names do not indicate the structure of the compound. They are common for complex molecules, which includes most natural products. Thus, the informally named lysergic acid diethylamide is systematically named (6aR,9R)-N,N-diethyl-7-methyl-4,6,6a,7,8,9-hexahydroindolo-[4,3-*fg*] quinoline-9-carboxamide.

With the increased use of computing, other naming methods have evolved that are intended to be interpreted by machines. Two popular formats are SMILES and InChI.

Structural Drawings

Organic molecules are described more commonly by drawings or structural formulas, combinations of drawings and chemical symbols. The line-angle formula is simple and unambiguous. In this system, the endpoints and intersections of each line represent one carbon, and hydrogen atoms can either be notated explicitly or assumed to be present as implied by tetravalent carbon. The depiction of organic compounds with drawings is greatly simplified by the fact that carbon in almost all organic compounds has four bonds, nitrogen three, oxygen two, and hydrogen one.

Classification of Organic Compounds

Functional Groups

The concept of functional groups is central in organic chemistry, both as a means to classify structures and for predicting properties. A functional group is a molecular module, and the reactivity of that functional group is assumed, within limits, to be the same in a variety of molecules. Functional groups can have decisive influence on the chemical and physical properties of organic compounds. Molecules are classified on the basis of their functional groups. Alcohols, for example, all have the subunit C-O-H. All alcohols tend to be somewhat hydrophilic, usually form esters, and usually can be converted to the corresponding halides. Most functional groups feature heteroatoms (atoms other than C and H). Organic compounds are classified according to functional groups, alcohols, carboxylic acids, amines, etc.

The family of carboxylic acids contains a carboxyl (-COOH) functional group. Acetic acid, shown here, is an example.

Aliphatic Compounds

The aliphatic hydrocarbons are subdivided into three groups of homologous series according to their state of saturation:

- paraffins, which are alkanes without any double or triple bonds,

- olefins or alkenes which contain one or more double bonds, i.e. di-olefins (dienes) or poly-olefins.

- alkynes, which have one or more triple bonds.

The rest of the group is classed according to the functional groups present. Such compounds can be "straight-chain", branched-chain or cyclic. The degree of branching affects characteristics, such as the octane number or cetane number in petroleum chemistry.

Both saturated (alicyclic) compounds and unsaturated compounds exist as cyclic derivatives. The most stable rings contain five or six carbon atoms, but large rings (macrocycles) and smaller rings are common. The smallest cycloalkane family is the three-membered cyclopropane ($(CH_2)_3$). Saturated cyclic compounds contain single bonds only, whereas aromatic rings have an alternating (or conjugated) double bond. Cycloalkanes do not contain multiple bonds, whereas the cycloalkenes and the cycloalkynes do.

Aromatic Compounds

Aromatic hydrocarbons contain conjugated double bonds. This means that every carbon atom in the ring is sp2 hybridized, allowing for added stability. The most important example is benzene, the structure of which was formulated by Kekulé who first proposed the delocalization or resonance principle for explaining its structure. For "conventional" cyclic compounds, aromaticity is conferred by the presence of $4n + 2$ delocalized pi electrons, where n is an integer. Particular instability (antiaromaticity) is conferred by the presence of $4n$ conjugated pi electrons.

Benzene is one of the best-known aromatic compounds as it is one of the simplest and most stable aromatics.

Heterocyclic Compounds

The characteristics of the cyclic hydrocarbons are again altered if heteroatoms are present, which can exist as either substituents attached externally to the ring (exocyclic) or as a member of the ring itself (endocyclic). In the case of the latter, the ring is termed a heterocycle. Pyridine and furan are examples of aromatic heterocycles while piperidine and tetrahydrofuran are the corresponding alicyclic heterocycles. The heteroatom of heterocyclic molecules is generally oxygen, sulfur, or nitrogen, with the latter being particularly common in biochemical systems.

Heterocycles are commonly found in a wide range of products including aniline dyes and medicines. Additionally, they are prevalent in a wide range of biochemical compounds such as alkaloids, vitamins, steroids, and nucleic acids (e.g. DNA, RNA).

Rings can fuse with other rings on an edge to give polycyclic compounds. The purine nucleoside bases are notable polycyclic aromatic heterocycles. Rings can also fuse on a "corner" such that one atom (almost always carbon) has two bonds going to one ring and two to another. Such compounds are termed spiro and are important in a number of natural products.

Polymers

One important property of carbon is that it readily forms chains, or networks, that are linked by carbon-carbon (carbon-to-carbon) bonds. The linking process is called polymerization, while the chains, or networks, are called polymers. The source compound is called a monomer.

This swimming board is made of polystyrene, an example of a polymer.

Two main groups of polymers exist: synthetic polymers and biopolymers. Synthetic polymers are artificially manufactured, and are commonly referred to as industrial polymers. Biopolymers occur within a respectfully natural environment, or without human intervention.

Since the invention of the first synthetic polymer product, bakelite, synthetic polymer products have frequently been invented.

Common synthetic organic polymers are polyethylene (polythene), polypropylene, nylon, teflon (PTFE), polystyrene, polyesters, polymethylmethacrylate (called perspex and plexiglas), and polyvinylchloride (PVC).

Both synthetic and natural rubber are polymers.

Varieties of each synthetic polymer product may exist, for purposes of a specific use. Changing the conditions of polymerization alters the chemical composition of the product and its properties. These alterations include the chain length, or branching, or the tacticity.

With a single monomer as a start, the product is a homopolymer.

Secondary component(s) may be added to create a heteropolymer (co-polymer) and the degree of clustering of the different components can also be controlled.

Physical characteristics, such as hardness, density, mechanical or tensile strength, abrasion resistance, heat resistance, transparency, colour, etc. will depend on the final composition.

Biomolecules

Maitotoxin, a complex organic biological toxin.

Biomolecular chemistry is a major category within organic chemistry which is frequently studied by biochemists. Many complex multi-functional group molecules are important in living organisms. Some are long-chain biopolymers, and these include peptides, DNA, RNA and the polysaccharides such as starches in animals and celluloses in plants. The other main classes are amino acids (monomer building blocks of peptides and proteins), carbohydrates (which includes the polysaccharides), the nucleic acids (which include DNA and RNA as polymers), and the lipids. In addition, animal biochemistry contains many small molecule intermediates which assist in energy production through the Krebs cycle, and produces isoprene, the most common hydrocarbon in animals. Isoprenes in animals form the important steroid structural (cholesterol) and steroid hormone compounds; and in plants form terpenes, terpenoids, some alkaloids, and a class of hydrocarbons called biopolymer polyisoprenoids present in the latex of various species of plants, which is the basis for making rubber.

Small Molecules

In pharmacology, an important group of organic compounds is small molecules, also referred to as 'small organic compounds'. In this context, a small molecule is a small organic compound that is biologically active, but is not a polymer. In practice, small molecules have a molar mass less than approximately 1000 g/mol.

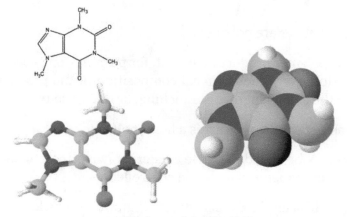

Molecular models of caffeine.

Fullerenes

Fullerenes and carbon nanotubes, carbon compounds with spheroidal and tubular structures, have stimulated much research into the related field of materials science. The first fullerene was discovered in 1985 by Sir Harold W. Kroto (one of the authors of this article) of the United Kingdom and by Richard E. Smalley and Robert F. Curl, Jr., of the United States. Using a laser to vaporize graphite rods in an atmosphere of helium gas, these chemists and their assistants obtained cagelike molecules composed of 60 carbon atoms (C60) joined together by single and double bonds to form a hollow sphere with 12 pentagonal and 20 hexagonal faces—a design that resembles a football, or soccer ball. In 1996 the trio was awarded the Nobel Prize for their pioneering efforts. The C60 molecule was named buckminsterfullerene (or, more simply, the buckyball) after the American architect R. Buckminster Fuller, whose geodesic dome is constructed on the same structural principles.

Others

Organic compounds containing bonds of carbon to nitrogen, oxygen and the halogens are not normally grouped separately. Others are sometimes put into major groups within organic chemistry and discussed under titles such as organosulfur chemistry, organometallic chemistry, organophosphorus chemistry and organosilicon chemistry.

Organic Reactions

Organic reactions are chemical reactions involving organic compounds. Many of these reactions are associated with functional groups. The general theory of these reactions involves careful analysis of such properties as the electron affinity of key atoms, bond strengths and steric hindrance. These factors can determine the relative stability of short-lived reactive intermediates, which usually directly determine the path of the reaction.

The basic reaction types are: addition reactions, elimination reactions, substitution reactions, pericyclic reactions, rearrangement reactions and redox reactions. An example of a common reaction is a substitution reaction written as:

$$Nu^- + C\text{-}X \rightarrow C\text{-}Nu + X^-$$

where X is some functional group and Nu is a nucleophile.

The number of possible organic reactions is basically infinite. However, certain general patterns are observed that can be used to describe many common or useful reactions. Each reaction has a stepwise reaction mechanism that explains how it happens in sequence—although the detailed description of steps is not always clear from a list of reactants alone.

The stepwise course of any given reaction mechanism can be represented using arrow pushing techniques in which curved arrows are used to track the movement of electrons as starting materials transition through intermediates to final products.

Organic Synthesis

Synthetic organic chemistry is an applied science as it borders engineering, the "design, analysis, and/or construction of works for practical purposes". Organic synthesis of a novel compound is a problem solving task, where a synthesis is designed for a target molecule by selecting optimal reactions from optimal starting materials. Complex compounds can have tens of reaction steps that sequentially build the desired molecule. The synthesis proceeds by utilizing the reactivity of the functional groups in the molecule. For example, a carbonyl compound can be used as a nucleophile by converting it into an enolate, or as an electrophile; the combination of the two is called the aldol reaction. Designing practically useful syntheses always requires conducting the actual synthesis in the laboratory. The scientific practice of creating novel synthetic routes for complex molecules is called total synthesis.

A synthesis designed by E.J. Corey for oseltamivir (Tamiflu). This synthesis has 11 distinct reactions.

Strategies to design a synthesis include retrosynthesis, popularized by E.J. Corey, starts with the target molecule and splices it to pieces according to known reactions. The pieces, or the proposed precursors, receive the same treatment, until available and ideally inexpensive starting materials are reached. Then, the retrosynthesis is written in the opposite direction to give the synthesis. A "synthetic tree" can be constructed, because each compound and also each precursor has multiple syntheses.

References

- Greenwood, Norman N.; Earnshaw, Alan (1997). Chemistry of the Elements (2nd ed.). Butterworth-Heinemann. ISBN 0-08-037941-9.

- Nicolaou, K. C.; Sorensen, E. J. (1996). Classics in Total Synthesis: Targets, Strategies, Methods. Wiley. ISBN 978-3-527-29231-8.

- Allan, Barbara. Livesey, Brian (1994). How to Use Biological Abstracts, Chemical Abstracts and Index Chemicus. Gower. ISBN 978-0566075568

- Shriner, R.L.; Hermann, C.K.F.; Morrill, T.C.; Curtin, D.Y. and Fuson, R.C. (1997) The Systematic Identification of Organic Compounds. John Wiley & Sons, ISBN 0-471-59748-1

- "What Is Organic Chemistry?". Science Alive! The Life and Science of Percy Julian. Chemical Heritage Foundation. Retrieved 27 January 2015.

2

Key Concepts of Organic Chemistry

The key concepts explained in this chapter are organic compounds, organic reactions, organic matter and the structural formula of a compound. This chapter explains to the reader the basic characteristics of organic chemistry which is the science concerned with all aspects of organic compounds.

Organic Compound

An organic compound is any member of a large class of gaseous, liquid, or solid chemical compounds whose molecules contain carbon. For historical reasons discussed below, a few types of carbon-containing compounds, such as carbides, carbonates, simple oxides of carbon (such as CO and CO_2), and cyanides are considered inorganic. The distinction between *organic* and *inorganic* carbon compounds, while "useful in organizing the vast subject of chemistry... is somewhat arbitrary".

Methane is one of the simplest organic compounds.

Organic chemistry is the science concerned with all aspects of organic compounds. Organic synthesis is the methodology of their preparation.

History

Vitalism

The word *organic* is historical, dating to the 1st century. For many centuries, Western alchemists believed in vitalism. This is the theory that certain compounds could be synthesized only from their classical elements—earth, water, air, and fire—by the action of a "life-force" (*vis vitalis*) that only organisms possessed. Vitalism taught that these "organic" compounds were fundamentally different from the "inorganic" compounds that could be obtained from the elements by chemical manipulation.

Vitalism survived for a while even after the rise of modern atomic theory and the replacement of the Aristotelian elements by those we know today. It first came under question in 1824, when Friedrich Wöhler synthesized oxalic acid, a compound known to occur only in living organisms, from cyanogen. A more decisive experiment was Wöhler's 1828 synthesis of urea from the inorganic salts potassium cyanate and ammonium sulfate. Urea had long been considered an "organic" compound, as it was known to occur only in the urine of living organisms. Wöhler's experiments were followed by many others, where increasingly complex "organic" substances were produced from "inorganic" ones without the involvement of any living organism.

Modern Classification

Even though vitalism has been discredited, scientific nomenclature retains the distinction between *organic* and *inorganic* compounds. The modern meaning of *organic compound* is any compound that contains a significant amount of carbon—even though many of the organic compounds known today have no connection to any substance found in living organisms.

There is no single "official" definition of an organic compound. Some textbooks define an organic compound as one that contains one or more C-H bonds. Others include C-C bonds in the definition. Others state that if a molecule contains carbon—it is organic.

Even the broader definition of "carbon-containing molecules" requires the exclusion of carbon-containing alloys (including steel), a relatively small number of carbon-containing compounds, such as metal carbonates and carbonyls, simple oxides of carbon and cyanides, as well as the allotropes of carbon and simple carbon halides and sulfides, which are usually considered inorganic.

The "C-H" definition excludes compounds that are historically and practically considered organic. Neither urea nor oxalic acid is organic by this definition, yet they were two key compounds in the vitalism debate. The IUPAC Blue Book on organic nomenclature specifically mentions urea and oxalic acid. Other compounds lacking C-H bonds that are also traditionally considered organic include benzenehexol, mesoxalic acid, and carbon tetrachloride. Mellitic acid, which contains no C-H bonds, is considered a possible organic substance in Martian soil. C-C bonds are found in most organic compounds, except some small molecules like methane and methanol, which have only one carbon atom in their structure.

The "C-H bond-only" rule also leads to somewhat arbitrary divisions in sets of carbon-fluorine compounds, as, for example, Teflon is considered by this rule to be "inorganic", whereas Tefzel is considered to be organic. Likewise, many Halons are considered inorganic, whereas the rest are considered organic. For these and other reasons, most sources believe that C-H compounds are only a subset of "organic" compounds.

In summary, most carbon-containing compounds are organic, and almost all organic compounds contain at least a C-H bond or a C-C bond. A compound does not need to contain C-H bonds to be considered organic (e.g., urea), but most organic compounds do.

Classification

Organic compounds may be classified in a variety of ways. One major distinction is between natural and synthetic compounds. Organic compounds can also be classified or subdivided by the presence of heteroatoms, e.g., organometallic compounds, which feature bonds between carbon and a metal, and organophosphorus compounds, which feature bonds between carbon and a phosphorus.

Another distinction, based on the size of organic compounds, distinguishes between small molecules and polymers.

Natural Compounds

Natural compounds refer to those that are produced by plants or animals. Many of these are still extracted from natural sources because they would be more expensive to produce artificially. Examples include most sugars, some alkaloids and terpenoids, certain nutrients such as vitamin B_{12}, and, in general, those natural products with large or stereoisometrically complicated molecules present in reasonable concentrations in living organisms.

Further compounds of prime importance in biochemistry are antigens, carbohydrates, enzymes, hormones, lipids and fatty acids, neurotransmitters, nucleic acids, proteins, peptides and amino acids, lectins, vitamins, and fats and oils.

Synthetic Compounds

Compounds that are prepared by reaction of other compounds are known as "synthetic". They may be either compounds that already are found in plants or animals or those that do not occur naturally.

Most polymers (a category that includes all plastics and rubbers), are organic synthetic or semi-synthetic compounds.

Biotechnology

Several compounds are industrially manufactured utilizing the biochemistry of organisms such as bacteria and yeast. Two examples are ethanol and insulin. Regularly, the DNA of the organism is altered to express desired compounds that are often not ordinarily produced by that organism. Sometimes the biotechnologically engineered compounds were never present in nature in the first place.

Nomenclature

The IUPAC nomenclature of organic compounds slightly differs from the CAS nomenclature.

Databases

- The *CAS* database is the most comprehensive repository for data on organic compounds. The search tool *SciFinder* is offered.

- The *Beilstein database* contains information on 9.8 million substances, covers the scientific literature from 1771 to the present, and is today accessible via Reaxys. Structures and a large diversity of physical and chemical properties is available for each substance, with reference to original literature.

- *PubChem* contains 18.4 million entries on compounds and especially covers the field of medicinal chemistry.

There is a great number of more specialized databases for diverse branches of organic chemistry.

Structure Determination

Today, the main tools are proton and carbon-13 NMR spectroscopy, IR Spectroscopy, Mass spectrometry, UV/Vis Spectroscopy and X-ray crystallography.

Organic Reaction

Organic reactions are chemical reactions involving organic compounds. The basic organic chemistry reaction types are addition reactions, elimination reactions, substitution reactions, pericyclic reactions, rearrangement reactions, photochemical reactions and redox reactions. In organic synthesis, organic reactions are used in the construction of new organic molecules. The production of many man-made chemicals such as drugs, plastics, food additives, fabrics depend on organic reactions.

The oldest organic reactions are combustion of organic fuels and saponification of fats to make soap. Modern organic chemistry starts with the Wöhler synthesis in 1828. In the history of the Nobel Prize in Chemistry awards have been given for the invention of specific organic reactions such as the Grignard reaction in 1912, the Diels-Alder reaction in 1950, the Wittig reaction in 1979 and olefin metathesis in 2005.

Classifications

Organic chemistry has a strong tradition of naming a specific reaction to its inventor or inventors and a long list of so-called named reactions exists, conservatively estimated at 1000. A very old named reaction is the Claisen rearrangement (1912) and a recent named reaction is the Bingel reaction (1993). When the named reaction is difficult to pronounce or very long as in the Corey-House-Posner-Whitesides reaction it helps to use the abbreviation as in the CBS reduction. The number of reactions hinting at the actual process taking place is much smaller, for example the ene reaction or aldol reaction.

Another approach to organic reactions is by type of organic reagent, many of them inorganic, required in a specific transformation. The major types are oxidizing agents such as osmium tetroxide, reducing agents such as Lithium aluminium hydride, bases such as lithium diisopropylamide and acids such as sulfuric acid.

Fundamentals

Factors governing organic reactions are essentially the same as that of any chemical reaction. Factors specific to organic reactions are those that determine the stability of reactants and products such as conjugation, hyperconjugation and aromaticity and the presence and stability of reactive intermediates such as free radicals, carbocations and carbanions.

An organic compound may consist of many isomers. Selectivity in terms of regioselectivity, diastereoselectivity and enantioselectivity is therefore an important criterion for many organic reactions. The stereochemistry of pericyclic reactions is governed by the Woodward–Hoffmann rules and that of many elimination reactions by the Zaitsev's rule.

Organic reactions are important in the production of pharmaceuticals. In a 2006 review it was estimated that 20% of chemical conversions involved alkylations on nitrogen and oxygen atoms, another 20% involved placement and removal of protective groups, 11% involved formation of new carbon-carbon bond and 10% involved functional group interconversions.

Organic Reactions by Mechanism

There is no limit to the number of possible organic reactions and mechanisms. However, certain general patterns are observed that can be used to describe many common or useful reactions. Each reaction has a stepwise reaction mechanism that explains how it happens, although this detailed description of steps is not always clear from a list of reactants alone. Organic reactions can be organized into several basic types. Some reactions fit into more than one category. For example, some substitution reactions follow an addition-elimination pathway. This overview isn't intended to include every single organic reaction. Rather, it is intended to cover the basic reactions.

Reaction type	Subtype	Comment
Addition reactions	electrophilic addition	include such reactions as halogenation, hydrohalogenation and hydration.
	nucleophilic addition	
	radical addition	
Elimination reaction		include processes such as dehydration and are found to follow an E1, E2 or E1cB reaction mechanism
Substitution reactions	nucleophilic aliphatic substitution	with S_N1, S_N2 and S_Ni reaction mechanisms
	nucleophilic aromatic substitution	
	nucleophilic acyl substitution	
	electrophilic substitution	
	electrophilic aromatic substitution	
	radical substitution	
Organic redox reactions		are redox reactions specific to organic compounds and are very common.
Rearrangement reactions	1,2-rearrangements	
	pericyclic reactions	
	metathesis	

In condensation reactions a small molecule, usually water, is split off when two reactants combine in a chemical reaction. The opposite reaction, when water is consumed in a reaction, is called hydrolysis. Many Polymerization reactions are derived from organic reactions. They are divided into addition polymerizations and step-growth polymerizations.

In general the stepwise progression of reaction mechanisms can be represented using arrow pushing techniques in which curved arrows are used to track the movement of electrons as starting materials transition to intermediates and products.

Organic Reactions by Functional Groups

Organic reactions can be categorized based on the type of functional group involved in the reaction as a reactant and the functional group that is formed as a result of this reaction. For example in the Fries rearrangement the reactant is an ester and the reaction product an alcohol.

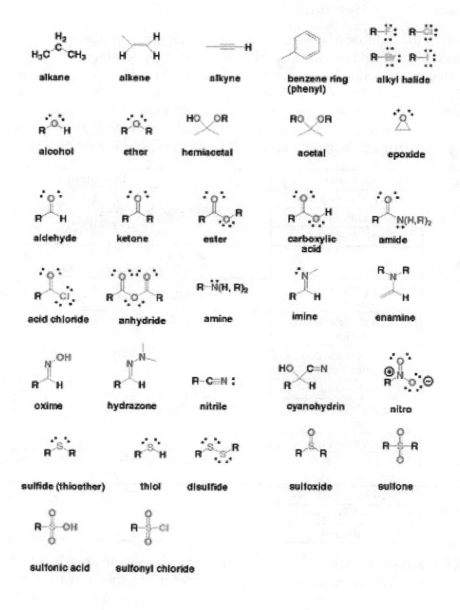

Other Organic Reaction Classification

In heterocyclic chemistry, organic reactions are classified by the type of heterocycle formed with respect to ring-size and type of heteroatom. See for instance the chemistry of indoles. Reactions are also categorized by the change in the carbon framework. Examples are ring expansion and ring contraction, homologation reactions, polymerization reactions, insertion reactions, ring-opening reactions and ring-closing reactions.

Organic reactions can also be classified by the type of bond to carbon with respect to the element involved. More reactions are found in organosilicon chemistry, organosulfur chemistry, organophosphorus chemistry and organofluorine chemistry. With the introduction of carbon-metal bonds the field crosses over to organometallic chemistry.

CH																	He
CLi	CBe											CB	CC	CN	CO	CF	Ne
CNa	CMg											CAl	CSi	CP	CS	CCl	CAr
CK	CCa	CSc	CTi	CV	CCr	CMn	CFe	CCo	CNi	CCu	CZn	CGa	CGe	CAs	CSe	CBr	CKr
CRb	CSr	CY	CZr	CNb	CMo	CTc	CRu	CRh	CPd	CAg	CCd	CIn	CSn	CSb	CTe	CI	CXe
CCs	CBa		CHf	CTa	CW	CRe	COs	CIr	CPt	CAu	CHg	CTl	CPb	CBi	CPo	CAt	Rn
Fr	CRa		Rf	Db	CSg	Bh	Hs	Mt	Ds	Rg	Cn	Nh	Fl	Mc	Lv	Ts	Og

↓

CLa	CCe	CPr	CNd	CPm	CSm	CEu	CGd	CTb	CDy	CHo	CEr	CTm	CYb	CLu
Ac	CTh	CPa	CU	CNp	CPu	CAm	CCm	CBk	CCf	CEs	Fm	Md	No	Lr

Chemical bonds to carbon	
Core organic chemistry	Many uses in chemistry
Academic research, but no widespread use	Bond unknown

Organic Matter

Organic matter or organic material, natural organic matter, NOM refers to the large pool of carbon-based compounds found within natural and engineered, terrestrial and aquatic environments. It is matter composed of organic compounds that has come from the remains of organisms such as plants and animals and their waste products in the environment. Organic molecules can also be made by chemical reactions that don't involve life. Basic structures are created from cellulose, tannin, cutin, and lignin, along with other various proteins, lipids, and carbohydrates. Organic matter is very important in the movement of nutrients in the environment and plays a role in water retention on the surface of the planet.

Formation

Living organisms are composed of organic compounds. In life they secrete or excrete organic materials into their environment, shed body parts such as leaves and roots and after the organism dies, its body is broken down by bacterial and fungal action. Larger molecules of organic matter can be formed from the polymerization of different parts of already broken down matter. The composition of natural organic matter depends on its origin, transformation mode, age, and existing environment, thus its bio-physico-chemical functions vary with different environments.

Natural Ecosystem Functions

Organic matter is present throughout the ecosystem. After degrading and reacting, it can move into soil and mainstream water via waterflow. Organic matter provides nutrition to living organisms. Organic matter acts as a buffer in aqueous solution to maintain a neutral pH in the environment. The buffer acting component has been proposed to be relevant for neutralizing acid rain.

Source Cycle

A majority of organic matter not already in the soil comes from groundwater. When the groundwater saturates the soil or sediment around it, organic matter can freely move between the phases. Groundwater has its own sources of natural organic matter also:

- "organic matter deposits, such as kerogen and coal

- soil and sediment organic matter

- organic matter infiltrating into the subsurface from rivers, lakes, and marine systems"

Note that one source of groundwater organic matter is soil organic matter and sedimentary organic matter. The major method of movement into soil is from groundwater, but organic matter from soil moves into groundwater as well. Most of the organic matter in lakes, rivers, and surface water areas comes from deteriorated material in the water and surrounding shores. However, organic matter can pass into or out of water to soil and sediment in the same respect as with the soil.

Importance of The Cycle

Organic matter can migrate through soil, sediment, and water. This movement enables a cycle. Organisms decompose into organic matter, which can then be transported and recycled. Not all biomass migrates, some is rather stationary, turning over only over the course of millions of years.

Soil Organic Matter

The organic matter in soil derives from plants and animals. In a forest, for example, leaf litter and woody material falls to the forest floor. This is sometimes referred to as organic material. When it decays to the point in which it is no longer recognizable it is called soil organic matter. When the organic matter has broken down into a stable substance that resist further decomposition it is called humus. Thus soil organic matter comprises all of the organic matter in the soil exclusive of the material that has not decayed.

One of the advantages of humus is that it is able to withhold water and nutrients, therefore giving plants the capacity for growth. Another advantage of humus is that it helps the soil to stick together which allows nematodes, or microscopic bacteria, to easily decay the nutrients in the soil.

There are several ways to quickly increase the amount of humus. Combining compost, plant or animal materials/waste, or green manure with soil will increase the amount of humus in the soil.

1. Compost: decomposed organic material.

2. Plant and animal material and waste: dead plants or plant waste such as leaves or bush and tree trimmings, or animal manure.

3. Green manure: plants or plant material that is grown for the sole purpose of being incorporated with soil.

These three materials supply nematodes and bacteria with nutrients for them to thrive and produce more humus, which will give plants enough nutrients to survive and grow.

Factors Controlling Rates of Decomposition

- o Environmental factors
  - Aeration
  - Temperature
  - Soil Moisture
  - Soil pH
- o Quality of added residues
  - Size of organic residues
  - C/N of organic residues

- Rate of decomposition of plant residues, in order from fastest to slowest decomposition rates:

    o   Sugars, starches, simple proteins

    o   Hemicellulose

    o   Cellulose

    o   Fats, waxes, oils, resins

    o   Lignin, phenolic compounds

Priming Effect

The *priming effect* is characterized by intense changes in the natural process of soil organic matter (SOM) turnover, resulting from relatively moderate intervention with the soil. The phenomenon is generally caused by either pulsed or continuous changes to inputs of fresh organic matter (FOM). Priming effects usually result in an acceleration of mineralization due to a *trigger* such as the FOM inputs. The cause of this increase in decomposition has often been attributed to an increase in microbial activity resulting from higher energy and nutrient availability released from the FOM. After the input of FOM, specialized microorganisms are believed to grow quickly and only decompose this newly added organic matter. The turnover rate of SOM in these areas is at least one order of magnitude higher than the bulk soil.

Other soil treatments, besides organic matter inputs, which lead to this short-term change in turnover rates, include "input of mineral fertilizer, exudation of organic substances by roots, mere mechanical treatment of soil or its drying and rewetting."

Priming effects can be either *positive* or *negative* depending on the reaction of the soil with the added substance. A positive priming effect results in the acceleration of mineralization while a negative priming effect results in immobilization, leading to N unavailability. Although most changes have been documented in C and N pools, the priming effect can also be found in phosphorus and sulfur, as well as other nutrients.

Löhnis was the first to discover the priming effect phenomenon in 1926 through his studies of green manure decomposition and its effects on legume plants in soil. He noticed that when adding fresh organic residues to the soil, it resulted in intensified mineralization by the humus N. It was not until 1953, though, that the term *priming effect* was given by Bingeman in his paper titled, *The effect of the addition of organic materials on the decomposition of an organic soil.* Several other terms had been used before *priming effect* was coined, including priming action, added nitrogen interaction (ANI), extra N and additional N. Despite these early contributions, the concept of the priming effect was widely disregarded until about the 1980s-1990s.

The priming effect has been found in many different studies and is regarded as a common occurrence, appearing in most plant soil systems. However, the mechanisms which lead to the priming effect are more complex then originally thought, and still remain generally misunderstood.

Although there is a lot of uncertainty surrounding the reason for the priming effect, a few *undisputed facts* have emerged from the collection of recent research:

1. The priming effect can arise either instantaneously or very shortly (potentially days or weeks) after the addition of a substance is made to the soil.

2. The priming effect is larger in soils that are rich in C and N as compared to those poor in these nutrients.

3. Real priming effects have not been observed in sterile environments.

4. The size of the priming effect increases as the amount of added treatment to the soil increases.

Recent findings suggest that the same priming effect mechanisms acting in soil systems may also be present in aquatic environments, which suggests a need for broader considerations of this phenomenon in the future.

Decomposition

One suitable definition of organic matter is biological material in the process of decaying or decomposing, such as humus. A closer look at the biological material in the process of decaying reveals so-called organic compounds (biological molecules) in the process of breaking up (disintegrating).

The main processes by which soil molecules disintegrates are by bacterial or fungal enzymatic catalysis. If bacteria or fungi were not present on Earth, the process of decomposition would have proceeded much slower.

Organic Chemistry

Measurements of organic matter generally measure only organic compounds or carbon, and so are only an approximation of the level of once-living or decomposed matter. Some definitions of organic matter likewise only consider "organic matter" to refer to only the carbon content, or organic compounds, and do not consider the origins or decomposition of the matter. In this sense, not all organic compounds are created by living organisms, and living organisms do not only leave behind organic material. A clam's shell, for example, while biotic, does not contain much organic carbon, so may not be considered organic matter in this sense. Conversely, urea is one of many organic compounds that can be synthesized without any biological activity.

Organic matter is heterogeneous and very complex. Generally, organic matter, in terms of weight, is:

- 45-55% carbon

- 35-45% oxygen

- 3-5% hydrogen

- 1-4% nitrogen

The molecular weights of these compounds can vary drastically, depending on if they repolymerize or not, from 200-20,000 amu. Up to one third of the carbon present is in aromatic compounds in which the carbon atoms form usually 6 membered rings. These rings are very stable due to resonance stabilization, so they are difficult to break down. The aromatic rings are also susceptible to electrophilic and nucleophilic attack from other electron-donating or electron-accepting material, which explains the possible polymerization to create larger molecules of organic matter.

There are also reactions that occur with organic matter and other material in the soil to create compounds never seen before. Unfortunately, it is very difficult to characterize these because so little is known about natural organic matter in the first place. Research is currently being done to figure out more about these new compounds and how many of them are being formed.

Organic Matter in Water (Aquatic)

Aquatic organic matter can be further divided into two subsections: dissolved organic matter (DOM) and particulate organic matter (POM). They are typically differentiated by that which can pass through a 0.45 micrometre filter (DOM), and that which cannot (POM).

Detection of Aquatic Organic Matter

Organic matter plays an important role in drinking water and wastewater treatment and recycling, natural aquatic ecosystems, aquaculture, and environmental rehabilitation. It is therefore important to have reliable methods of detection and characterisation, for both short- and long-term monitoring. A variety of analytical detection methods for organic matter have existed for up to decades, to describe and characterise organic matter. These include, but are not limited to: total and dissolved organic carbon, mass spectrometry, nuclear magnetic resonance (NMR) spectroscopy, infrared (IR) spectroscopy, UV-Visible spectroscopy, and fluorescence spectroscopy. Each of these methods has its own advantages and limitations.

Water Purification

The same capability of natural organic matter that helped with water retention in soil creates problems for current water purification methods. In water, organic matter can still bind to metal ions and minerals. These bound molecules are not necessarily stopped by the purification process, but do not cause harm to any humans, animals, or plants. However, because of the high level of reactivity of organic matter, by-products that do not contain nutrients can be made. These by-products can induce biofouling, which essentially clogs water filtration systems in water purification facilities, as the by-products are larger than membrane pore sizes. This clogging problem can be treated by chlorine disinfection (chlorination), which can break down residual material that clogs systems. However, chlorination can form disinfection by-products.

Potential Solutions

Water with organic matter can be disinfected with ozone-initiated radical reactions. The ozone (three oxygens) has very strong oxidation characteristics. It can form hydroxyl radicals (OH) when it decomposes, which will react with the organic matter to shut down the problem of biofouling.

False Positives

Many water quality groups, such as the North Carolina State University Water Quality Group, believe that having too much organic material will cause deoxygenation and essentially remove oxygen from the water. Although organic material, which consists of many hydrocarbon and cyclic carbon chains, is susceptible to attack by oxygen, it would be sterically unfavorable to attach oxygens to every single carbon.

Of course, there are exceptions, such as varying the temperature at which these reactions occur. As the temperature becomes much higher, there is a better chance that an unfavorable reaction will occur because molecules move around faster increasing the randomness of the system (entropy).

Vitalism

The equation of "organic" with living organisms comes from the now-abandoned idea of vitalism that attributed a special force to life that alone could create organic substances. This idea was first questioned after the artificial synthesis of urea by Friedrich Wöhler in 1828.

Hydrocarbon

Ball-and-stick model of the methane molecule, CH_4. Methane is part of a homologous series known as the alkanes, which contain single bonds only.

In organic chemistry, a hydrocarbon is an organic compound consisting entirely of hydrogen and carbon, and thus are group 14 hydrides. Hydrocarbons from which one hydrogen atom has been removed are functional groups, called hydrocarbyls. Aromatic hydrocarbons (arenes), alkanes, alkenes, cycloalkanes and alkyne-based compounds are different types of hydrocarbons.

The majority of hydrocarbons found on Earth naturally occur in crude oil, where decomposed organic matter provides an abundance of carbon and hydrogen which, when bonded, can catenate to form seemingly limitless chains.

Types of Hydrocarbons

The classifications for hydrocarbons, defined by IUPAC nomenclature of organic chemistry are as follows:

1. Saturated hydrocarbons are the simplest of the hydrocarbon species. They are composed entirely of single bonds and are saturated with hydrogen. The formula for acyclic saturated hydrocarbons (i.e., alkanes) is C_nH_{2n+2}. The most general form of saturated hydrocarbons is $C_nH_{2n+2(1-r)}$, where r is the number of rings. Those with exactly one ring are the cycloalkanes. Saturated hydrocarbons are the basis of petroleum fuels and are found as either linear or branched species. Substitution reaction is their characteristics property (like chlorination reaction to form chloroform). Hydrocarbons with the same molecular formula but different structural formulae are called structural isomers. As given in the example of 3-methylhexane and its higher homologues, branched hydrocarbons can be chiral. Chiral saturated hydrocarbons constitute the side chains of biomolecules such as chlorophyll and tocopherol.

2. Unsaturated hydrocarbons have one or more double or triple bonds between carbon atoms. Those with double bond are called alkenes. Those with one double bond have the

formula C_nH_{2n} (assuming non-cyclic structures). Those containing triple bonds are called alkynes, with general formula C_nH_{2n-2}.

3. Aromatic hydrocarbons, also known as arenes, are hydrocarbons that have at least one aromatic ring.

Hydrocarbons can be gases (e.g. methane and propane), liquids (e.g. hexane and benzene), waxes or low melting solids (e.g. paraffin wax and naphthalene) or polymers (e.g. polyethylene, polypropylene and polystyrene).

General Properties

Because of differences in molecular structure, the empirical formula remains different between hydrocarbons; in linear, or "straight-run" alkanes, alkenes and alkynes, the amount of bonded hydrogen lessens in alkenes and alkynes due to the "self-bonding" or catenation of carbon preventing entire saturation of the hydrocarbon by the formation of double or triple bonds.

This inherent ability of hydrocarbons to bond to themselves is known as catenation, and allows hydrocarbons to form more complex molecules, such as cyclohexane, and in rarer cases, arenes such as benzene. This ability comes from the fact that the bond character between carbon atoms is entirely non-polar, in that the distribution of electrons between the two elements is somewhat even due to the same electronegativity values of the elements (~0.30), and does not result in the formation of an electrophile.

Generally, with catenation comes the loss of the total amount of bonded hydrocarbons and an increase in the amount of energy required for bond cleavage due to strain exerted upon the molecule; in molecules such as cyclohexane, this is referred to as ring strain, and occurs due to the "destabilized" spatial electron configuration of the atom.

In simple chemistry, as per valence bond theory, the carbon atom must follow the "*4-hydrogen rule*", which states that the maximum number of atoms available to bond with carbon is equal to the number of electrons that are attracted into the outer shell of carbon. In terms of shells, carbon consists of an incomplete outer shell, which comprises 4 electrons, and thus has 4 electrons available for covalent or dative bonding.

Hydrocarbons are hydrophobic like lipids.

Some hydrocarbons also are abundant in the solar system. Lakes of liquid methane and ethane have been found on Titan, Saturn's largest moon, confirmed by the Cassini-Huygens Mission. Hydrocarbons are also abundant in nebulae forming polycyclic aromatic hydrocarbon (PAH) compounds.

Simple Hydrocarbons and Their Variations

Number of carbon atoms	Alkane (single bond)	Alkene (double bond)	Alkyne (triple bond)	Cycloalkane	Alkadiene
1	Methane	-	-	-	-
2	Ethane	Ethene (ethylene)	Ethyne (acetylene)	–	–

3	Propane	Propene (propylene)	Propyne (methylacetylene)	Cyclopropane	Propadiene (allene)
4	Butane	Butene (butylene)	Butyne	Cyclobutane	Butadiene
5	Pentane	Pentene	Pentyne	Cyclopentane	Pentadiene (piperylene)
6	Hexane	Hexene	Hexyne	Cyclohexane	Hexadiene
7	Heptane	Heptene	Heptyne	Cycloheptane	Heptadiene
8	Octane	Octene	Octyne	Cyclooctane	Octadiene
9	Nonane	Nonene	Nonyne	Cyclononane	Nonadiene
10	Decane	Decene	Decyne	Cyclodecane	Decadiene

Usage

Oil refineries are one way hydrocarbons are processed for use. Crude oil is processed in several stages to form desired hydrocarbons, used as fuel and in other products.

Hydrocarbons are a primary energy source for current civilizations. The predominant use of hydrocarbons is as a combustible fuel source. In their solid form, hydrocarbons take the form of asphalt (bitumen).

Mixtures of volatile hydrocarbons are now used in preference to the chlorofluorocarbons as a propellant for aerosol sprays, due to chlorofluorocarbons' impact on the ozone layer.

Methane (CH_4) and ethane (C_2H_6) are gaseous at ambient temperatures and cannot be readily liquefied by pressure alone. Propane (C_3H_8) is however easily liquefied, and exists in 'propane bottles' mostly as a liquid. Butane (C_4H_{10}) is so easily liquefied that it provides a safe, volatile fuel for small pocket lighters. Pentane (C_5H_{12}) is a clear liquid at room temperature, commonly used in chemistry and industry as a powerful nearly odorless solvent of waxes and high molecular weight organic compounds, including greases. Hexane (C_6H_{14}) is also a widely used non-polar, non-aromatic solvent, as well as a significant fraction of common gasoline. The C^6 through C^{10} alkanes, alkenes and isomeric cycloalkanes are the top components of gasoline, naphtha, jet fuel and specialized industrial solvent mixtures. With the progressive addition of carbon units, the simple non-ring structured hydrocarbons have higher viscosities, lubricating indices, boiling points, solidification temperatures, and deeper color. At the opposite extreme from methane lie the heavy tars that remain as the *lowest fraction* in a crude oil refining retort. They are collected and widely utilized as roofing compounds, pavement composition, wood preservatives (the creosote series) and as extremely high viscosity shear-resisting liquids.

Hydrocarbon use is also prevalent in nature. Some eusocial arthropods, such as the Brazilian stingless bee *Schwarziana quadripunctata*, use unique hydrocarbon "scents" in order to determine kin from non-kin. The chemical hydrocarbon composition varies between age, sex, nest location, and hierarchal position.

Poisoning

Hydrocarbon poisoning such as that of benzene and petroleum usually occurs accidentally by inhalation or ingestion of these cytotoxic chemical compounds. Intravenous or subcutaneous injection of petroleum compounds with intent of suicide or abuse is an extraordinary event that can result in local damage or systemic toxicity such as tissue necrosis, abscess formation, respiratory system failure and partial damage to the kidneys, the brain and the nervous system. Moaddab and Eskandarlou report a case of chest wall necrosis and empyema resulting from attempting suicide by injection of petroleum into the pleural cavity.

Reactions

There are three main types of reactions:

- Substitution reaction

- Addition reaction

- Combustion

Substitution Reactions

Substitution reactions only occur in saturated hydrocarbons (single carbon–carbon bonds). In this reaction, an alkane reacts with a chlorine molecule. One of the chlorine atoms displace an hydrogen atom. This forms hydrochloric acid as well as the hydrocarbon with one chlorine atom.

$$CH_4 + Cl_2 \rightarrow CH_3Cl + HCl$$

$$CH_3Cl + Cl_2 \rightarrow CH_2Cl_2 + HCl$$

all the way to CCl_4 (carbon tetrachloride)

$$C_2H_6 + Cl_2 \rightarrow C_2H_5Cl + HCl$$

$$C_2H_4Cl_2 + Cl_2 \rightarrow C_2H_4Cl_3 + HCl$$

all the way to C_2Cl_6 (hexachloroethane)

Addition Reactions

Addition reactions involve alkenes and alkynes. In this reaction a halogen molecule breaks the double or triple bond in the hydrocarbon and forms a bond.

Combustion

Hydrocarbons are currently the main source of the world's electric energy and heat sources (such as home heating) because of the energy produced when burnt. Often this energy is used directly as heat such as in home heaters, which use either petroleum or natural gas. The hydrocarbon is burnt and the heat is used to heat water, which is then circulated. A similar principle is used to create electric energy in power plants.

Common properties of hydrocarbons are the facts that they produce steam, carbon dioxide and heat during combustion and that oxygen is required for combustion to take place. The simplest hydrocarbon, methane, burns as follows:

$$CH_4 + 2\ O_2 \rightarrow 2\ H_2O + CO_2 + energy$$

In inadequate supply of air, carbon monoxide gas and water vapour are formed:

$$2\ CH_4 + 3\ O_2 \rightarrow 2\ CO + 4\ H_2O$$

Another example of this reaction is propane:

$$C_3H_8 + 5\ O_2 \rightarrow 4\ H_2O + 3\ CO_2 + energy$$

$$C_nH_{2n+2} + 3n + 1/2\ O_2 \rightarrow (n + 1)\ H_2O + n\ CO_2 + energy$$

Burning of hydrocarbons is an example of an exothermic chemical reaction.

Hydrocarbons can also be burned with elemental fluorine, resulting in carbon tetrafluoride and hydrogen fluoride products.

Petroleum

Extracted hydrocarbons in a liquid form are referred to as petroleum (literally "rock oil") or mineral oil, whereas hydrocarbons in a gaseous form are referred to as natural gas. Petroleum and natural gas are found in the Earth's subsurface with the tools of petroleum geology and are a significant source of fuel and raw materials for the production of organic chemicals.

The extraction of liquid hydrocarbon fuel from sedimentary basins is integral to modern energy development. Hydrocarbons are mined from oil sands and oil shale, and potentially extracted from sedimentary methane hydrates. These reserves require distillation and upgrading to produce synthetic crude and petroleum.

Oil reserves in sedimentary rocks are the source of hydrocarbons for the energy, transport and petrochemical industries.

Economically important hydrocarbons include fossil fuels such as coal, petroleum and natural gas, and its derivatives such as plastics, paraffin, waxes, solvents and oils. Hydrocarbons – along with NO_x and sunlight – contribute to the formation of tropospheric ozone and greenhouse gases.

Bioremediation

Bacteria in the gabbroic layer of the ocean's crust can degrade hydrocarbons; but the extreme environment makes research difficult. Other bacteria such as *Lutibacterium anuloederans* can also degrade hydrocarbons. Mycoremediation or breaking down of hydrocarbon by mycellium and mushroom is possible.

Safety

Many hydrocarbons are highly flammable, therefore, care should be taken to prevent injury. Benzene and many aromatic compounds are possible carcinogens, and proper safety equipment must be worn to prevent these harmful compounds from entering the body. If hydrocarbons undergo combustion in tight areas, toxic carbon monoxide can form. Hydrocarbons should be kept away from fluorine compounds due to the high probability of forming toxic hydrofluoric acid.

Molecule

A molecule is an electrically neutral group of two or more atoms held together by chemical bonds. Molecules are distinguished from ions by their lack of electrical charge. However, in quantum physics, organic chemistry, and biochemistry, the term *molecule* is often used less strictly, also being applied to polyatomic ions.

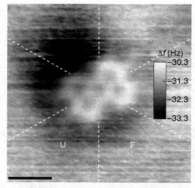

Atomic force microscopy image of a PTCDA molecule, which contains five carbon rings in a non-linear arrangement.

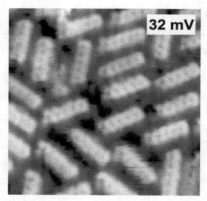

A scanning tunneling microscopy image of pentacene molecules, which consist of linear chains of five carbon rings.

In the kinetic theory of gases, the term *molecule* is often used for any gaseous particle regardless of its composition. According to this definition, noble gas atoms are considered molecules as they are in fact monoatomic molecules.

A molecule may be homonuclear, that is, it consists of atoms of one chemical element, as with oxygen (O_2); or it may be heteronuclear, a chemical compound composed of more than one element, as with water (H_2O). Atoms and complexes connected by non-covalent interactions, such as hydrogen bonds or ionic bonds, are generally not considered single molecules.

Molecules as components of matter are common in organic substances (and therefore biochemistry). They also make up most of the oceans and atmosphere. However, the majority of familiar solid substances on Earth, including most of the minerals that make up the crust, mantle, and core of the Earth, contain many chemical bonds, but are *not* made of identifiable molecules. Also, no typical molecule can be defined for ionic crystals (salts) and covalent crystals (network solids), although these are often composed of repeating unit cells that extend either in a plane (such as in graphene) or three-dimensionally (such as in diamond, quartz, or sodium chloride). The theme of repeated unit-cellular-structure also holds for most condensed phases with metallic bonding, which means that solid metals are also not made of molecules. In glasses (solids that exist in a vitreous disordered state), atoms may also be held together by chemical bonds with no presence of any definable molecule, nor any of the regularity of repeating units that characterizes crystals.

Molecular Science

The science of molecules is called *molecular chemistry* or *molecular physics*, depending on whether the focus is on chemistry or physics. Molecular chemistry deals with the laws governing the interaction between molecules that results in the formation and breakage of chemical bonds, while molecular physics deals with the laws governing their structure and properties. In practice, however, this distinction is vague. In molecular sciences, a molecule consists of a stable system (bound state) composed of two or more atoms. Polyatomic ions may sometimes be usefully thought of as electrically charged molecules. The term *unstable molecule* is used for very reactive species, i.e., short-lived assemblies (resonances) of electrons and nuclei, such as radicals, molecular ions, Rydberg molecules, transition states, van der Waals complexes, or systems of colliding atoms as in Bose–Einstein condensate.

History and Etymology

According to Merriam-Webster and the Online Etymology Dictionary, the word "molecule" derives from the Latin "moles" or small unit of mass.

- Molecule (1794) – "extremely minute particle", from French *molécule* (1678), from New Latin *molecula*, diminutive of Latin *moles* "mass, barrier". A vague meaning at first; the vogue for the word (used until the late 18th century only in Latin form) can be traced to the philosophy of Descartes.

The definition of the molecule has evolved as knowledge of the structure of molecules has increased. Earlier definitions were less precise, defining molecules as the smallest particles of pure chemical substances that still retain their composition and chemical properties. This definition often breaks down since many substances in ordinary experience, such as rocks, salts, and metals, are composed of large crystalline networks of chemically bonded atoms or ions, but are not made of discrete molecules.

Bonding

Molecules are held together by either covalent bonding or ionic bonding. Several types of non-metal elements exist only as molecules in the environment. For example, hydrogen only exists as hydrogen molecule. A molecule of a compound is made out of two or more elements.

Covalent

A covalent bond is a chemical bond that involves the sharing of electron pairs between atoms. These electron pairs are termed *shared pairs* or *bonding pairs*, and the stable balance of attractive and repulsive forces between atoms, when they share electrons, is termed *covalent bonding*.

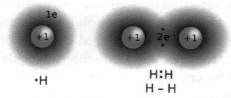

A covalent bond forming H_2 (right) where two hydrogen atoms share the two electrons

Ionic

Ionic bonding is a type of chemical bond that involves the electrostatic attraction between oppositely charged ions, and is the primary interaction occurring in ionic compounds. The ions are atoms that have lost one or more electrons (termed cations) and atoms that have gained one or more electrons (termed anions). This transfer of electrons is termed *electrovalence* in contrast to covalence. In the simplest case, the cation is a metal atom and the anion is a nonmetal atom, but these ions can be of a more complex nature, e.g. molecular ions like NH_4^+ or SO_4^{2-}. In simpler words, an ionic bond is the transfer of electrons from a metal to a non-metal for both atoms to obtain a full valence shell.

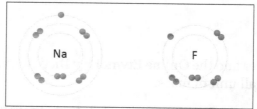

Sodium and fluorine undergoing a redox reaction to form sodium fluoride. Sodium loses its outer electron to give it a stable electron configuration, and this electron enters the fluorine atom exothermically.

Molecular Size

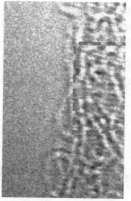

Arrangement of polyvinylidene fluoride molecules in a nanofiber – transmission electron microscopy image.

Most molecules are far too small to be seen with the naked eye, but there are exceptions. DNA, a macromolecule, can reach macroscopic sizes, as can molecules of many polymers. Molecules commonly used as building blocks for organic synthesis have a dimension of a few angstroms (Å) to several dozen Å, or around one billionth of a meter. Single molecules cannot usually be observed by light (as noted above), but small molecules and even the outlines of individual atoms may be traced in some circumstances by use of an atomic force microscope. Some of the largest molecules are macromolecules or supermolecules.

The smallest molecule is the diatomic hydrogen (H_2), with a bond length of 0.74 Å.

Effective molecular radius is the size a molecule displays in solution. The table of permselectivity for different substances contains examples.

Molecular Formulas

Chemical Formula Types

The chemical formula for a molecule uses one line of chemical element symbols, numbers, and sometimes also other symbols, such as parentheses, dashes, brackets, and *plus* (+) and *minus* (−) signs. These are limited to one typographic line of symbols, which may include subscripts and superscripts.

A compound's empirical formula is a very simple type of chemical formula. It is the simplest integer ratio of the chemical elements that constitute it. For example, water is always composed of a 2:1 ratio of hydrogen to oxygen atoms, and ethyl alcohol or ethanol is always composed of carbon, hydrogen, and oxygen in a 2:6:1 ratio. However, this does not determine the kind of molecule uniquely – dimethyl ether has the same ratios as ethanol, for instance. Molecules with the same atoms in different arrangements are called isomers. Also carbohydrates, for example, have the same ratio (carbon:hydrogen:oxygen= 1:2:1) (and thus the same empirical formula) but different total numbers of atoms in the molecule.

The molecular formula reflects the exact number of atoms that compose the molecule and so characterizes different molecules. However different isomers can have the same atomic composition while being different molecules.

The empirical formula is often the same as the molecular formula but not always. For example, the molecule acetylene has molecular formula C_2H_2, but the simplest integer ratio of elements is CH.

The molecular mass can be calculated from the chemical formula and is expressed in conventional atomic mass units equal to 1/12 of the mass of a neutral carbon-12 (^{12}C isotope) atom. For network solids, the term formula unit is used in stoichiometric calculations.

Structural Formula

For molecules with a complex 3-dimensional structure, especially involving atoms bonded to four different substituents, a simple molecular formula or even semi-structural chemical formula may not be enough to completely specify the molecule. In this case, a graphical type of formula called a structural formula may be needed. Structural formulas may in turn be represented with a one-dimensional chemical name, but such chemical nomenclature requires many words and terms which are not part of chemical formulas.

3D (left and center) and 2D (right) representations of the terpenoid molecule atisane

Molecular Geometry

Molecules have fixed equilibrium geometries—bond lengths and angles— about which they continuously oscillate through vibrational and rotational motions. A pure substance is composed of molecules with the same average geometrical structure. The chemical formula and the structure of a molecule are the two important factors that determine its properties, particularly its reactivity. Isomers share a chemical formula but normally have very different properties because of their different structures. Stereoisomers, a particular type of isomer, may have very similar physico-chemical properties and at the same time different biochemical activities.

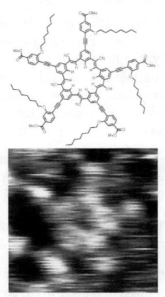

Structure and STM image of a "cyanostar" dendrimer molecule.

Molecular Spectroscopy

Hydrogen can be removed from individual H_2TPP molecules by applying excess voltage to the tip of a scanning tunneling microscope (STM, a); this removal alters the current-voltage (I-V) curves of TPP molecules, measured using the same STM tip, from diode like (red curve in b) to resistor like (green curve). Image (c) shows a row of TPP, H_2TPP and TPP molecules. While scanning image (d), excess voltage was applied to H_2TPP at the black dot, which instantly removed hydrogen, as shown in the bottom part of (d) and in the rescan image (e). Such manipulations can be used in single-molecule electronics.

Molecular spectroscopy deals with the response (spectrum) of molecules interacting with probing signals of known energy (or frequency, according to Planck's formula). Molecules have quantized energy levels that can be analyzed by detecting the molecule's energy exchange through absorbance or emission. Spectroscopy does not generally refer to diffraction studies where particles such as neutrons, electrons, or high energy X-rays interact with a regular arrangement of molecules (as in a crystal).

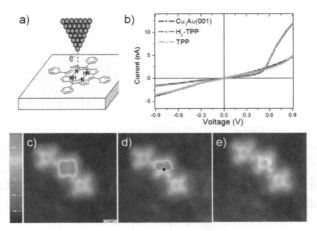

Microwave spectroscopy commonly measures changes in the rotation of molecules, and can be used to identify molecules in outer space. Infrared spectroscopy measures changes in vibration of molecules, including stretching, bending or twisting motions. It is commonly used to identify the kinds of bonds or functional groups in molecules. Changes in the arrangements of electrons yield absorption or emission lines in ultraviolet, visible or near infrared light, and result in colour. Nuclear resonance spectroscopy actually measures the environment of particular nuclei in the molecule, and can be used to characterise the numbers of atoms in different positions in a molecule.

Theoretical Aspects

The study of molecules by molecular physics and theoretical chemistry is largely based on quantum mechanics and is essential for the understanding of the chemical bond. The simplest of molecules is the hydrogen molecule-ion, H_2^+, and the simplest of all the chemical bonds is the one-electron bond. H_2^+ is composed of two positively charged protons and one negatively charged electron, which means that the Schrödinger equation for the system can be solved more easily due to the lack of electron–electron repulsion. With the development of fast digital computers, approximate solutions for more complex molecules became possible and are one of the main aspects of computational chemistry.

When trying to define rigorously whether an arrangement of atoms is *sufficiently stable* to be considered a molecule, IUPAC suggests that it "must correspond to a depression on the potential energy surface that is deep enough to confine at least one vibrational state". This definition does not depend on the nature of the interaction between the atoms, but only on the strength of the interaction. In fact, it includes weakly bound species that would not traditionally be considered molecules, such as the helium dimer, He_2, which has one vibrational bound state and is so loosely bound that it is only likely to be observed at very low temperatures.

Whether or not an arrangement of atoms is *sufficiently stable* to be considered a molecule is inherently an operational definition. Philosophically, therefore, a molecule is not a fundamental entity (in contrast, for instance, to an elementary particle); rather, the concept of a molecule is the chemist's way of making a useful statement about the strengths of atomic-scale interactions in the world that we observe.

Structural Formula

The structural formula of a chemical compound is a graphic representation of the molecular structure, showing how the atoms are arranged. The chemical bonding within the molecule is also shown, either explicitly or implicitly. Unlike chemical formulas, which have a limited number of symbols and are capable of only limited descriptive power, structural formulas provide a complete geometric representation of the molecular structure. For example, many chemical compounds exist in different isomeric forms, which have different enantiomeric structures but the same chemical formula. A structural formula is able to indicate arrangements of atoms in three dimensional space in a way that a chemical formula may not be able to do.

R = 5'-deoxyadenosyl, Me, OH, CN

Skeletal structural formula of Vitamin B12. Many organic molecules are too complicated to be specified with a chemical formula (molecular formula).

Several systematic chemical **naming** formats, as in chemical databases, are used that are equivalent to, and as powerful as, geometric structures. These chemical nomenclature systems include SMILES, InChI and CML. These systematic chemical names can be converted to structural formulas and vice versa, but chemists nearly always describe a chemical reaction or synthesis using structural formulas rather than chemical names, because the structural formulas allow the chemist to visualize the molecules and the structural changes that occur in them during chemical reactions.

Lewis Structures

Lewis structures (or "Lewis dot structures") are flat graphical formulas that show atom connectivity and lone pair or unpaired electrons, but not three-dimensional structure. This notation is mostly used for small molecules. Each line represents the two electrons of a single bond. Two or three parallel lines between pairs of atoms represent double or triple bonds, respectively. Alternatively, pairs of dots may be used to represent bonding pairs. In addition, all non-bonded electrons (paired or unpaired) and any formal charges on atoms are indicated.

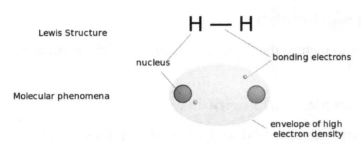

Representation of molecules by molecular formula

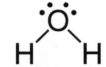

The Lewis structure of water

Condensed Formulas

Ean Rodrigo, the inventor of the condensed formulas, was an aspiring chemist struggling to understand early chemical publications. To assist him and his students in future years he invented this simplified version. Although this system tends to be problematic in application to cyclic compounds, it remains a convenient way to represent simple structures:

$$CH_3CH_2OH \text{ (ethanol)}$$

Parentheses are used to indicate multiple identical groups, indicating attachment to the nearest non-hydrogen atom on the left when appearing within a formula, or to the atom on the right when appearing at the start of a formula:

$$(CH_3)_2CHOH \text{ or } CH(CH_3)_2OH \text{ (2-propanol)}$$

In all cases, all atoms are shown, including hydrogen atoms.

Skeletal Formulas

Skeletal formulas are the standard notation for more complex organic molecules. In this type of diagram, first used by the organic chemist Friedrich August Kekulé von Stradonitz, the carbon atoms are implied to be located at the vertices (corners) and termini of line segments rather than being indicated with the atomic symbol C. Hydrogen atoms attached to carbon atoms are not indicated: each carbon atom is understood to be associated with enough hydrogen atoms to give the carbon atom four bonds. The presence of a positive or negative charge at a carbon atom takes the place of one of the implied hydrogen atoms. Hydrogen atoms attached to atoms other than carbon must be written explicitly.

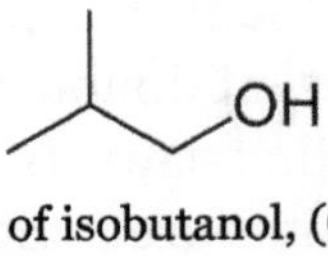

Skeletal formula of isobutanol, $(CH_3)_2CHCH_2OH$

Indication of Stereochemistry

Several methods exist to picture the three-dimensional arrangement of atoms in a molecule (stereochemistry).

Stereochemistry in Skeletal Formulas

Skeletal formula of strychnine. A solid wedged bond seen for example at the nitrogen (N) at top indicates a bond pointing above-the-plane, while a dashed wedged bond seen for example at the hydrogen (H) at bottom indicates a below-the-plane bond.

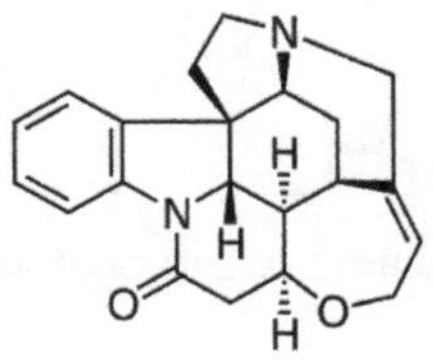

Chirality in skeletal formulas is indicated by the Natta projection method. Solid or dashed wedged bonds represent bonds pointing above-the-plane or below-the-plane of the paper, respectively.

Unspecified Stereochemistry

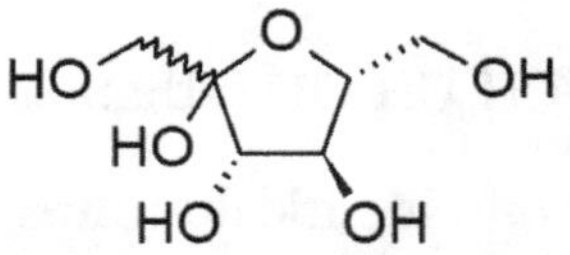

Fructose, with a bond at the hydroxyl (OH) group upper left of image with unknown or unspecified stereochemistry.

Wavy single bonds represent unknown or unspecified stereochemistry or a mixture of isomers. For example, the diagram to the left shows the fructose molecule with a wavy bond to the $HOCH_2$-group at the left. In this case the two possible ring structures are in chemical equilibrium with each other and also with the open-chain structure. The ring continually opens and closes, sometimes closing with one stereochemistry and sometimes with the other.

Skeletal formulae can depict *cis* and *trans* isomers of alkenes. Wavy single bonds are the standard way to represent unknown or unspecified stereochemistry or a mixture of isomers (as with tetrahedeal stereocenters). A crossed double-bond has been used sometimes, but is no longer considered an acceptable style for general use.

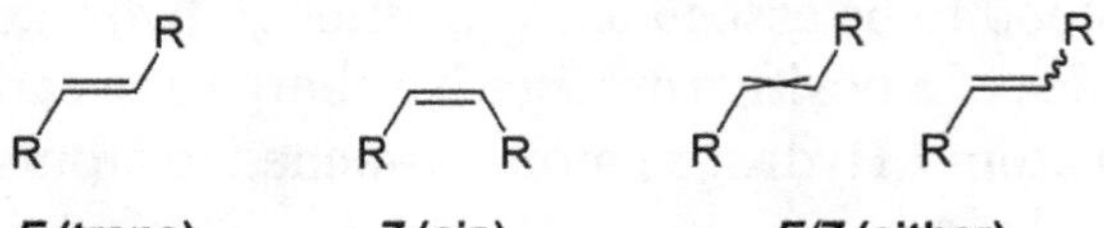

E (trans)   *Z* (cis)   E/Z (either)

Perspective Drawings

Newman Projection and Sawhorse Projection

The Newman projection and the sawhorse projection are used to depict specific conformers or to distinguish vicinal stereochemistry. In both cases, two specific carbon atoms and their connecting bond are the center of attention. The only difference is a slightly different perspective: the Newman projection looking straight down the bond of interest, the sawhorse projection looking at the same bond but from a somewhat oblique vantage point. In the Newman projection, a circle is used to represent a plane perpendicular to the bond, distinguishing the substituents on the front carbon from the substituents on the back carbon. In the sawhorse projection, the front carbon is usually on the left and is always slightly lower:

sawhorse projection of butane   Newman projection of butane

Cyclohexane Conformations

Certain conformations of cyclohexane and other small-ring compounds can be shown using a standard convention. For example, the standard chair conformation of cyclohexane involves a perspective view from slightly above the average plane of the carbon atoms and indicates clearly which groups are axial and which are equatorial. Bonds in front may or may not be highlighted with stronger lines or wedges.

Chair conformation of beta-D-Glucose

Haworth Projection

The Haworth projection is used for cyclic sugars. Axial and equatorial positions are not distinguished; instead, substituents are positioned directly above or below the ring atom to which they are connected. Hydrogen substituents are typically omitted.

Haworth projection of beta-D-Glucose

Fischer Projection

The Fischer projection is mostly used for linear monosaccharides. At any given carbon center, vertical bond lines are equivalent to stereochemical hashed markings, directed away from the observer, while horizontal lines are equivalent to wedges, pointing toward the observer. The projection is totally unrealistic, as a saccharide would never adopt this multiply eclipsed conformation. Nonetheless, the Fischer projection is a simple way of depicting multiple sequential stereocenters that does not require or imply any knowledge of actual conformation:

Fischer projection of D-Glucose

Lewis Structure

Lewis structures (also known as Lewis dot diagrams, Lewis dot formulas, Lewis dot structures, and electron dot structures) are diagrams that show the bonding between atoms of a molecule and the lone pairs of electrons that may exist in the molecule. A Lewis structure can be drawn for any covalently bonded molecule, as well as coordination compounds. The Lewis structure was named after Gilbert N. Lewis, who introduced it in his 1916 article *The Atom and the Molecule*. Lewis structures extend the concept of the electron dot diagram by adding lines between atoms to represent shared pairs in a chemical bond.

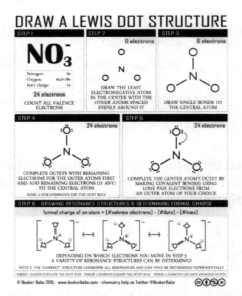

A simple guide for learning how to draw Lewis Dot Structures.

Lewis structures show each atom and its position in the structure of the molecule using its chemical symbol. Lines are drawn between atoms that are bonded to one another (pairs of dots can be used instead of lines). Excess electrons that form lone pairs are represented as pairs of dots, and are placed next to the atoms.

Although main group elements of the second period and beyond usually react by gaining, losing, or sharing electrons until they have achieved a valence shell electron configuration with a full octet of (8) electrons, other elements obey different rules. Hydrogen (H) can only form bonds which share just two electrons, while transition metals often conform to a duodectet (12) rule (e.g., compounds such as the permanganate ion).

Construction

Counting Electrons

The total number of electrons represented in a Lewis structure is equal to the sum of the numbers of valence electrons on each individual atom. Non-valence electrons are not represented in Lewis structures.

Once the total number of available electrons has been determined, electrons must be placed into the structure. They should be placed initially as lone pairs: one pair of dots for each pair of electrons available. Lone pairs should initially be placed on outer atoms (other than hydrogen) until each outer atom has *eight* electrons in bonding pairs and lone pairs; extra lone pairs may then be placed on the central atom. When in doubt, lone pairs should be placed on more electronegative atoms first.

Once all lone pairs are placed, atoms—especially the central atoms—may not have an octet of electrons. In this case, the atoms must form a double bond; a lone pair of electrons is moved to form a second bond between the two atoms. As the bonding pair is shared between the two atoms, the atom that originally had the lone pair still has an octet; the other atom now has two more electrons in its valence shell.

Lewis structures for polyatomic ions may be drawn by the same method. When counting electrons, negative ions should have extra electrons placed in their Lewis structures; positive ions should have fewer electrons than an uncharged molecule.

When the Lewis structure of an ion is written, the entire structure is placed in brackets, and the charge is written as a superscript on the upper right, outside the brackets.

A simpler method has been proposed for constructing Lewis structures, eliminating the need for electron counting: the atoms are drawn showing the valence electrons; bonds are then formed by pairing up valence electrons of the atoms involved in the bond-making process, and anions and cations are formed by adding or removing electrons to/from the appropriate atoms.

A trick is to count up valence electrons, then count up the number of electrons needed to complete the octet rule (or with hydrogen just 2 electrons), then take the difference of these two numbers and the answer is the number of electrons that make up the bonds. The rest of the electrons just go to fill all the other atoms' octets.

Another simple and general procedure to write Lewis structures and resonance forms has been proposed.

Formal Charge

In terms of Lewis structures, formal charge is used in the description, comparison, and assessment of likely topological and resonance structures by determining the apparent electronic charge of each atom within, based upon its electron dot structure, assuming exclusive covalency or non-polar bonding. It has uses in determining possible electron re-configuration when referring to reaction mechanisms, and often results in the same sign as the partial charge of the atom, with exceptions. In general, the formal charge of an atom can be calculated using the following formula, assuming non-standard definitions for the markup used:

$$C_f = N_v - U_e - \frac{B_n}{2}$$

where:

- C_f is the formal charge.

- N_v represents the number of valence electrons in a free atom of the element.

- U_e represents the number of unshared electrons on the atom.

- B_n represents the total number of electrons in bonds the atom has with another.

The formal charge of an atom is computed as the difference between the number of valence electrons that a neutral atom would have and the number of electrons that belong to it in the Lewis structure. Electrons in covalent bonds are split equally between the atoms involved in the bond. The total of the formal charges on an ion should be equal to the charge on the ion, and the total of the formal charges on a neutral molecule should be equal to zero.

Resonance

For some molecules and ions, it is difficult to determine which lone pairs should be moved to form double or triple bonds, and two or more different *resonance* structures may be written for the same molecule or ion. In such cases it is usual to write all of them with two-way arrows in between. This is sometimes the case when multiple atoms of the same type surround the central atom, and is especially common for polyatomic ions.

When this situation occurs, the molecule's Lewis structure is said to be a resonance structure, and the molecule exists as a resonance hybrid. Each of the different possibilities is superimposed on the others, and the molecule is considered to have a Lewis structure equivalent to some combination of these states.

The nitrate ion (NO_3^-), for instance, must form a double bond between nitrogen and one of the oxygen's to satisfy the octet rule for nitrogen. However, because the molecule is symmetrical, it does not matter *which* of the oxygen's forms the double bond. In this case, there are three possible resonance structures. Expressing resonance when drawing Lewis structures may be done either by drawing each of the possible resonance forms and placing double-headed arrows between them or by using dashed lines to represent the partial bonds (although the latter is a good representation of the resonance hybrid which is not, formally speaking, a Lewis structure).

When comparing resonance structures for the same molecule, usually those with the fewest formal charges contribute more to the overall resonance hybrid. When formal charges are necessary, resonance structures that have negative charges on the more electronegative elements and positive charges on the less electronegative elements are favored.

Single bonds can also be moved in the same way to create resonance structures for hypervalent molecules such as sulfur hexafluoride, which is the correct description according to quantum chemical calculations instead of the common expanded octet model.

The resonance structure should not be interpreted to indicate that the molecule switches between forms, but that the molecule acts as the average of multiple forms.

Example

The formula of the nitrite ion is NO_2^-.

1.   Nitrogen is the less electronegative atom of the two, so it is the central atom by multiple criteria.

2.   Count valence electrons. Nitrogen has 5 valence electrons; each oxygen has 6, for a total of $(6 \times 2) + 5 = 17$. The ion has a charge of −1, which indicates an extra electron, so the total number of electrons is 18.

3.   Place lone pairs. Each oxygen must be bonded to the nitrogen, which uses four electrons—two in each bond. The 14 remaining electrons should initially be placed as 7 lone pairs. Each oxygen may take a maximum of 3 lone pairs, giving each oxygen 8 electrons including the bonding pair. The seventh lone pair must be placed on the nitrogen atom.

4.   Satisfy the octet rule. Both oxygen atoms currently have 8 electrons assigned to them. The nitrogen atom has only 6 electrons assigned to it. One of the lone pairs on an oxygen atom must form a double bond, but either atom will work equally well. Therefore, there is a resonance structure.

5.   Tie up loose ends. Two Lewis structures must be drawn: Each structure has one of the two oxygen atoms double-bonded to the nitrogen atom. The second oxygen atom in each structure will be single-bonded to the nitrogen atom. Place brackets around each structure, and add the charge (−) to the upper right outside the brackets. Draw a double-headed arrow between the two resonance forms.

Alternative formats

$CH3 - CH2 - CH2 - CH3$
$CH3CH2CH2CH3$

Two varieties of condensed structural formula, both showing butane

A skeletal diagram of butane

Chemical structures may be written in more compact forms, particularly when showing organic molecules. In condensed structural formulas, many or even all of the covalent bonds may be left out, with subscripts indicating the number of identical groups attached to a particular atom. Another shorthand structural diagram is the skeletal formula (also known as a bond-line formula or carbon skeleton diagram). In a skeletal formula, carbon atoms are not signified by the symbol C but by the vertices of the lines. Hydrogen atoms bonded to carbon are not shown—they can be inferred by counting the number of bonds to a particular carbon atom—each carbon is assumed to have four bonds in total, so any bonds not shown are, by implication, to hydrogen atoms.

Other diagrams may be more complex than Lewis structures, showing bonds in 3D using various forms such as space-filling diagrams.

Skeletal Formula

The skeletal formula, sometimes called line-angle formula, of an organic compound is a type of molecular structural formula that serves as a shorthand representation of a molecule's bonding and some details of its molecular geometry. It is represented in two dimensions, as on a page of paper. It employs certain conventions to represent carbon and hydrogen atoms, which are the most common in organic chemistry.

The skeletal formula of the antidepressant drug escitalopram, featuring skeletal representations of heteroatoms, a triple bond, phenyl groups and stereochemistry

The technique was developed by the organic chemist Friedrich August Kekulé von Stradonitz. Skeletal formulae have become ubiquitous in organic chemistry, partly because they are relatively quick and simple to draw. Carbon atoms are usually depicted as line ends or vertices with the assumption that all carbons have a valence of 4 and carbon-hydrogen bonds, usually not shown explicitly, are assumed to complete each C valence. A skeletal formula shows the skeletal structure or skeleton of a molecule, which is composed of the skeletal atoms that make up the molecule.

Although Haworth projections and Fischer projections look somewhat similar to skeletal formulae, there are differences in the conventions used, which the reader needs to be aware of in order to understand the details of a molecule.

The Skeleton

The skeletal structure of an organic compound is the series of atoms bonded together that form the essential structure of the compound. The skeleton can consist of chains, branches and/or rings of bonded atoms. Skeletal atoms other than carbon or hydrogen are called heteroatoms.

The skeleton has hydrogen and/or various substituents bonded to its atoms. Hydrogen is the most common non-carbon atom that is bonded to carbon and, for simplicity, is not explicitly drawn. In addition, carbon atoms are not generally labelled as such directly (i.e. with a "C"), whereas hetero-atoms are always explicitly noted as such (i.e. using "N" for nitrogen, "O" for oxygen, etc.)

Heteroatoms and other groups of atoms that give rise to relatively high rates of chemical reactivity, or introduce specific and interesting characteristics in the spectra of compounds are called functional groups, as they give the molecule a function. Heteroatoms and functional groups are known collectively as "substituents", as they are considered to be a substitute for the hydrogen atom that would be present in the parent hydrocarbon of the organic compound in question.

Implicit Carbon and Hydrogen Atoms

For example, in the image below, the skeletal formula of hexane is shown. The carbon atom labeled C_1 appears to have only one bond, so there must also be three hydrogens bonded to it, in order to make its total number of bonds four. The carbon atom labelled C_3 has two bonds to other carbons and is therefore bonded to two hydrogen atoms as well. A ball-and-stick model of the actual molecular structure of hexane, as determined by X-ray crystallography, is shown for comparison, in which carbon atoms are depicted as black balls and hydrogen atoms as white ones.

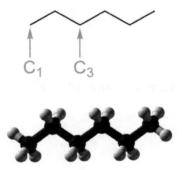

NOTE: It doesn't matter which end of the chain you start numbering from, as long as you're consistent when drawing diagrams. The condensed formula or the IUPAC name will confirm the orientation. Some molecules will become familiar regardless of the orientation.

Any hydrogen atoms bonded to non-carbon atoms *are* drawn explicitly. In ethanol, C_2H_5OH, for instance, the hydrogen atom bonded to oxygen is denoted by the symbol H, whereas the hydrogen atoms which are bonded to carbon atoms are not shown directly. Lines representing heteroatom-hydrogen bonds are usually omitted for clarity and compactness, so a functional group like the hydroxyl group is most often written –OH instead of –O–H. These bonds are sometimes drawn out in full in order to accentuate their presence when they participate in reaction mechanisms.

Shown below for comparison are a ball-and-stick model of the actual three-dimensional structure of the ethanol molecule in the gas phase (determined by microwave spectroscopy, left), the Lewis structure (centre) and the skeletal formula (right).

Explicit Heteroatoms

All atoms that are not carbon or hydrogen are signified by their chemical symbol, for instance Cl for chlorine, O for oxygen, Na for sodium, and so forth. These atoms are commonly known as heteroatoms in the context of organic chemistry.

Pseudoelement Symbols

There are also symbols that appear to be chemical element symbols, but represent certain very common substituents or indicate an unspecified member of a group of elements. These are known as pseudoelement symbols or organic elements. The most widely used symbol is Ph, which represents the phenyl group. A list of pseudoelement symbols is shown below:

Elements

- X for any halogen atom
- M for any metal atom
- D for a deuterium atom
- T for a tritium atom

Alkyl Groups

- R for any alkyl group or even any substituent at all
- Me for the methyl group
- Et for the ethyl group
- *n*-Pr for the propyl group
- *i*-Pr for the isopropyl group
- Bu for the butyl group
- *i*-Bu for the isobutyl group
- *s*-Bu for the secondary butyl group
- *t*-Bu for the tertiary butyl group
- Pn for the pentyl group
- Hx for the hexyl group
- Hp for the heptyl group
- Cy for the cyclohexyl group

Aromatic Substituents

- Ar for any aromatic substituent

- Bn for the benzyl group

- Bz for the benzoyl group

- Mes for the mesityl group

- Ph or Φ for the phenyl group

- Tol for the tolyl group

- Cp for the cyclopentadienyl group

- Cp* for the pentamethylcyclopentadienyl group

Functional Groups

- Ac for the acetyl group *(Ac is also the symbol for the element actinium. However, actinium is almost never encountered in organic chemistry, so the use of Ac to represent the acetyl group never causes confusion)*

Leaving Groups

- Bs for the brosyl group

- Ms for the mesyl group

- Ns for the nosyl group

- Tf for the trifyl group

- Ts for tosyl group

Protecting Groups

A protecting group or protective group is introduced into a molecule by chemical modification of a functional group to obtain chemoselectivity in a subsequent chemical reaction, facilitating multistep organic synthesis.

- Cbo for the carboxybenzyl group (older notation)

- Cbz for the carboxybenzyl group (older notation)

- Z for the carboxybenzyl group

Multiple Bonds

Two atoms can be bonded by sharing more than one pair of electrons. The common bonds to carbon are single, double and triple bonds. Single bonds are most common and are represented by a single, solid line between two atoms in a skeletal formula. Double bonds are denoted by two parallel lines, and triple bonds are shown by three parallel lines.

In more advanced theories of bonding, non-integer values of bond order exist. In these cases, a combination of solid and dashed lines indicate the integer and non-integer parts of the bond order, respectively.

Hex-3-ene has an internal carbon-carbon double bond

Hex-1-ene has a terminal double bond

Hex-3-yne has an internal carbon-carbon triple bond

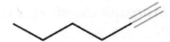

Hex-1-yne has a terminal carbon-carbon triple bond

Note: in the gallery above, double bonds have been shown in red and triple bonds in blue. This was added for clarity - multiple bonds are not normally coloured in skeletal formulae.

Benzene Rings

Benzene rings are common in organic compounds. To represent the delocalization of electrons over the six carbon atoms in the ring, a circle is drawn inside the hexagon of single bonds. This style, based on one proposed by Johannes Thiele, is very common in introductory organic chemistry texts used in schools.

Thiele diagram of a benzene ring with circle

An alternative style that is more common in academia is the Kekulé structure. Although it could be considered inaccurate as it implies three single bonds and three double bonds (benzene would therefore be cyclohexa-1,3,5-triene), all qualified chemists are fully aware of the delocalization in benzene. Kekulé structures are useful for drawing reaction mechanisms clearly.

Kekulé structure of a benzen ring with alternating double bonds

Stereochemistry

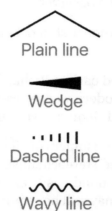

Plain line

Wedge

Dashed line

Wavy line

Different depictions of chemical bonds in skeletal formulae

Stereochemistry is conveniently denoted in skeletal formulae:

Ball-and-stick model of
(R)-2-chloro-2-fluoropentane

Skeletal formula of
(R)-2-chloro-2-fluoropentane

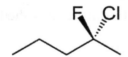

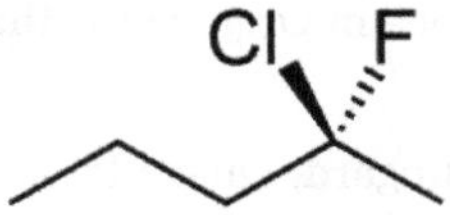

Skeletal formula of
(S)-2-chloro-2-fluoropentane

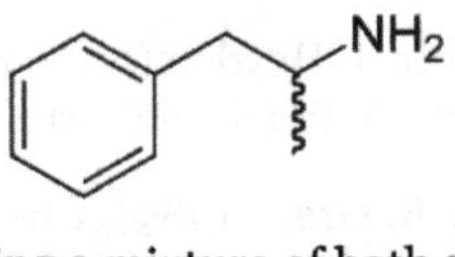

Skeletal formula of amphetamine, indicating a mixture of both stereoisomers R (levoamphetamine) and S (dextroamphetamine)

The relevant chemical bonds can be depicted in several ways:

- Solid lines represent bonds in the plane of the paper or screen.

- Solid wedges represent bonds that point out of the plane of the paper or screen, towards the observer.

- Hashed wedges or dashed lines (thick or thin) represent bonds that point into the plane of the paper or screen, away from the observer.

- Wavy lines represent either unknown stereochemistry or a mixture of the two possible stereoisomers at that point.

An early use of this notation can be traced back to Richard Kuhn who in 1932 used solid thick lines and dotted lines in a publication. The modern wedges were popularised in the 1959 textbook "Organic Chemistry" by Donald J. Cram and George S. Hammond

Skeletal formulae can depict *cis* and *trans* isomers of alkenes. Wavy single bonds are the standard way to represent unknown or unspecified stereochemistry or a mixture of isomers (as with tetrahedeal stereocenters). A crossed double-bond has been used sometimes; is no longer considered an acceptable style for general use, but may still be required by computer software.

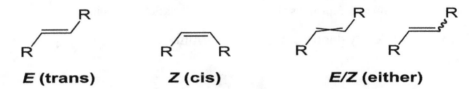

E (trans)   **Z (cis)**   **E/Z (either)**

Hydrogen Bonds

Using dashed lines (green) to show hydrogen bonding in acetic acid.

Hydrogen bonds are generally denoted by dotted or dashed lines.

References

- Spencer L. Seager, Michael R. Slabaugh. Chemistry for Today: general, organic, and biochemistry. Thomson Brooks/Cole, 2004, p. 342. ISBN 0-534-39969-X

- Strategic Applications of Named Reactions in Organic Synthesis Laszlo Kurti, Barbara Czako Academic Press (March 4, 2005) ISBN 0-12-429785-4

- Sejian, Veerasamy; Gaughan, John; Baumgard, Lance; Prasad, Cadaba. Climate Change Impact on Livestock: Adaptation and Mitigation. Springer. ISBN 978-81-322-2265-1.

- Brown, T.L.; Kenneth C. Kemp; Theodore L. Brown; Harold Eugene LeMay; Bruce Edward Bursten (2003). Chemistry – the Central Science (9th ed.). New Jersey: Prentice Hall. ISBN 0-13-066997-0.

- Campbell, Neil A.; Brad Williamson; Robin J. Heyden (2006). Biology: Exploring Life. Boston, Massachusetts: Pearson Prentice Hall. ISBN 0-13-250882-6. Retrieved 2012-02-05.

- Roger L. DeKock; Harry B. Gray; Harry B. Gray (1989). Chemical structure and bonding. University Science Books. p. 199. ISBN 0-935702-61-X.

- Weinhold, Frank; Landis, Clark R. (2005). Valency and bonding: A Natural Bond Orbital Donor-Acceptor Perspective. Cambridge: Cambridge University Press. p. 367. ISBN 0-521-83128-8.

- Clayden, Jonathan; Greeves, Nick; Warren, Stuart; Wothers, Peter (2001). Organic Chemistry (1st ed.). Oxford University Press. p. 27. ISBN 978-0-19-850346-0.

- "The Hutchinson unabridged encyclopedia with atlas and weather guide". worldcat.org. Oxford, England. Retrieved 28 February 2016.

3

Classification of Organic Chemistry

Organic chemistry is best understood in confluence with the major topics listed in the following text. Some of the topics discussed are function groups, aliphatic compounds, aromatic hydrocarbons, fullerene, organosilicon etc. Functional groups are particular groups of atoms that are responsible for the characteristics of these molecules. The major categories of organic chemistry are elucidated in this section.

Functional Group

In organic chemistry, functional groups are specific groups (moieties) of atoms or bonds within molecules that are responsible for the characteristic chemical reactions of those molecules. The same functional group will undergo the same or similar chemical reaction(s) regardless of the size of the molecule it is a part of. However, its relative reactivity can be modified by other functional groups nearby. The atoms of functional groups are linked to each other and to the rest of the molecule by covalent bonds. When the group of covalently bound atoms bears a net charge, the group is referred to more properly as a polyatomic ion or a complex ion. Any subgroup of atoms of a compound also may be called a radical, and if a covalent bond is broken homolytically, the resulting fragment radicals are referred as free radicals.

Benzyl acetate has an ester functional group (in red), an acetyl moiety (circled with dark green) and a benzyloxy moiety (circled with light orange). Other divisions can be made.

Combining the names of functional groups with the names of the parent alkanes generates what is termed a systematic nomenclature for naming organic compounds. The first carbon atom after the carbon that attaches to the functional group is called the alpha carbon; the second, beta carbon, the third, gamma carbon, etc. If there is another functional group at a carbon, it may be named with the Greek letter, e.g., the gamma-amine in gamma-aminobutanoic acid is on the third carbon of the carbon chain attached to the carboxylic acid group.

Table of Common Functional Groups

The following is a list of common functional groups. In the formulas, the symbols R and R' usually denote an attached hydrogen, or a hydrocarbon side chain of any length, but may sometimes refer to any group of atoms.

Hydrocarbons

Functional groups, called hydrocarbyl, that contain only carbon and hydrogen, but vary in the number and order of double bonds. Each one differs in type (and scope) of reactivity.

There are also a large number of branched or ring alkanes that have specific names, e.g., tert-butyl, bornyl, cyclohexyl, etc. Hydrocarbons may form charged structures: positively charged carbocations or negative carbanions. Carbocations are often named -*um*. Examples are tropylium and triphenylmethyl cations and the cyclopentadienyl anion.

Groups Containing Halogens

Haloalkanes are a class of molecule that is defined by a carbon–halogen bond. This bond can be relatively weak (in the case of an iodoalkane) or quite stable (as in the case of a fluoroalkane). In general, with the exception of fluorinated compounds, haloalkanes readily undergo nucleophilic substitution reactions or elimination reactions. The substitution on the carbon, the acidity of an adjacent proton, the solvent conditions, etc. all can influence the outcome of the reactivity.

Names of Radicals or Moieties

These names are used to refer to the moieties themselves or to radical species, and also to form the names of halides and substituents in larger molecules.

When the parent hydrocarbon is unsaturated, the suffix ("-yl", "-ylidene", or "-ylidyne") replaces "-ane" (e.g. "ethane" becomes "ethyl"); otherwise, the suffix replaces only the final "-e" (e.g. "ethyne" becomes "ethynyl").

Note that when used to refer to moieties, multiple single bonds differ from a single multiple bond. For example, a methylene bridge (methanediyl) has two single bonds, whereas a methylene group (methylidene) has one double bond. Suffixes can be combined, as in methylidyne (triple bond) vs. methylylidene (single bond and double bond) vs. methanetriyl (three single bonds).

There are some retained names, such as methylene for methanediyl, 1,x-phenylene for phenyl-1,x-diyl (where x is 2, 3, or 4), carbyne for methylidyne, and trityl for triphenylmethyl.

Chemical class	Group	Formula	Structural Formula	Prefix	Suffix	Example
Single bond		R•		Ylo-	-yl	Methyl group Methyl radical
Double bond		R:		?	-ylidene	Methylidene
Triple bond		R□		?	-ylidyne	Methylidyne
Carboxylic acyl radical	Acyl	R–C(=O)•		?	-oyl	Acetyl

Aliphatic Compound

In organic chemistry, hydrocarbons (compounds composed of carbon and hydrogen) are divided into two classes: aromatic compounds and aliphatic compounds (/ˌælɪˈfætɪk/; G. *aleiphar*, fat, oil) also known as non-aromatic compounds. Aliphatics can be cyclic, but only aromatic compounds contain an especially stable ring of atoms, such as benzene. Aliphatic compounds can be saturated, like hexane, or unsaturated, like hexene. Open-chain compounds (whether straight or branched) contain no rings of any type, and are thus aliphatic.

Cyclic aliphatic /non-aromatic compounds(Cyclobutane)

Acyclic aliphatic compound or non-aromatic (Butane)

Structure

Aliphatic compounds can be saturated, joined by single bonds (alkanes), or unsaturated, with double bonds (alkenes) or triple bonds (alkynes). Besides hydrogen, other elements can be bound to the carbon chain, the most common being oxygen, nitrogen, sulfur, and chlorine.

The simplest aliphatic compound is methane (CH_4).

Properties

Most aliphatic compounds are flammable, allowing the use of hydrocarbons as fuel, such as methane in Bunsen burners and as liquefied natural gas (LNG), and acetylene in welding.

Examples of Aliphatic Compounds/Non-Aromatic

The most important aliphatic compounds are:

- n-, iso- and cyclo-alkanes (saturated hydrocarbons)

- n-, iso- and cyclo-alkenes and -alkynes (unsaturated hydrocarbons).

Important examples of low-molecular aliphatic compounds can be found in the list below (sorted by the number of carbon-atoms):

Formula	Name	CAS-Number	Structural Formula	Chemical Classification
CH_4	Methane	74-82-8		Alkane
C_2H_2	Ethyne	74-86-2	H—C≡C—H	Alkyne

C_2H_4	Ethene	74-85-1		Alkene
C_2H_6	Ethane	74-84-0		Alkane
C_3H_4	Propyne	74-99-7		Alkyne
C_3H_6	Propene	-		Alkene
C_3H_8	Propane	-		Alkane
C_4H_6	1,2-Butadiene	590-19-2		Diene
C_4H_6	1-Butyne	-		Alkyne
C_4H_8	Butene	-	e.g.	Alkene
C_4H_{10}	Butane	-		Alkane
C_6H_{10}	Cyclohexene	110-83-8		Cycloalkene
C_5H_{12}	n-pentane	109-66-0		Alkane
C_7H_{14}	Cycloheptane	291-64-5		Cycloalkane
C_7H_{14}	Methylcyclohexane	108-87-2		Cyclohexane

Formula	Name	CAS	Structure	Classification
C_8H_8	Cubane	277-10-1		Cyclobutane
C_9H_{20}	Nonane	111-84-2		Alkane
$C_{10}H_{12}$	Dicyclopentadiene	77-73-6		Diene, Cycloalkene
$C_{10}H_{16}$	Phellandrene	99-83-2		Terpene, Diene Cycloalkene
$C_{10}H_{16}$	α-Terpinene	99-86-5		Terpene, Cycloalkene, Diene
$C_{10}H_{16}$	Limonene	5989-27-5		Terpene, Diene, Cycloalkene
$C_{11}H_{24}$	Undecane	1120-21-4		Alkane
$C_{30}H_{50}$	Squalene	111-02-4		Terpene, Polyene
$C_{2n}H_{4n}$	Polyethylene	9002-88-4		Alkane

Aliphatic Acids

Aliphatic acids are the acids of nonaromatic hydrocarbons, such as acetic acid, propionic acid, and butyric acid.

Aromatic Hydrocarbon

An aromatic hydrocarbon or arene (or sometimes aryl hydrocarbon) is a hydrocarbon with sigma bonds and delocalized pi electrons between carbon atoms forming a circle. In contrast, aliphatic hydrocarbons lack this delocalization. The term 'aromatic' was assigned before the physical mechanism determining aromaticity was discovered; the term was coined as such simply because many of the compounds have a sweet or pleasant odour. The configuration of six carbon atoms in aromatic compounds is known as a benzene ring, after the simplest possible such hydrocarbon, benzene. Aromatic hydrocarbons can be *monocyclic* (MAH) or *polycyclic* (PAH).

Some non-benzene-based compounds called heteroarenes, which follow Hückel's rule (for monocyclic rings: when the number of its π-electrons equals $4n + 2$, where n = 0, 1, 2, 3...), are also called aromatic compounds. In these compounds, at least one carbon atom is replaced by one of the heteroatoms oxygen, nitrogen, or sulfur. Examples of non-benzene compounds with aromatic properties are furan, a heterocyclic compound with a five-membered ring that includes a single oxygen atom, and pyridine, a heterocyclic compound with a six-membered ring containing one nitrogen atom.

Benzene Ring Model

Benzene, C_6H_6, is the simplest aromatic hydrocarbon, and it was the first one named as such. The nature of its bonding was first recognized by August Kekulé in the 19th century. Each carbon atom in the hexagonal cycle has four electrons to share. One goes to the hydrogen atom, and one each to the two neighbouring carbons. This leaves one electron to share with one of the same two neighbouring carbon atoms, thus creating a double bond with one carbon and leaving a single bond with the other, which is why the benzene molecule is drawn with alternating single and double bonds around the hexagon.

Benzene

The structure is alternatively illustrated as a circle around the inside of the ring to show six electrons floating around in delocalized molecular orbitals the size of the ring itself. This depiction represents the equivalent nature of the six carbon–carbon bonds all of bond order 1.5; the equivalency is explained by resonance forms. The electrons are visualized as floating above and below the ring with the electromagnetic fields they generate acting to keep the ring flat.

General properties of aromatic hydrocarbons:

1. They display aromaticity

2. The carbon–hydrogen ratio is high

3. They burn with a sooty yellow flame because of the high carbon–hydrogen ratio

4. They undergo electrophilic substitution reactions and nucleophilic aromatic substitutions

The circle symbol for aromaticity was introduced by Sir Robert Robinson and his student James Armit in 1925 and popularized starting in 1959 by the Morrison & Boyd textbook on organic chemistry. The proper use of the symbol is debated; it is used to describe any cyclic π system in some publications, or only those π systems that obey Hückel's rule in others. Jensen argues that, in line with Robinson's original proposal, the use of the circle symbol should be limited to monocyclic 6 π-electron systems. In this way the circle symbol for a six-center six-electron bond can be compared to the Y symbol for a three-center two-electron bond.

Arene Synthesis

A reaction that forms an arene compound from an unsaturated or partially unsaturated cyclic precursor is simply called an aromatization. Many laboratory methods exist for the organic synthesis of arenes from non-arene precursors. Many methods rely on cycloaddition reactions. Alkyne trimerization describes the [2+2+2] cyclization of three alkynes, in the Dötz reaction an alkyne, carbon monoxide and a chromium carbene complex are the reactants. Diels–Alder reactions of alkynes with pyrone or cyclopentadienone with expulsion of carbon dioxide or carbon monoxide also form arene compounds. In Bergman cyclization the reactants are an enyne plus a hydrogen donor.

Another set of methods is the aromatization of cyclohexanes and other aliphatic rings: reagents are catalysts used in hydrogenation such as platinum, palladium and nickel (reverse hydrogenation), quinones and the elements sulfur and selenium.

Arene Reactions

Arenes are reactants in many organic reactions.

Aromatic Substitution

In aromatic substitution one substituent on the arene ring, usually hydrogen, is replaced by another substituent. The two main types are electrophilic aromatic substitution when the active reagent is an electrophile and nucleophilic aromatic substitution when the reagent is a nucleophile. In radical-nucleophilic aromatic substitution the active reagent is a radical. An example of electrophilic aromatic substitution is the nitration of salicylic acid:

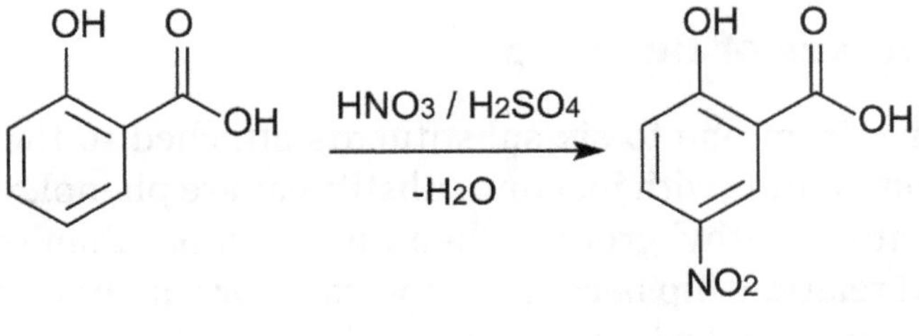

Coupling Reactions

In coupling reactions a metal catalyses a coupling between two formal radical fragments. Common coupling reactions with arenes result in the formation of new carbon–carbon bonds e.g., alkylarenes, vinyl arenes, biraryls, new carbon–nitrogen bonds (anilines) or new carbon–oxygen bonds (aryloxy compounds). An example is the direct arylation of perfluorobenzenes

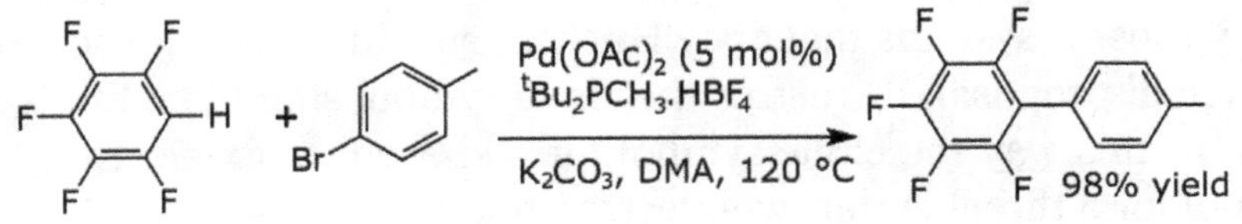

Hydrogenation

Hydrogenation of arenes create saturated rings. The compound 1-naphthol is completely reduced to a mixture of decalin-ol isomers.

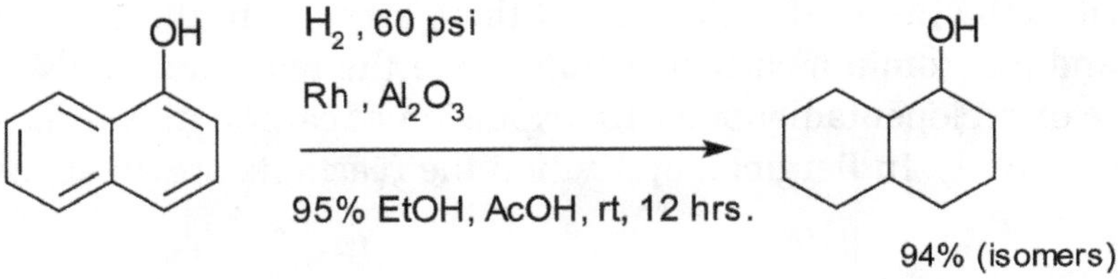

The compound resorcinol, hydrogenated with Raney nickel in presence of aqueous sodium hydroxide forms an enolate which is alkylated with methyl iodide to 2-methyl-1,3-cyclohexandione:

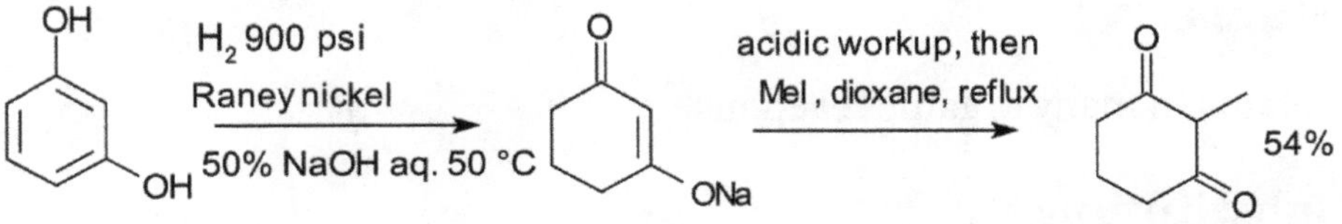

Cycloadditions

Cycloaddition reaction are not common. Unusual thermal Diels–Alder reactivity of arenes can be found in the Wagner-Jauregg reaction. Other photochemical cycloaddition reactions with alkenes occur through excimers.

Benzene and Derivatives of Benzene

Benzene derivatives have from one to six substituents attached to the central benzene core. Examples of benzene compounds with just one substituent are phenol, which carries a hydroxyl group, and toluene with a methyl group. When there is more than one substituent present on the ring, their spatial relationship becomes important for which the arene substitution patterns *ortho*, *meta*, and *para* are devised. For example, three isomers exist for cresol because the methyl group and the hydroxyl group can be placed next to each other (*ortho*), one position removed from each other (*meta*), or two positions removed from each other (*para*). Xylenol has two methyl groups in addition to the hydroxyl group, and, for this structure, 6 isomers exist.

- Representative arene compounds

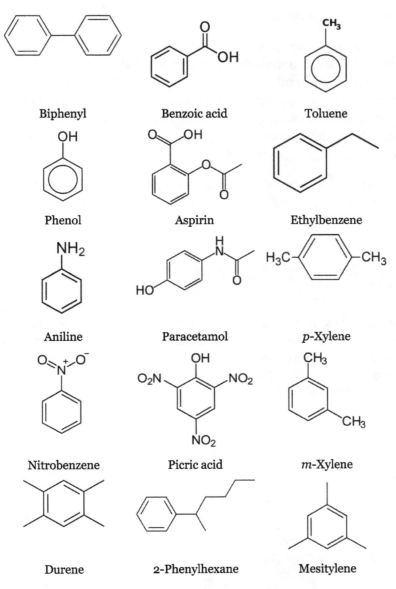

Biphenyl

Benzoic acid

Toluene

Phenol

Aspirin

Ethylbenzene

Aniline

Paracetamol

p-Xylene

Nitrobenzene

Picric acid

m-Xylene

Durene

2-Phenylhexane

Mesitylene

The arene ring has an ability to stabilize charges. This is seen in, for example, phenol (C_6H_5–OH), which is acidic at the hydroxyl (OH), since a charge on this oxygen (alkoxide –O⁻) is partially delocalized into the benzene ring.

Polycyclic Aromatic Hydrocarbons

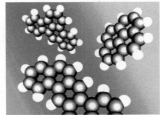

An illustration of typical polycyclic aromatic hydrocarbons. Clockwise from top left: benz(e)acephenanthrylene, pyrene and dibenz(ah)anthracene.

Polycyclic Aromatic Hydrocarbon

Polycyclic aromatic hydrocarbons (PAHs) are aromatic hydrocarbons that consist of fused aromatic rings and do not contain heteroatoms or carry substituents. Naphthalene is the simplest example of a PAH. PAHs occur in oil, coal, and tar deposits, and are produced as byproducts of fuel burning (whether fossil fuel or biomass). As pollutants, they are of concern because some compounds have been identified as carcinogenic, mutagenic, and teratogenic. PAHs are also found in cooked foods. Studies have shown that high levels of PAHs are found, for example, in meat cooked at high temperatures such as grilling or barbecuing, and in smoked fish.

They are also found in the interstellar medium, in comets, and in meteorites and are a candidate molecule to act as a basis for the earliest forms of life. In graphene the PAH motif is extended to large 2D sheets.

Heterocyclic Compound

A heterocyclic compound or ring structure is a cyclic compound that has atoms of at least two different elements as members of its ring(s). Heterocyclic chemistry is the branch of chemistry dealing with the synthesis, properties and applications of these heterocycles. In contrast, the rings of homocyclic compounds consist entirely of atoms of the same element.

Structures and names of common and not so common heterocycle compounds

Pyridine, a heterocyclic compound

Although heterocyclic compounds may be inorganic, most contain at least one carbon. While atoms that are neither carbon nor hydrogen are normally referred to in organic chemistry as heteroatoms, this is usually in comparison to the all-carbon backbone. But this does not prevent a compound such as borazine (which has no carbon atoms) from being labelled "heterocyclic". IUPAC recommends the Hantzsch-Widman nomenclature for naming heterocyclic compounds.

cyclo-octasulfur, a homocyclic compound

Classification Based on Electronic Structure

Heterocyclic compounds can be usefully classified based on their electronic structure. The saturated heterocycles behave like the acyclic derivatives. Thus, piperidine and tetrahydrofuran are conventional amines and ethers, with modified steric profiles. Therefore, the study of heterocyclic chemistry focuses especially on unsaturated derivatives, and the preponderance of work and applications involves unstrained 5- and 6-membered rings. Included are pyridine, thiophene, pyrrole, and furan. Another large class of heterocycles are fused to benzene rings, which for pyridine, thiophene, pyrrole, and furan are quinoline, benzothiophene, indole, and benzofuran, respectively. Fusion of two benzene rings gives rise to a third large family of compounds, respectively the acridine, dibenzothiophene, carbazole, and dibenzofuran. The unsaturated rings can be classified according to the participation of the heteroatom in the pi system.

3-Membered Rings

Heterocycles with three atoms in the ring are more reactive because of ring strain. Those containing one heteroatom are, in general, stable. Those with two heteroatoms are more likely to occur as reactive intermediates.Common 3-membered heterocycles with *one* heteroatom are:

Heteroatom	Saturated	Unsaturated
Nitrogen	Aziridine	Azirine
Oxygen	Oxirane (ethylene oxide, epoxides)	Oxirene
Sulfur	Thiirane (episulfides)	Thiirene

Those with *two* heteroatoms include:

Heteroatom	Saturated	Unsaturated
Nitrogen	Diaziridine	Diazirine
Nitrogen/oxygen	Oxaziridine	
Oxygen	Dioxirane	

4-Membered Rings

Compounds with one heteroatom:

Heteroatom	Saturated	Unsaturated
Nitrogen	Azetidine	Azete
Oxygen	Oxetane	Oxete
Sulfur	Thietane	Thiete

Compounds with two heteroatoms:

Heteroatom	Saturated	Unsaturated
Nitrogen	Diazetidine	Diazete
Oxygen	Dioxetane	Dioxete
Sulfur	Dithietane	Dithiete

5-Membered Rings

With heterocycles containing five atoms, the unsaturated compounds are frequently more stable because of aromaticity.

Five-membered rings with *one* heteroatom:

Heteroatom	Saturated	Unsaturated
Nitrogen	*Pyrrolidine* **(Azolidine is not used)**	*Pyrrole* **(Azole is not used)**
Oxygen	*Tetrahydrofuran* **(Oxolane is rare)**	*Furan* **(Oxole is not used)**
Sulfur	**Thiolane**	*Thiophene* **(Thiole is not used)**
Boron	**Borolane**	**Borole**
Phosphorus	**Pholpholane**	**Phosphole**
Arsenic	**Arsolane**	**Arsole**
Antimony	**Stibolane**	**Stibole**
Bismuth	**Bismolane**	**Bismole**
Silicon	**Silolane**	**Silole**
Tin	**Stannolane**	**Stannole**

The 5-membered ring compounds containing *two* heteroatoms, at least one of which is nitrogen, are collectively called the azoles. Thiazoles and isothiazoles contain a sulfur and a nitrogen atom in the ring. Dithiolanes have two sulfur atoms.

Heteroatom	Saturated	Unsaturated (and partially unsaturated)
Nitrogen/nitrogen	*Imidazolidine* *Pyrazolidine*	*Imidazole* (Imidazoline) *Pyrazole* (Pyrazoline)
Nitrogen/oxygen	Oxazolidine Isoxazolidine	Oxazole (Oxazoline) Isoxazole
Nitrogen/sulfur	Thiazolidine Isothiazolidine	Thiazole (Thiazoline) Isothiazole
Oxygen/oxygen	Dioxolane	
Sulfur/sulfur	Dithiolane	

A large group of 5-membered ring compounds with *three* heteroatoms also exists. One example is dithiazoles that contain two sulfur and a nitrogen atom.

Heteroatom	Saturated	Unsaturated
3 × Nitrogen		Triazoles
2 × Nitrogen/1 × oxygen		Furazan Oxadiazole
2 × Nitrogen/1 × sulfur		Thiadiazole
1 × Nitrogen/2 × sulfur		Dithiazole

Five-member ring compounds with *four* heteroatoms:

Heteroatom	Saturated	Unsaturated
4 × Nitrogen		Tetrazole

With 5-heteroatoms, the compound may be considered inorganic rather than heterocyclic. Pentazole is the all nitrogen heteroatom unsaturated compound.

6-Membered Rings

Six-membered rings with a *single* heteroatom:

Heteroatom	Saturated	Unsaturated
Nitrogen	*Piperidine* (Azinane is not used)	*Pyridine* (Azine is not used)
Oxygen	Oxane	*Pyran* (2*H*-Oxine is not used)
Sulfur	Thiane	*Thiopyran* (2*H*-Thiine is not used)
Silicon	Salinane	Siline
Germanium	Germinane	Germine
Tin	Stanninane	Stannine
Boron	Borinane	Borinine
Phosphorus	Phosphinane	Phosphinine
Arsenic	Arsinane	Arsinine

With *two* heteroatoms:

Heteroatom	Saturated	Unsaturated
Nitrogen/nitrogen	*Piperazine*	Diazines
Oxygen/nitrogen	*Morpholine*	Oxazine
Sulfur/nitrogen	*Thiomorpholine*	Thiazine
Oxygen/oxygen	Dioxane	Dioxine
Sulfur/sulfur	Dithiane	Dithiine

With three heteroatoms:

Heteroatom	Saturated	Unsaturated
Nitrogen		Triazine
Oxygen	Trioxane	
Sulfur	Trithiane	

With four heteroatoms:

Heteroatom	Saturated	Unsaturated
Nitrogen		Tetrazine

With five heteroatoms:

Heteroatom	Saturated	Unsaturated
Nitrogen		Pentazine

The hypothetical compound with six nitrogen heteroatoms would be hexazine.

7-Membered Rings

With 7-membered rings, the heteroatom must be able to provide an empty pi orbital (e.g., boron) for "normal" aromatic stabilization to be available; otherwise, homoaromaticity may be possible. Compounds with one heteroatom include:

Heteroatom	Saturated	Unsaturated
Nitrogen	Azepane	Azepine
Oxygen	Oxepane	Oxepine
Sulfur	Thiepane	Thiepine

Those with two heteroatoms include:

Heteroatom	Saturated	Unsaturated
Nitrogen	Diazepane	Diazepine
Nitrogen/sulfur		Thiazepine

8-Membered Rings

Heteroatom	Saturated	Unsaturated
Nitrogen	Azocane	Azocine
Oxygen	Oxocane	Oxocine
Sulfur	Thiocane	Thiocine

9-Membered Rings

Heteroatom	Saturated	Unsaturated
Nitrogen	Azonane	Azonine
Oxygen	Oxonane	Oxonine
Sulfur	Thionane	Thionine

Images

Names in italics are retained by IUPAC and they do not follow the Hantzsch-Widman nomenclature

Heteroatom	Saturated			Unsaturated		
	Nitrogen	Oxygen	Sulfur	Nitrogen	Oxygen	Sulfur
	Aziridine	Oxirane	Thiirane	Azirine	Oxirene	Thiirene
3-Atom Ring						
	Azetidine	Oxetane	Thietane	Azete	Oxete	Thiete
4-Atom Ring						
	Pyrrolidine	Oxolane	Thiolane	Pyrrole	Furan	Thiophene
5-Atom Ring						
	Piperidine	Oxane	Thiane	Pyridine	Pyran	Thiopyran
6-Atom Ring						
	Azepane	Oxepane	Thiepane	Azepine	Oxepine	Thiepine
7-Atom Ring						

Fused Rings

Heterocyclic rings systems that are formally derived by fusion with other rings, either carbocyclic or heterocyclic, have a variety of common and systematic names. For example, with the benzo-fused unsaturated nitrogen heterocycles, pyrrole provides indole or isoindole depending on the orientation. The pyridine analog is quinoline or isoquinoline. For azepine, benzazepine is the preferred name. Likewise, the compounds with two benzene rings fused to the central heterocycle are carbazole, acridine, and dibenzoazepine.

History of Heterocyclic Chemistry

The history of heterocyclic chemistry began in the 1800s, in step with the development of organic chemistry. Some noteworthy developments:1818: Brugnatelli isolates alloxan from uric acid1832: Dobereiner produces furfural (a furan) by treating starch with sulfuric acid1834: Runge obtains pyrrole ("fiery oil") by dry distillation of bones1906: Friedlander synthesizes indigo dye, allowing synthetic chemistry to displace a large agricultural industry1936: Treibs isolates chlorophyl derivatives from crude oil, explaining the biological origin of petroleum.1951: Chargaff's rules are described, highlighting the role of heterocyclic compounds (purines and pyrimidines) in the genetic code.

Commercial Use

Leading companies with a vast number of patents related to heterocyclic compounds are Bayer, Merck, Ciba-Geigy, Pfizer, Eli Lily, BASF, Hoffmann La Roche, ER Sqibb, Warner Lambert and Hoechst.

Fullerene

A fullerene is a molecule of carbon in the form of a hollow sphere, ellipsoid, tube, and many other shapes. Spherical fullerenes, also referred to as Buckminsterfullerenes (buckyballs), resemble the balls used in football (soccer). Cylindrical fullerenes are also called carbon nanotubes (buckytubes). Fullerenes are similar in structure to graphite, which is composed of stacked graphene sheets of linked hexagonal rings; they may also contain pentagonal (or sometimes heptagonal) rings.

 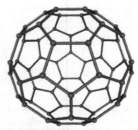

Buckminsterfullerene C_{60} (left) and carbon nanotubes (right) are two examples of structures in the fullerene family.

The first fullerene molecule to be discovered, and the family's namesake, buckminsterfullerene (C_{60}), was manufactured in 1985 by Richard Smalley, Robert Curl, James Heath, Sean O'Brien, and Harold Kroto at Rice University. The name was an homage to Buckminster Fuller, whose geodesic domes it resembles. The structure was also identified some five years earlier by Sumio Iijima, from an electron microscope image, where it formed the core of a "bucky onion". Fullerenes have since been found to occur in nature. More recently, fullerenes have been detected in outer space. According to astronomer Letizia Stanghellini, "It's possible that buckyballs from outer space provided seeds for life on Earth."

The discovery of fullerenes greatly expanded the number of known carbon allotropes, which until recently were limited to graphite, graphene, diamond, and amorphous carbon such as soot and charcoal. Buckyballs and buckytubes have been the subject of intense research, both for their unique chemistry and for their technological applications, especially in materials science, electronics, and nanotechnology.

History

The icosahedral $C_{60}H_{60}$ cage was mentioned in 1965 as a possible topological structure. Eiji Osawa of Toyohashi University of Technology predicted the existence of C_{60} in 1970. He noticed that the structure of a corannulene molecule was a subset of an Association football shape, and he hypothesised that a full ball shape could also exist. Japanese scientific journals reported his idea, but neither it nor any translations of it reached Europe or the Americas.

The icosahedral fullerene C_{540}, another member of the family of fullerenes.

Also in 1970, R. W. Henson (then of the Atomic Energy Research Establishment) proposed the structure and made a model of C_{60}. Unfortunately, the evidence for this new form of carbon was very weak and was not accepted, even by his colleagues. The results were never published but were acknowledged in *Carbon* in 1999.

In 1973 independently from Henson, a group of scientists from the USSR, directed by Prof. Bochvar, made a quantum-chemical analysis of the stability of C_{60} and calculated its electronic structure. As in the previous cases, the scientific community did not accept the theoretical prediction. The paper was published in 1973 in *Proceedings of the USSR Academy of Sciences* (in Russian).

In mass spectrometry discrete peaks appeared corresponding to molecules with the exact mass of sixty or seventy or more carbon atoms. In 1985 Harold Kroto of the University of Sussex, James R. Heath, Sean O'Brien, Robert Curl and Richard Smalley from Rice University, discovered C_{60}, and shortly thereafter came to discover the fullerenes. Kroto, Curl, and Smalley were awarded the 1996 Nobel Prize in Chemistry for their roles in the discovery of this class of molecules. C_{60} and other fullerenes were later noticed occurring outside the laboratory (for example, in normal candle-soot). By 1990 it was relatively easy to produce gram-sized samples of fullerene powder using the techniques of Donald Huffman, Wolfgang Krätschmer, Lowell D. Lamb, and Konstantinos Fostiropoulos. Fullerene purification remains a challenge to chemists and to a large extent determines fullerene prices. So-called endohedral fullerenes have ions or small molecules incorporated inside the cage atoms. Fullerene is an unusual reactant in many organic reactions such as the Bingel reaction discovered in 1993. Carbon nanotubes were first discovered and synthesized in 1991.

Minute quantities of the fullerenes, in the form of C_{60}, C_{70}, C_{76}, C_{82} and C_{84} molecules, are produced in nature, hidden in soot and formed by lightning discharges in the atmosphere. In 1992, fullerenes were found in a family of minerals known as Shungites in Karelia, Russia.

In 2010, fullerenes (C_{60}) have been discovered in a cloud of cosmic dust surrounding a distant star 6500 light years away. Using NASA's Spitzer infrared telescope the scientists spotted the molecules' unmistakable infrared signature. Sir Harry Kroto, who shared the 1996 Nobel Prize in Chemistry for the discovery of buckyballs commented: "This most exciting breakthrough provides convincing evidence that the buckyball has, as I long suspected, existed since time immemorial in the dark recesses of our galaxy."

Naming

The discoverers of the Buckminsterfullerene (C_{60}) allotrope of carbon named it after Richard Buckminster Fuller, a noted architectural modeler who popularized the geodesic dome. Since buckminsterfullerenes have a similar shape to those of such domes, they thought the name appropriate. As the discovery of the fullerene family came *after* buckminsterfullerene, the shortened name 'fullerene' is used to refer to the family of fullerenes. The suffix "-ene" indicates that each C atom is covalently bonded to three others (instead of the maximum of four), a situation that classically would correspond to the existence of bonds involving two pairs of electrons ("double bonds").

Types of Fullerene

Since the discovery of fullerenes in 1985, structural variations on fullerenes have evolved well beyond the individual clusters themselves. Examples include:

- Buckyball clusters: smallest member is C20 (unsaturated version of dodecahedrane) and the most common is C60;

- Nanotubes: hollow tubes of very small dimensions, having single or multiple walls; potential applications in electronics industry;

- Megatubes: larger in diameter than nanotubes and prepared with walls of different thickness; potentially used for the transport of a variety of molecules of different sizes;

- polymers: chain, two-dimensional and three-dimensional polymers are formed under high-pressure high-temperature conditions; single-strand polymers are formed using the Atom Transfer Radical Addition Polymerization (ATRAP) route;

- nano"onions": spherical particles based on multiple carbon layers surrounding a buckyball core; proposed for lubricants;

- linked "ball-and-chain" dimers: two buckyballs linked by a carbon chain;

- fullerene rings.

Buckyballs

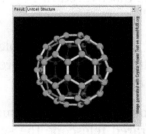   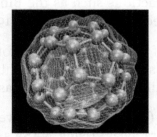

rotating view of C_{60}, one kind of fullerene

C_{60} with isosurface of ground state electron density as calculated with DFT

Buckminsterfullerene

Buckminsterfullerene is the smallest fullerene molecule containing pentagonal and hexagonal rings in which no two pentagons share an edge (which can be destabilizing, as in pentalene). It is also most common in terms of natural occurrence, as it can often be found in soot.

The structure of C_{60} is a truncated icosahedron, which resembles an association football ball of the type made of twenty hexagons and twelve pentagons, with a carbon atom at the vertices of each polygon and a bond along each polygon edge.

The van der Waals diameter of a C_{60} molecule is about 1.1 nanometers (nm). The nucleus to nucleus diameter of a C_{60} molecule is about 0.71 nm.

The C_{60} molecule has two bond lengths. The 6:6 ring bonds (between two hexagons) can be considered "double bonds" and are shorter than the 6:5 bonds (between a hexagon and a pentagon). Its average bond length is 1.4 angstroms.

Silicon buckyballs have been created around metal ions.

Boron Buckyball

A type of buckyball which uses boron atoms, instead of the usual carbon, was predicted and described in 2007. The B_{80} structure, with each atom forming 5 or 6 bonds, is predicted to be more stable than the C_{60} buckyball. One reason for this given by the researchers is that the B-80 is actually more like the original geodesic dome structure popularized by Buckminster Fuller, which uses triangles rather than hexagons. However, this work has been subject to much criticism by quantum chemists as it was concluded that the predicted I_h symmetric structure was vibrationally unstable and the resulting cage undergoes a spontaneous symmetry break, yielding a puckered cage with rare T_h symmetry (symmetry of a volleyball). The number of six-member rings in this molecule is 20 and number of five-member rings is 12. There is an additional atom in the center of each six-member ring, bonded to each atom surrounding it. By employing a systematic global search algorithm, later it was found that the previously proposed B80 fullerene is not global minimum for 80 atom boron clusters and hence can not be found in nature. In the same paper by Sandip De et al., it was concluded that boron's energy landscape is significantly different from other fullerenes already found in nature hence pure boron fullerenes are unlikely to exist in nature.

Other Buckyballs

Another fairly common fullerene is C_{70}, but fullerenes with 72, 76, 84 and even up to 100 carbon atoms are commonly obtained.

In mathematical terms, the structure of a fullerene is a trivalent convex polyhedron with pentagonal and hexagonal faces. In graph theory, the term fullerene refers to any 3-regular, planar graph with all faces of size 5 or 6 (including the external face). It follows from Euler's polyhedron formula, $V - E + F = 2$ (where V, E, F are the numbers of vertices, edges, and faces), that there are exactly 12 pentagons in a fullerene and $V/2 - 10$ hexagons.

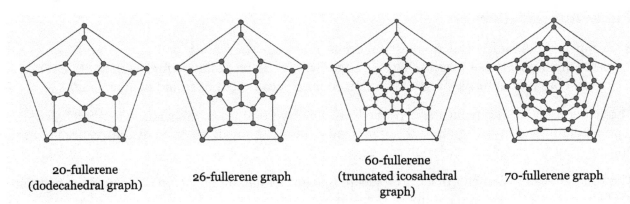

| 20-fullerene
(dodecahedral graph) | 26-fullerene graph | 60-fullerene
(truncated icosahedral
graph) | 70-fullerene graph |

The smallest fullerene is the dodecahedral C_{20}. There are no fullerenes with 22 vertices. The number of fullerenes C_{2n} grows with increasing $n = 12, 13, 14, ...$, roughly in proportion to n^9 (sequence A007894 in the OEIS). For instance, there are 1812 non-isomorphic fullerenes C_{60}. Note that only one form of C_{60}, the buckminsterfullerene alias truncated icosahedron, has no pair of adjacent pentagons (the smallest such fullerene). To further illustrate the growth, there are 214,127,713 non-isomorphic fullerenes C_{200}, 15,655,672 of which have no adjacent pentagons. Optimized structures of many fullerene isomers are published and listed on the web.

Heterofullerenes have heteroatoms substituting carbons in cage or tube-shaped structures. They were discovered in 1993 and greatly expand the overall fullerene class of compounds. Notable examples include boron, nitrogen (azafullerene), oxygen, and phosphorus derivatives.

Trimetasphere carbon nanomaterials were discovered by researchers at Virginia Tech and licensed exclusively to Luna Innovations. This class of novel molecules comprises 80 carbon atoms (C 80) forming a sphere which encloses a complex of three metal atoms and one nitrogen atom. These fullerenes encapsulate metals which puts them in the subset referred to as metallofullerenes. Trimetaspheres have the potential for use in diagnostics (as safe imaging agents), therapeutics and in organic solar cells.

Carbon Nanotubes

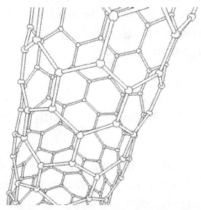

This rotating model of a carbon nanotube shows its 3D structure.

Nanotubes are cylindrical fullerenes. These tubes of carbon are usually only a few nanometres wide, but they can range from less than a micrometer to several millimeters in length. They often have closed ends, but can be open-ended as well. There are also cases in which the tube reduces in diameter before closing off. Their unique molecular structure results in extraordinary macroscopic properties, including high tensile strength, high electrical conductivity, high ductility, high heat conductivity, and relative chemical inactivity (as it is cylindrical and "planar" — that is, it has no "exposed" atoms that can be easily displaced). One proposed use of carbon nanotubes is in paper batteries, developed in 2007 by researchers at Rensselaer Polytechnic Institute. Another highly speculative proposed use in the field of space technologies is to produce high-tensile carbon cables required by a space elevator.

Carbon Nanobuds

Nanobuds have been obtained by adding buckminsterfullerenes to carbon nanotubes.

Fullerite

Fullerites are the solid-state manifestation of fullerenes and related compounds and materials.

The C_{60} fullerene in crystalline form

"Ultrahard fullerite" is a coined term frequently used to describe material produced by high-pressure high-temperature (HPHT) processing of fullerite. Such treatment converts fullerite into a nanocrystalline form of diamond which has been reported to exhibit remarkable mechanical properties.

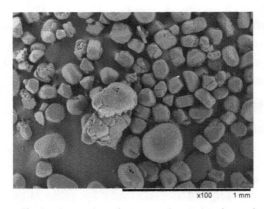

Fullerite (scanning electron microscope image)

Inorganic Fullerenes

Materials with fullerene-like molecular structures but lacking carbon include MoS_2, WS_2, TiS_2 and NbS_2. Prof. J. M. Martin from Ecole Centrale de Lyon in France tested the new material under isostatic pressure and found it to be stable up to at least 350 tons/cm^2.

Properties

For the past decade, the chemical and physical properties of fullerenes have been a hot topic in the field of research and development, and are likely to continue to be for a long time. *Popular Science* has discussed possible uses of fullerenes (graphene) in armor. In April 2003, fullerenes were under study for potential medicinal use: binding specific antibiotics to the structure to target resistant bacteria and even target certain cancer cells such as melanoma. The October 2005 issue of *Chemistry & Biology* contains an article describing the use of fullerenes as light-activated antimicrobial agents.

In the field of nanotechnology, heat resistance and superconductivity are some of the more heavily studied properties.

A common method used to produce fullerenes is to send a large current between two nearby graphite electrodes in an inert atmosphere. The resulting carbon plasma arc between the electrodes cools into sooty residue from which many fullerenes can be isolated.

There are many calculations that have been done using ab-initio quantum methods applied to fullerenes. By DFT and TD-DFT methods one can obtain IR, Raman and UV spectra. Results of such calculations can be compared with experimental results.

Aromaticity

Researchers have been able to increase the reactivity of fullerenes by attaching active groups to their surfaces. Buckminsterfullerene does not exhibit "superaromaticity": that is, the electrons in the hexagonal rings do not delocalize over the whole molecule.

A spherical fullerene of n carbon atoms has n pi-bonding electrons, free to delocalize. These should try to delocalize over the whole molecule. The quantum mechanics of such an arrangement should be like one shell only of the well-known quantum mechanical structure of a single atom, with a stable filled shell for n = 2, 8, 18, 32, 50, 72, 98, 128, etc.; i.e. twice a perfect square number; but this series does not include 60. This $2(N + 1)^2$ rule (with N integer) for spherical aromaticity is the three-dimensional analogue of Hückel's rule. The 10+ cation would satisfy this rule, and should be aromatic. This has been shown to be the case using quantum chemical modelling, which showed the existence of strong diamagnetic sphere currents in the cation.

As a result, C_{60} in water tends to pick up two more electrons and become an anion. The nC_{60} described below may be the result of C_{60} trying to form a loose metallic bond.

Chemistry

Fullerenes are stable, but not totally unreactive. The sp^2-hybridized carbon atoms, which are at their energy minimum in planar graphite, must be bent to form the closed sphere or tube, which produces angle strain. The characteristic reaction of fullerenes is electrophilic addition at 6,6-double bonds, which reduces angle strain by changing sp²-hybridized carbons into sp³-hybridized ones. The change in hybridized orbitals causes the bond angles to decrease from about 120° in the sp² orbitals to about 109.5° in the sp³ orbitals. This decrease in bond angles allows for the bonds to bend less when closing the sphere or tube, and thus, the molecule becomes more stable.

Other atoms can be trapped inside fullerenes to form inclusion compounds known as endohedral fullerenes. An unusual example is the egg-shaped fullerene $Tb_3N@C_{84}$, which violates the isolated pentagon rule. Recent evidence for a meteor impact at the end of the Permian period was found by analyzing noble gases so preserved. Metallofullerene-based inoculates using the rhonditic steel process are beginning production as one of the first commercially viable uses of buckyballs.

Solubility

Fullerenes are sparingly soluble in many solvents. Common solvents for the fullerenes include aromatics, such as toluene, and others like carbon disulfide. Solutions of pure buckminsterfullerene have a deep purple color. Solutions of C_{70} are a reddish brown. The higher fullerenes C_{76} to C_{84} have a variety of colors. C_{76} has two optical forms, while other higher fullerenes have several structural isomers. Fullerenes are the only known allotrope of carbon that can be dissolved in common solvents at room temperature.

C_{60} in solution

C_{60} in extra virgin olive oil showing the characteristic purple color of pristine C_{60} solutions.

Solvent	C_{60} mg/mL	C_{70} mg/mL
1-chloronaphthalene	51	ND
1-methylnaphthalene	33	ND
1,2-dichlorobenzene	24	36.2
1,2,4-trimethylbenzene	18	ND
tetrahydronaphthalene	16	ND

carbon disulfide	8	9.875
1,2,3-tribromopropane	8	ND
chlorobenzene	7	ND
p-xylene	5	3.985
bromoform	5	ND
cumene	4	ND
toluene	3	1.406
benzene	1.5	1.3
carbon tetrachloride	0.447	0.121
chloroform	0.25	ND
n-hexane	0.046	0.013
cyclohexane	0.035	0.08
tetrahydrofuran	0.006	ND
acetonitrile	0.004	ND
methanol	4.0×10^{-5}	ND
water	1.3×10^{-11}	ND
pentane	0.004	0.002
heptane	ND	0.047
octane	0.025	0.042
isooctane	0.026	ND
decane	0.070	0.053
dodecane	0.091	0.098
tetradecane	0.126	ND
acetone	ND	0.0019
isopropanol	ND	0.0021
dioxane	0.0041	ND
mesitylene	0.997	1.472
dichloromethane	0.254	0.080
ND, not determined		

Some fullerene structures are not soluble because they have a small band gap between the ground and excited states. These include the small fullerenes C_{28}, C_{36} and C_{50}. The C_{72} structure is also in this class, but the endohedral version with a trapped lanthanide-group atom is soluble due to the interaction of the metal atom and the electronic states of the fullerene. Researchers had originally been puzzled by C_{72} being absent in fullerene plasma-generated soot extract, but found in endohedral samples. Small band gap fullerenes are highly reactive and bind to other fullerenes or to soot particles.

Solvents that are able to dissolve buckminsterfullerene (C_{60} and C_{70}) are listed at left in order from highest solubility. The solubility value given is the approximate saturated concentration.

Solubility of C_{60} in some solvents shows unusual behaviour due to existence of solvate phases (analogues of crystallohydrates). For example, solubility of C_{60} in benzene solution shows maximum at about 313 K. Crystallization from benzene solution at temperatures below maximum results in formation of triclinic solid solvate with four benzene molecules $C_{60}\cdot 4C_6H_6$ which is rather unstable in air. Out of solution, this structure decomposes into usual face-centered cubic (fcc) C_{60} in few minutes' time. At temperatures above solubility maximum the solvate is not stable even when immersed in saturated solution and melts with formation of fcc C_{60}. Crystallization at temperatures above the solubility maximum results in formation of pure fcc C_{60}. Millimeter-sized crystals of C_{60} and C_{70} can be grown from solution both for solvates and for pure fullerenes.

Quantum Mechanics

In 1999, researchers from the University of Vienna demonstrated that wave-particle duality applied to molecules such as fullerene.

Superconductivity

Chirality

Some fullerenes (e.g. C_{76}, C_{78}, C_{80}, and C_{84}) are inherently chiral because they are D_2-symmetric, and have been successfully resolved. Research efforts are ongoing to develop specific sensors for their enantiomers.

Construction

Two theories have been proposed to describe the molecular mechanisms that make fullerenes. The older, "bottom-up" theory proposes that they are built atom-by-atom. The alternative "top-down" approach claims that fullerenes form when much larger structures break into constituent parts.

In 2013 researchers discovered that asymmetrical fullerenes formed from larger structures settle into stable fullerenes. The synthesized substance was a particular metallofullerene consisting of 84 carbon atoms with two additional carbon atoms and two yttrium atoms inside the cage. The process produced approximately 100 micrograms.

However, they found that the asymmetrical molecule could theoretically collapse to form nearly every known fullerene and metallofullerene. Minor perturbations involving the breaking of a few molecular bonds cause the cage to become highly symmetrical and stable. This insight supports the theory that fullerenes can be formed from graphene when the appropriate molecular bonds are severed.

Production Technology

Fullerene production processes comprise the following five subprocesses: (i) synthesis of fullerenes or fullerene-containing soot; (ii) extraction; (iii) separation (purification) for each fullerene molecule, yielding pure fullerenes such as C_{60}; (iv) synthesis of derivatives (mostly using the techniques of organic synthesis); (v) other post-processing such as dispersion into a matrix. The two synthesis methods used in practice are the arc method, and the combustion method. The latter, discovered at the Massachusetts Institute of Technology, is preferred for large scale industrial production.

Applications

Fullerenes have been extensively used for several biomedical applications including the design of high-performance MRI contrast agents, X-Ray imaging contrast agents, photodynamic therapy and drug and gene delivery, summarized in several comprehensive reviews.

Tumor Research

While past cancer research has involved radiation therapy, photodynamic therapy is important to study because breakthroughs in treatments for tumor cells will give more options to patients with different conditions. More recent experiments using HeLa cells in cancer research involves the development of new photosensitizers with increased ability to be absorbed by cancer cells and still trigger cell death. It is also important that a new photosensitizer does not stay in the body for a long time to prevent unwanted cell damage.

Fullerenes can be made to be absorbed by HeLa cells. The C_{60} derivatives can be delivered to the cells by using the functional groups L-phenylalanine, folic acid, and L-arginine among others. The purpose for functionalizing the fullerenes is to increase the solubility of the molecule by the cancer cells. Cancer cells take up these molecules at an increased rate because of an upregulation of transporters in the cancer cell, in this case amino acid transporters will bring in the L-arginine and L-phenylalanine functional groups of the fullerenes.

Once absorbed by the cells, the C_{60} derivatives would react to light radiation by turning molecular oxygen into reactive oxygen which triggers apoptosis in the HeLa cells and other cancer cells that can absorb the fullerene molecule. This research shows that a reactive substance can target cancer cells and then be triggered by light radiation, minimizing damage to surrounding tissues while undergoing treatment.

When absorbed by cancer cells and exposed to light radiation, the reaction that creates reactive oxygen damages the DNA, proteins, and lipids that make up the cancer cell. This cellular damage forces the cancerous cell to go through apoptosis, which can lead to the reduction in size of a tumor. Once the light radiation treatment is finished the fullerene will reabsorb the free radicals to prevent damage of other tissues. Since this treatment focuses on cancer cells, it is a good option for patients whose cancer cells are within reach of light radiation. As this research continues into the future, it will be able to penetrate deeper into the body and be absorbed by cancer cells more effectively.

Safety and Toxicity

A comprehensive and recent review on fullerene toxicity is given by Lalwani et al. These authors review the works on fullerene toxicity beginning in the early 1990s to present, and conclude that very little evidence gathered since the discovery of fullerenes indicate that C_{60} is toxic. The toxicity of these carbon nanoparticles is not only dose and time-dependent, but also depends on a number of other factors such as: type (e.g., C_{60}, C_{70}, $M@C_{60}$, $M@C_{82}$, functional groups used to water solubilize these nanoparticles (e.g., OH, COOH), and method of administration (e.g., intravenous, intraperitoneal). The authors therefore recommend that pharmacology of every new fullerene- or metallofullerene-based complex must be assessed individually as a different compound.

Popular Culture

Examples of fullerenes in popular culture are numerous. Fullerenes appeared in fiction well before scientists took serious interest in them. In a humorously speculative 1966 column for *New Scientist*, David Jones suggested that it may be possible to create giant hollow carbon molecules by distorting a plane hexagonal net by the addition of impurity atoms.

On 4 September 2010, Google used an interactively rotatable fullerene C_{60} as the second 'o' in their logo to celebrate the 25th anniversary of the discovery of the fullerenes.

Organosulfur Compounds

Organosulfur compounds are organic compounds that contain sulfur. They are often associated with foul odors, but many of the sweetest compounds known are organosulfur derivatives, e.g., saccharin. Nature abounds with organosulfur compounds—sulfur is essential for life. Of the 20 common amino acids, two (cysteine and methionine) are organosulfur compounds, and the antibiotics penicillin (pictured below) and sulfa drugs both contain sulfur. While sulfur-containing antibiotics save many lives, sulfur mustard is a deadly chemical warfare agent. Fossil fuels, coal, petroleum, and natural gas, which are derived from ancient organisms, necessarily contain organosulfur compounds, the removal of which is a major focus of oil refineries.

Sulfur shares the chalcogen group with oxygen, selenium and tellurium, and it is expected that organosulfur compounds have similarities with carbon–oxygen, carbon–selenium and carbon–tellurium compounds, which is true to some extent.

A classical chemical test for the detection of sulfur compounds is the Carius halogen method.

Classes of Organosulfur Compounds

Organosulfur compounds can be classified according to the sulfur-containing functional groups, which are listed (approximately) in decreasing order of their occurrence.

- Illustrative organosulfur compounds

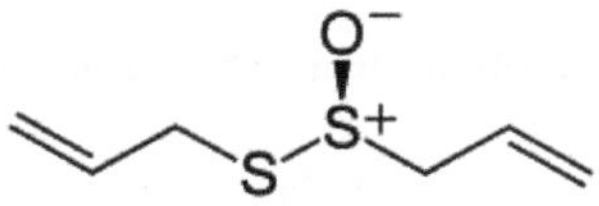

Allicin, the active flavor compound in crushed garlic

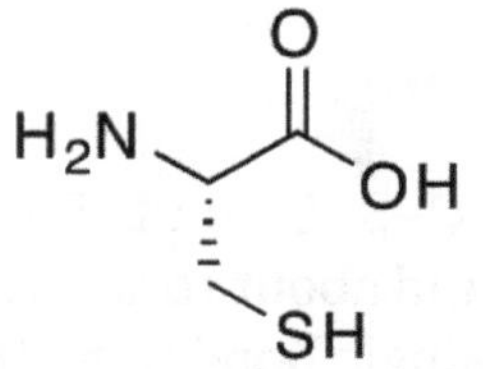

(*R*)-Cysteine, an amino acid containing a thiol group

Methionine, an amino acid containing a thioether

Diphenyl disulfide, a representative disulfide

Dibenzothiophene, a component of crude oil

Perfluorooctanesulfonic acid, a controversial surfactant

Lipoic acid, an essential cofactor of four mitochondrial enzyme complexes.

Penicillin core structure, where "R" is the variable group.

Sulfanilamide, a sulfonamide antibacterial, called a sulfa drug.

Sulfur mustard, a chemical warfare agent.

Thioethers, Thioesters, Thioacetals

These compounds are characterized by C–S–C bonds Relative to C–C bonds, C–S bond are both longer, because S is larger than carbon, and about 10% weaker. Representative bond lengths in sulfur compounds are 183 pm for the S–C single bond in methanethiol and 173 pm in thiophene. The C–S bond dissociation energy for thiomethane is 89 kcal/mol (370 kJ/mol) compared to methane's 100 kcal/mol (420 kJ/mol) and when hydrogen is replaced by a methyl group the energy decreases to 73 kcal/mol (305 kJ/mol). The single carbon to oxygen bond is shorter than that of the C–C bond. The bond dissociation energies for dimethyl sulfide and dimethyl ether are respectively 73 and 77 kcal/mol (305 and 322 kJ/mol).

Thioethers are typically prepared by alkylation of thiols. They can also be prepared via the Pummerer rearrangement. In one named reaction called the Ferrario reaction phenyl ether is converted to *phenoxathiin* by action of elemental sulfur and aluminium chloride.

87%

Thioacetals and thioketals feature C–S–C–S–C bond sequence. They represent a subclass of thioethers. The thioacetals are useful in "umpolung" of carbonyl groups. Thioacetals and thioketals can also be used to protect a carbonyl group in organic syntheses.

Thioesters have general structure R–CO–S–R. They are related to regular esters but are more reactive.

The above classes of sulfur compounds also exist in saturated and unsaturated heterocyclic structures, often in combination with other heteroatoms, as illustrated by thiiranes, thiirenes, thietanes, thietes, dithietanes, thiolanes, thianes, dithianes, thiepanes, thiepines, thiazoles, isothiazoles, and thiophenes, among others. The latter three compounds represent a special class of sulfur-containing heterocycles that are aromatic. The resonance stabilization of thiophene is 29 kcal/mol (121 kJ/mol) compared to 20 kcal/mol (84 kJ/mol) for the oxygen analogue furan. The reason for this difference is the higher electronegativity for oxygen drawing away electrons to itself at the expense of the aromatic ring current. Yet as an aromatic substituent the thio group is less effective as an activating group than the alkoxy group. Dibenzothiophene, a tricyclic heterocycle consisting of two benzene rings fused to a central thiophene ring occurs widely in heavier fractions of petroleum, along with its alkyl substituted derivatives.

Thiols, Disulfides, Polysulfides

Thiol group contain the functionality R–SH. Thiols are structurally similar to the alcohol group, but these functionalities are very different in their chemical properties. Thiols are more nucleophilic, more acidic, and more readily oxidized. This acidity can differ by 5 pK_a units.

The difference in electronegativity between sulfur (2.58) and hydrogen (2.20) is small and therefore hydrogen bonding in thiols is not prominent. Aliphatic thiols form monolayers on gold, which are topical in nanotechnology.

Certain aromatic thiols can be accessed through a Herz reaction.

Disulfides R–S–S–R with a covalent sulfur to sulfur bond are important for crosslinking: in biochemistry for the folding and stability of some proteins and in polymer chemistry for the crosslinking of rubber.

Longer sulfur chains are also known, such as in the natural product varacin which contains an unusual pentathiepin ring (5-sulfur chain cyclised onto a benzene ring).

Sulfoxides, Sulfones and Thiosulfinates

A sulfoxide, R–S(O)–R, is the S-oxide of a thioether, a sulfone, R–(O)$_2$–R, is the S,S-dioxide of a thioether, a thiosulfinate, R–S(O)–S–R, is the S-oxide of a disulfide, and a thiosulfonate, R–S(O)$_2$–S–R, is the S,S-dioxide of a disulfide. All of these compounds are well known with extensive chemistry, e.g., dimethyl sulfoxide, dimethyl sulfone, and allicin.

Sulfimides, Sulfoximides, Sulfonediimines

Sulfimides (also called a sulfilimines) are sulfur–nitrogen compounds of structure R$_2$S=NR′, the nitrogen analog of sulfoxides. They are of interest in part due to their pharmacological properties. When two different R groups are attached to sulfur, sulfimides are chiral. Sulfimides form stable α-carbanions.

Sulfoximides (also called sulfoximines) are tetracoordinate sulfur–nitrogen compounds, isoelectronic with sulfones, in which one oxygen atom of the sulfone is replaced by a substituted nitrogen atom, e.g., R$_2$S(O)=NR′. When two different R groups are attached to sulfur, sulfoximides are chiral. Much of the interest in this class of compounds is derived from the discovery that methionine sulfoximide (methionine sulfoximine) is an inhibitor of glutamine synthetase.

Sulfonediimines (also called sulfodiimines, sulfodiimides or sulfonediimides) are tetracoordinate sulfur–nitrogen compounds, isoelectronic with sulfones, in which both oxygen atoms of the sulfone are replaced by a substituted nitrogen atom, e.g., R$_2$S(=NR′)$_2$. They are of interest because of their biological activity and as building blocks for heterocycle synthesis.

S-Nitrosothiols

S-Nitrosothiols, also known as thionitrites, are compounds containing a nitroso group attached to the sulfur atom of a thiol, e.g. R–S–N=O. They have received considerable attention in biochemistry because they serve as donors of the nitrosonium ion, NO$^+$, and nitric oxide, NO, which may serve as signaling molecules in living systems, especially related to vasodilation.

Sulfur Halides

A wide range of organosulfur compounds are known which contain one or more halogen atom ("X" in the chemical formulas that follow) bonded to a single sulfur atom, e.g.: sulfenyl halides, RSX; sulfinyl halides, RS(O)X; sulfonyl halides, RSO$_2$X; alkyl and arylsulfur trichlorides, RSCl$_3$ and trifluorides, RSF$_3$; and alkyl and arylsulfur pentafluorides, RSF$_5$. Less well known are dialkylsulfur tetrahalides, mainly represented by the tetrafluorides, e.g., R$_2$SF$_4$.

Thioketones, Thioaldehydes, and Related Compounds

Compounds with double bonds between carbon and sulfur are relatively uncommon, but include the important compounds carbon disulfide, carbonyl sulfide, and thiophosgene. Thioketones (RC(=S)R′) are uncommon with alkyl substituents, but one example is thiobenzophenone.

Thioaldehydes are rarer still, reflecting their lack of steric protection ("thioformaldehyde" exists as a cyclic trimer). Thioamides, with the formula $R_1C(=S)N(R_2)R_3$ are more common. They are typically prepared by the reaction of amides with Lawesson's reagent. Isothiocyanates, with formula R–N=C=S, are found naturally. Vegetable foods with characteristic flavors due to isothiocyanates include wasabi, horseradish, mustard, radish, Brussels sprouts, watercress, nasturtiums, and capers.

S-Oxides and S,S-Dioxides of Thiocarbonyl Compounds

The S-oxides of thiocarbonyl compounds are known as thiocarbonyl S-oxides or sulfines, $R_2C=S=O$, and thiocarbonyl S,S-dioxides or sulfenes, $R_2C=SO_2$. These compounds are well known with extensive chemistry.

Triple Bonds Between Carbon and Sulfur

Triple bonds between sulfur and carbon in sulfaalkynes are rare and can be found in carbon monosulfide (CS) and have been suggested for the compounds F_3CCSF_3 and F_5SCSF_3. The compound HCSOH is also presented as having a formal triple bond.

Thiocarboxylic Acids and Thioamides

Thiocarboxylic acids (RC(O)SH) and dithiocarboxylic acids (RC(S)SH) are well known. They are structurally similar to carboxylic acids but more acidic. Thioamides are analogous to amides.

Sulfonic, Sulfinic And Sulfenic Acids, Esters, Amides, and Related Compounds

Sulfonic acids have functionality $R-S(=O)_2-OH$. They are strong acids that are typically soluble in organic solvents. Sulfonic acids like trifluoromethanesulfonic acid is a frequently used reagent in organic chemistry. Sulfinic acids have functionality R–S(O)–H while sulfenic acids have functionality R–S–OH. In the series sulfonic—sulfinic—sulfenic acids, both the acid strength and stability diminish in that order. Sulfonamides, sulfinamides and sulfenamides, with formulas $R-SO_2NR'_2$, $R-S(O)NR'_2$, and $R-SNR'_2$, respectively, each have a rich chemistry. For example, sulfa drugs are sulfonamides derived from aromatic sulfonation. Chiral sulfinamides are used in asymmetric synthesis, while sulfenamides are used extensively in the vulcanization process to assist cross-linking. Thiocyanates, R–S–CN, are related to sulfenyl halides and esters in terms of reactivity.

Sulfonium, Oxosulfonium and Related Salts

A sulfonium ion is a positively charged ion featuring three organic substituents attached to sulfur, with the formula $[R_3S]^+$. Together with their negatively charged counterpart, the anion, the compounds are called sulfonium salts. An oxosulfonium ion is a positively charged ion featuring three organic substituents and an oxygen attached to sulfur, with the formula $[R_3S=O]^+$. Together with their negatively charged counterpart, the anion, the compounds are called oxosulfonium salts. Related species include alkoxysulfonium and chlorosulfonium ions, $[R_2SOR]^+$ and $[R_2SCl]^+$, respectively.

Sulfonium, Oxosulfonium and Thiocarbonyl Ylides

Deprotonation of sulfonium and oxosulfonium salts affords ylides, of structure $R_2S^+-C^--R'_2$ and $R_2S(O)^+-C^--R'_2$. While sulfonium ylides, for instance in the Johnson–Corey–Chaykovsky reaction used to synthesize oxiranes, are sometimes drawn with a C=S double bond, e.g., $R_2S=CR'_2$, the ylidic carbon–sulfur bond is highly polarized and is better described as being ionic. Sulfonium ylides are key intermediates in the synthetically useful Stevens rearrangement. Thiocarbonyl ylides ($RR'C=S^+-C^--RR'$) can form by ring-opening of thiiranes, photocyclization of aryl vinyl sulfides, as well as by other processes.

Sulfuranes and Persulfuranes

Sulfuranes are relatively specialized functional group that are tetravalent, hypervalent sulfur compounds, with the formula SR_4 and likewise persulfuranes are hexavalent SR_6. All-carbon hexavalent complexes have been known for the heavier representatives of the chalcogen group, for instance the compound hexamethylpertellurane ($Te(Me)_6$) was discovered in 1990 by reaction of tetramethyltellurium with xenon difluoride to $Te(Me)_2)F_2$ followed by reaction with diethylzinc. The sulfur analogue hexamethylpersulfurane ($S(CH_3)_6$) has been predicted to be stable but has not been synthesized yet.

The first ever all-carbon persulfurane actually synthesized in a laboratory has two methyl and two biphenyl ligands:

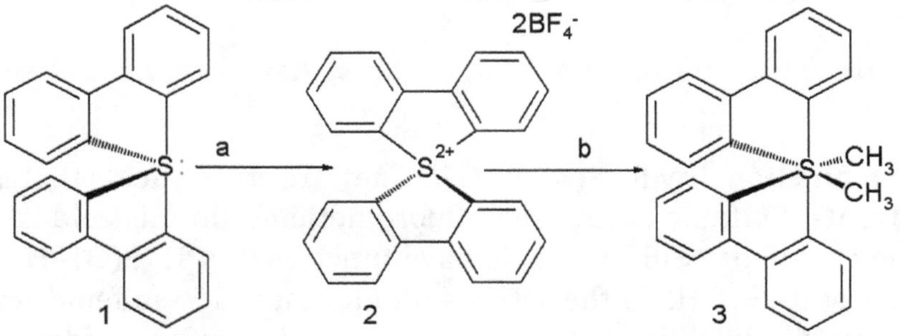

It is prepared from the corresponding sulfurane 1 with xenon difluoride/boron trifluoride in acetonitrile to the sulfuranyl dication 2 followed by reaction with methyllithium in tetrahydrofuran to (a stable) persulfurane 3 as the cis isomer. X-ray diffraction shows C–S bond lengths ranging between 189 and 193 pm (longer than the standard bond length) with the central sulfur atom in a distorted octahedral molecular geometry.

Computer simulation suggests that these bonds are very polar with the negative charges residing on carbon.

Naturally Occurring Organosulfur Compounds

Not all organosulfur compounds are foul-smelling pollutants. Penicillin and cephalosporin are life-saving antibiotics, derived from fungi. Gliotoxin is a sulfur-containing mycotoxin produced by several species of fungi under investigation as an antiviral agent. Compounds like allicin and ajoene are responsible for the odor of garlic, and lenthionine contributes to the flavor of shiitake mushrooms. Volatile organosulfur compounds also contribute subtle flavor characteristics to wine, nuts, cheddar cheese, chocolate, coffee, and tropical fruit flavors. Many of these natural products also have important medicinal properties such as preventing platelet aggregation or fighting cancer.

Organosulfur Compounds in Pollution

Most organic sulfur compounds in the environment are naturally occurring, as a consequence of the fact that sulfur is essential for life and two amino acids (cysteine and methionine) contain this element.

Some organosulfur compounds in the environment, are generated as minor by-products of industrial processes such as the manufacture of plastics and tires.

Selected smell-producing processes are organosulfur compounds produced by the coking of coal designed to drive out sulfurous compounds and other volatile impurities in order to produce 'clean carbon' (coke), which is primarily used for steel production.

Organosulfur Compounds in Fossil Fuels

Odours occur as well in chemical processing of coal or crude oil into precursor chemicals (feedstocks) for downstream industrial uses (e.g. plastics or pharmaceutical production) and the ubiquitous needs of petroleum distillation for gasolines, diesel, and other grades of fuel oils production.

Organosulfur compounds might be understood as aromatic contaminants that need to be removed from natural gas before commercial uses, from exhaust stacks and exhaust vents before discharge. In this latter context, organosulfur compounds may be said to account for the pollutants in sulfurous acid rain, or equivalently, said to be pollutants within most common fossil fuels, especially coal.

The most common organosulfur compound present in all petroleum fractions is thiophene (C_4H_4S), a cyclic and aromatic liquid. In addition, the heavy fractions of oil contain benzothiophene (C_8H_6S, thianaphtene) and dibenzothiophene. Most of the last compounds are solids and smell like naphthalene. Many methylated, dimethyl, diethyl benzothiophene derivatives are present in diesel and fuel oils which make fuel oils very difficult to clean.

All these heterocyclic thioethers account for 200–500 ppm of natural fuel, the heavily substituted dibenzothiophenes remain after HDS and account for 10–20 ppm. These molecules are also found in coals and they must be eliminated before consumption.

Reduced molybdenum together with nickel is currently used to eliminate thiophenes from petroleum (HDS) due to its great affinity towards sulfur. In addition tungsten together with nickel and cobalt is used for hydrodesulfurization (HDS) in large refineries. The adsorption mechanism of thiophene to transition metals is proposed to occur through the π system, where the organosulfur compound lies almost parallel to the metal surface. Many researchers focus their efforts in optimizing the oxidation state of the transition metals for HDS, like Cu(I) and Ag(II) which together with Pd(0) have proven to be more specific for π bonding with thiophenes of all kinds.

Basis for Odor of Organosulfur Compounds

Humans and other animals have an exquisitely sensitive sense of smell toward the odor of low-valent organosulfur compounds such as thiols, thioethers, and disulfides. Malodorous volatile thiols are protein-degradation products found in putrid food, so sensitive identification of these compounds is crucial to avoiding intoxication. Low-valent volatile sulfur compounds are also found in areas where oxygen levels in the air are low, posing a risk of suffocation. It has been found that copper is required for the highly sensitive detection of certain volatile thiols and related organosulfur compounds by olfactory receptors in mice. Whether humans, too, require copper for sensitive detection of thiols is not yet known.

Organophosphorus Compound

Organophosphorus compounds are organic compounds containing phosphorus. They are used primarily in pest control as an alternative to chlorinated hydrocarbons that persist in the environment. Organophosphorus chemistry is the corresponding science of the properties and reactivity of organophosphorus compounds. Phosphorus, like nitrogen, is in group 15 of the periodic table, and thus phosphorus compounds and nitrogen compounds have many similar properties.

These compounds are highly effective insecticides, though some are also lethal to humans at minuscule doses (nerve gas) and include some of the most toxic substances ever created by man.

The definition of organophosphorus compounds is variable, which can lead to confusion. In industrial and environmental chemistry, an organophosphorus compound need contain only an organic substituent, but need not have a direct phosphorus-carbon (P-C) bond. Thus a large proportion of pesticides (e.g., malathion), are often included in this class of compounds.

Phosphorus can adopt a variety of oxidation states, and it is general to classify organophosphorus compounds based on their being derivatives of phosphorus(V) vs phosphorus(III), which are the predominant classes of compounds. In a descriptive but only intermittently used nomenclature, phosphorus compounds are identified by their coordination number δ and their valency λ. In this system, a phosphine is a $\delta^3\lambda^3$ compound.

Organophosphorus(V) Compounds, Main Categories

Phosphate Esters and Amides

Phosphate esters have the general structure $P(=O)(OR)_3$ feature P(V). Such species are of technological importance as flame retardant agents, and plasticizers. Lacking a P–C bond, these compounds are in the technical sense not organophosphorus compounds but esters of phosphoric acid. Many derivatives are found in nature, such as phosphatidylcholine. Phosphate ester are synthesized by alcoholysis of phosphorus oxychloride. A variety of mixed amido-alkoxo derivatives are known, one medically significant example being the anti-cancer drug cyclophosphamide. Also derivatives containing the thiophosphoryl group (P=S) include the pesticide malathion. The organophosphates prepared on the largest scale are the zinc dithiophosphates, as additives for motor oil. Several million kilograms of this coordination complex are produced annually by the reaction of phosphorus pentasulfide with alcohols.

Illustrative organophosphates and related compounds: phosphatidylcholine, triphenylphosphate, cyclophosphamide, parathion, and zinc dithiophosphate.

In the environment, these compounds break down via hydrolysis to eventually afford phosphate and the organic alcohol or amine from which they are derived.

Phosphonic and Phosphinic Acids and Their Esters

Phosphonates are esters of phosphonic acid and have the general formula $RP(=O)(OR')_2$. Phosphonates have many technical applications, a well-known member being glyphosate, better known as Roundup. With the formula $(HO)_2P(O)CH_2NHCH_2CO_2H$, this derivative of glycine is one of the most widely used herbicides. Bisphosphonates are a class of drugs to treat osteoporosis. The nerve gas agent sarin, containing both C–P and F–P bonds, is a phosphonate.

Phosphinates feature *two* P–C bonds, with the general formula $R_2P(=O)(OR')$. A commercially significant member is the herbicide Glufosinate. Similar to glyphosate mentioned above, it has the structure $CH_3P(O)(OH)CH_2CH_2CH(NH_2)CO_2H$.

Illustrative examples of phosphonates and phosphinates in the order shown: Sarin (phosphonate), Glyphosate (phosphonate), fosfomycin (phosphonate), zoledronic acid (phosphonate), and Glufosinate (phosphinate). In aqueous solution, phosphonic acids ionize to give the corresponding organophosphonates.

The Michaelis–Arbuzov reaction is the main method for the synthesis of these compounds. For example, dimethylmethylphosphonate arises from the rearrangement of trime-thylphosphite, which is catalyzed by methyl iodide. In the Horner–Wadsworth–Emmons reaction and the Seyferth–Gilbert homologation, phosphonates are used in reactions with carbonyl compounds. The Kabachnik–Fields reaction is a method for the preparation of aminophosphonates. These compounds contain a very inert bond between phosphorus and carbon. Consequently, they hydrolyze to give phosphonic and phosphinic acid derivatives, but not phosphate.

Phosphine Oxides, Imides, and Chalcogenides

Phosphine oxides (designation $\delta^3\lambda^3$) have the general structure $R_3P=O$ with formal oxidation state V. Phosphine oxides form hydrogen bonds and some are therefore soluble in water. The P=O bond is very polar with a dipole moment of 4.51 D for triphenylphosphine oxide.

Compounds related to phosphine oxides are the imides (R_3PNR') and related chalcogenides (R_3PE, where E = S, Se, Te). These compounds are some of the most thermally stable organophosphorus compounds, but few are useful in significant amounts.

Phosphonium Salts and Phosphoranes

Compounds with the formula $[PR_4^+]X^-$ comprise the phosphonium salts. These species are tetrahedral phosphorus(V) compounds. From the commercial perspective, the most important member is tetrakis(hydroxymethyl)phosphonium chloride, $[P(CH_2OH)_4]Cl$, which is used as a fire retardant in textiles. Approximately 2M kg are produced annually of the chloride and the related sulfate. They are generated by the reaction of phosphine with formaldehyde in the presence of the mineral acid:

$$PH_3 + HX + 4\,CH_2O \rightarrow [P(CH_2OH)_4^+]X^-$$

A variety of phosphonium salts can be prepared by alkylation and arylation of organophosphines:

$$PR_3 + R'X \rightarrow [PR_3R'^+]X^-$$

The methylation of triphenylphosphine is the first step in the preparation of the Wittig reagent.

Illustrative phosphorus(V) compounds: the phosphonium ion $P(CH_2OH)_4^+$, two resonance structures for the Wittig reagent Ph_3PCH_2, and pentaphenylphosphorane, a rare pentaorganophophorus compound.

The parent phosphorane ($\delta^5\lambda^5$) is PH_5, which is unknown. Related compounds containing both halide and organic substituents on phosphorus are fairly common. Those with five organic substituents are rare, although $P(C_6H_5)_5$ is known, being derived from $P(C_6H_5)_4^+$ by reaction with phenyllithium.

Phosphorus ylides are unsaturated phosphoranes, known as Wittig reagents, e.g. $CH_2P(C_6H_5)_3$. These compounds feature tetrahedral phosphorus(V) and are considered relatives of phosphine oxides. They also are derived from phosphonium salts, but by deprotonation not alkylation.

Organophosphorus(III) Compounds, Main Categories

Phosphites, Phosphonites, and Phosphinites

Phosphites, sometimes called phosphite esters, have the general structure $P(OR)_3$ with oxidation state +3. Such species arise from the alcoholysis of phosphorus trichloride:

$$PCl_3 + 3\,ROH \rightarrow P(OR)_3 + 3\,HCl$$

The reaction is general, thus a vast number of such species are known. Phosphites are employed in the Perkow reaction and the Michaelis–Arbuzov reaction. They also serve as ligands in organometallic chemistry.

Intermediate between phosphites and phosphines are phosphonites ($P(OR)_2R'$) and phosphinite ($P(OR)R'_2$). Such species arise via alcoholysis reactions of the corresponding phosphinous and phosphonous chlorides (($PClR'_2$) and PCl_2R', respectively).

Phosphines

The parent compound of the phosphines is PH_3, called phosphine in the US and British Common-wealth, but phosphane elsewhere. Replacement of one or more hydrogen centers by an organic substituents (alkyl, aryl), gives $PH_{3-x}R_x$, an organophosphine, generally referred to as phosphines.

Various reduced organophosphorus compounds: a complex of an organophosphine pincer ligand, the chiral diphosphine used in homogeneous catalysis, the primary phosphine $PhPH_2$, and the phosphorus(I) compound $(PPh)_5$.

Comparison of Phosphines and Amines

The phosphorus atom in phosphines has a formal oxidation state −3 ($\delta^3\lambda^3$) and are the phosphorus analogues of amines. Like amines, phosphines have a trigonal pyramidal molecular geometry although often with smaller C-E-C angles (E = N, P), at least in the absence of steric effects. The C-P-C bond angle is 98.6° for trimethylphosphine increasing to 109.7° when the methyl groups are replaced by *tert*-butyl groups. When used as ligands, the steric bulk of tertiary phosphines is evaluated by their cone angle. The barrier to inversion is also much higher than in amines for a process like nitrogen inversion to occur, and therefore phosphines with three different substituents can be resolved into thermally stable optical isomers. Phosphines are often less basic than corresponding amines, for instance the phosphonium ion itself has a pK_a of −14 compared to 9.21 for the ammonium ion; trimethylphosphonium has a pK_a of 8.65 compared to 9.76 for trimethylammonium. However, triphenylphosphine (pK_a 2.73) is more basic than triphenylamine (pK_a −5), mainly because the lone pair of the nitrogen in NPh_3 is partially delocalized into the three phenyl rings. Whereas the lone pair on nitrogen is delocalized in pyrrole, the lone pair on phosphorus atom in the phosphorus equivalent of pyrrole (phosphole) is not. The reactivity of phosphines matches that of amines with regard to nucleophilicity in the formation of phosphonium salts with the general structure $PR_4^+X^-$. This property is used in the Appel reaction for converting alcohols to alkyl halides. Phosphines are easily oxidized to the corresponding phosphine oxides, whereas amine oxides are less readily generated. In part for this reason, phosphines are very rarely encountered in nature.

Synthetic Routes

From the commercial perspective, the most important phosphine is triphenylphosphine, several million kilograms being produced annually. It is prepared from the reaction of chlorobenzene, PCl_3, and sodium. Phosphines of a more specialized nature are usually prepared by other routes. Phosphorus halides undergo Nucleophilic displacement by organometallic reagents such as Grignard reagents. Conversely, some syntheses entail nucleophilic displacement of phosphide anion equivalents ("R_2P^-") by aryl- and alkyl halides. Primary (RPH_2) and secondary phosphines ($RRPH$ and R_2PH) add to alkenes in presence of a strong base (e.g., KOH in DMSO). Markovnikov's rules apply. Similar reactions occur involving alkynes. Base is not required for electron-deficient alkenes (e.g., derivatives of acrylonitrile) and alkynes.

Under free-radical conditions the P-H bonds of primary and secondary phosphines add across alkenes. Such reactions proceed with anti-Markovnikov regiochemistry. AIBN or organic peroxides are used as initiators. Tertiary phosphine oxides and sulfides can be reduced with chlorosilanes and other reagents.

Reactions

The main reaction types of phosphines are as nucleophiles and bases. Their nucleophilicity is evidence by their reactions with alkyl halides to phosphonium salts. Phosphines are nucleophilic catalysts in the dimerization of enones in various reactions in organic synthesis, e.g. the Rauhut–Currier reaction.

Phosphines are reducing agents, as illustrated in the Staudinger reduction converting azides to amines and in the Mitsunobu reaction for converting alcohols into esters. In these processes, the phosphine is oxidized to the phosphine oxide. Phosphines have also been found to reduce activated carbonyl groups for instance the reduction of an α-keto ester to an α-hydroxy ester in *scheme 2*. In the proposed reaction mechanism, the first proton is on loan from the methyl group in trimethylphosphine (triphenylphosphine does not react).

Reduction of activated carbonyl groups by alkyl phosphines

Phosphine Ligands

Phosphines such as trimethylphosphine are important ligands in inorganic chemistry. Mainly owing to the utility of asymmetric synthesis, a variety of chiral diphosphines have been popularized, such as BINAP and DIPAMP. A large number of phosphine ligands including diphosphines are simply called "phos ligands".

sPhos		SPANphos	
SEGphos		Triphos	
Xantphos		XPhos	
Chiraphos		duPhos	
A selection of phos ligands			

Primary and Secondary Phosphines

In addition to the other reactions associated with phosphines, those bearing P-H groups exhibit additional reactivity associated with the P-H bonds. They are readily deprotonated using strong bases to give phosphide anions. Primary and secondary phosphines are generally prepared by reduction of related phosphorus halides or esters. For example, phosphonates are reduced to primary phosphines:

Example of reduction of phosphonate ester to a primary phosphine.

The stability is attributed to conjugation between the aromatic ring and the phosphorus lone pair.

Phosphaalkenes and Phosphaalkynes

Compounds with carbon phosphorus(III) multiple bonds are called phosphaalkenes ($R_2C=PR$) and phosphaalkynes ($RC\equiv P$). They are similar in structure, but not in reactivity, to imines ($R_2C=NR$) and nitriles ($RC\equiv N$), respectively. In the compound phosphorine, one carbon atom in benzene is replaced by phosphorus. Species of this type are relatively rare but for that reason are of interest to researchers. A general method for the synthesis of phosphaalkenes is by 1,2-elimination of suitable precursors, initiated thermally or by base such as DBU, DABCO, or triethylamine:

Thermolysis of Me_2PH generates $CH_2=PMe$, an unstable species in the condensed phase.

Organophosphorus(0), (I), and (II) Compounds

Compounds where phosphorus exists in a formal oxidation state of less than III are uncommon, but examples are known for each class. Organophosphorus(0) species are illustrated by the carbene adducts, $[P(NHC)]_2$, where NHC is an N-heterocyclic carbene. With the formulae $(RP)_n$ and $(R_2P)_2$, respectively, compounds of phosphorus(I) and (II) are generated by reduction of the related organophosphorus(III) chlorides:

$$5\ PhPCl_2 + 5\ Mg \rightarrow (PhP)_5 + 5\ MgCl_2$$

$$2\ Ph_2PCl + Mg \rightarrow Ph_2P\text{-}PPh_2 + MgCl_2$$

Diphosphenes, with the formula R_2P_2, formally contain phosphorus-phosphorus double bonds. These phosphorus(I) species are rare but are stable provided that the organic substituents are large enough to prevent catenation. Many mixed-valence compounds are known, e.g. the cage $P_7(CH_3)_3$.

CH																	He
CLi	CBe											CB	CC	CN	CO	CF	Ne
CNa	CMg											CAl	CSi	CP	CS	CCl	CAr
CK	CCa	CSc	CTi	CV	CCr	CMn	CFe	CCo	CNi	CCu	CZn	CGa	CGe	CAs	CSe	CBr	CKr
CRb	CSr	CY	CZr	CNb	CMo	CTc	CRu	CRh	CPd	CAg	CCd	CIn	CSn	CSb	CTe	CI	CXe
CCs	CBa		CHf	CTa	CW	CRe	COs	CIr	CPt	CAu	CHg	CTl	CPb	CBi	CPo	CAt	Rn
Fr	CRa		Rf	Db	CSg	Bh	Hs	Mt	Ds	Rg	Cn	Nh	Fl	Mc	Lv	Ts	Og

↓

		CLa	CCe	CPr	CNd	CPm	CSm	CEu	CGd	CTb	CDy	CHo	CEr	CTm	CYb	CLu
		Ac	CTh	CPa	CU	CNp	CPu	CAm	CCm	CBk	CCf	CEs	Fm	Md	No	Lr

Chemical bonds to carbon	
Core organic chemistry	Many uses in chemistry
Academic research, but no widespread use	Bond unknown

Organosilicon

Organosilicon compounds are organic compounds containing carbon–silicon bonds. Organosilicon chemistry is the corresponding science exploring their properties and reactivity. Most organosilicon compounds are similar to the ordinary organic compounds, being colourless, flammable, hydrophobic, and stable. The first organosilicon compound, tetraethylsilane, was discovered by Charles Friedel and James Crafts in 1863 by reaction of tetrachlorosilane with diethylzinc. The carbosilicon silicon carbide is an *inorganic* compound.

C—Si

A carbon–silicon bond present in all organosilicon compounds

Occurrence and Applications

Organosilicon compounds are widely encountered in commercial products. Most common are sealants, caulks, adhesives, and coatings made from silicones.

Silicone caulk, commercial sealants, are mainly composed of organosilicon compounds.

Polydimethylsiloxane (PDMS) is the principal component of silicones.

Biology and Medicine

Carbon–silicon bonds are however generally absent in biochemical processes, although their fleeting existence has been reported in a freshwater alga. Silafluofen is an organosilicon compound that functions as a pyrethroid insecticide. Several organosilicon compounds are being investigated as pharmaceuticals.

Properties of Si–C, Si–O, and Si–F Bonds

In most organosilicon compounds, Si is tetravalent and tetrahedral. Carbon–silicon bonds compared to carbon–carbon bonds are longer (186 pm vs. 154 pm) and weaker with bond dissociation energy 451 kJ/mol vs. 607 kJ/mol. The C–Si bond is somewhat polarised towards carbon due to carbon's greater electronegativity (C 2.55 vs Si 1.90). The Si–C bond can be broken more readily than typical C–C bonds. One manifestation of bond polarization in organosilanes is found in the Sakurai reaction. Certain alkyl silanes can be oxidized to an alcohol in the Fleming–Tamao oxidation.

Another manifestation is the β-silicon effect describes the stabilizing effect of a β-silicon atom on a carbocation with many implications for reactivity.

Si–O bonds are much stronger (809 kJ/mol compared to 538 kJ/mol) than a typical C–O single bond. The favorable formation of Si–O bonds drive many organic reactions such as the Brook rearrangement and Peterson olefination. Compared to the strong Si–O bond, the Si–F bond is even stronger.

Production

The bulk of organosilicon compounds derive from organosilicon chlorides $(CH_3)_{4-x}SiCl_x$. These chlorides produced by the "Direct process", which entails the reaction of methyl chloride with a silicon-copper alloy. The main and most sought-after product is dimethyldichlorosilane:

$$2\ CH_3Cl + Si \rightarrow (CH_3)_2SiCl_2$$

A variety of other products are obtained, including trimethylsilyl chloride and methyltrichlorosilane. About 1 million tons of organosilicon compounds are prepared annually by this route. The method can also be used for phenyl chlorosilanes.

Hydrosilylation

Compounds with Si-H bonds add to unsaturated substrates in the process called hydrosilylation (also called hydrosilation). Commercially, the main substrates are alkenes. Other unsaturated functional groups—alkynes, imines, ketones, and aldehydes—also participate, although these uses are rather specialized. An example is the hydrosilation of phenylacetylene:

In the related silylmetalation, a metal replaces the hydrogen atom.

Functional Groups

Silanols, Siloxides, and Siloxanes

Silanols are analogues of alcohols. They are generally prepared by hydrolysis of silyl chlorides and oxidation of silyl hydrides:

$$R_3SiCl + H_2O \rightarrow R_3SiOH + HCl$$

Less frequently they are prepared by oxidation of silyl hydrides:

$$2\,R_3SiH + O_2 \rightarrow 2R_3SiOH$$

The parent R_3SiOH is too unstable for isolation, but the many organic derivatives are known including $(CH_3)_3SiOH$ and $(C_6H_5)_3SiOH$. They are about 500x more acidic than the corresponding alcohols. Siloxides (silanoates) are the deprotonated derivatives of silanols:

$$R_3SiOH + NaOH \rightarrow R_3SiONa + H_2O$$

Silanols tend to dehydrate to give siloxanes:

$$2\,R_3SiOH \rightarrow R_3Si\text{-}O\text{-}SiR_3 + H_2O$$

Polymers with repeating siloxane linkages are called silicones. Compounds with an Si=O double bond called silanones are extremely unstable.

Silyl Ethers

Silyl ethers have the connectivity Si-O-C. They are typically prepared by the reaction of alcohols with silyl chlorides:

$$(CH_3)_3SiCl + ROH \rightarrow (CH_3)_3Si\text{-}O\text{-}R + HCl$$

Silyl ethers are extensively used as protective groups for alcohols.

Exploiting the strength of the Si-F bond, fluoride sources such as tetra-n-butylammonium fluoride (TBAF) are used in deprotection of silyl ethers:

$$(CH_3)_3Si\text{-}O\text{-}R + F^- + H_2O \rightarrow (CH_3)_3Si\text{-}F + H\text{-}O\text{-}R + OH^-$$

Silyl Chlorides

Organosilyl chlorides are important commodity chemicals. They are mainly used to produce silicone polymers as described above. Especially important silyl chlorides are dimethyldichlorosilane (Me_2SiCl_2), methyltrichlorosilane ($MeSiCl_3$), and trimethylsilyl chloride (Me_3SiCl). More specialized derivatives that find commercial applications include dichloromethylphenylsilane, trichloro(chloromethyl)silane, trichloro(dichlorophenyl)silane, trichloroethylsilane, and phenyltrichlorosilane.

Although proportionately a minor outlet, organosilicon compounds are widely used in organic synthesis. Notably trimethylsilyl chloride Me_3SiCl is the main silylating agent. One classic method called the Flood reaction for the synthesis of this compound class is by heating hexaalkyldisiloxanes $R_3SiOSiR_3$ with concentrated sulfuric acid and a sodium halide.

Silyl hydrides

The silicon to hydrogen bond is longer than the C–H bond (148 compared to 105 pm) and weaker (299 compared to 338 kJ/mol). Hydrogen is more electronegative than silicon hence the naming convention of silyl *hydrides*. Commonly the presence of the hydride is not mentioned in the name of the compound. Triethylsilane has the formula Et_3SiH. Phenylsilane is $PhSiH_3$. The parent compound SiH_4 is called silane.

Silanylidenes

Silanylidenes are compounds containing a silicon based chain, joined by a double bond to the main molecule, such as silylidenemethanol. Where it is the main functional group, the molecule is named after the parent silane, with the -ylidene- infix, such as methylidenesilane.

Silenes

Organosilicon compounds, unlike their carbon counterparts, do not have a rich double bond chemistry due to the large difference in electronegativity. Existing compounds with silene Si=C bonds (also known as alkylidenesilanes) are laboratory curiosities such as the silicon benzene analogue silabenzene. In 1967, Gusel'nikov and Flowers provided the first evidence for silenes from pyrolysis of *dimethylsilacyclobutane*. The first stable (kinetically shielded) silene was reported in 1981 by Brook

Disilenes have Si=Si double bonds and disilynes are silicon analogues of an alkyne. The first Silyne (with a silicon to carbon triple bond) was reported in 2010

Siloles

Siloles, also called silacyclopentadienes, are members of a larger class of compounds called metalloles. They are the silicon analogs of cyclopentadienes and are of current academic interest due to their electroluminescence and other electronic properties. Siloles are efficient in electron transport. They owe their low lying LUMO to a favorable interaction between the antibonding sigma silicon orbital with an antibonding pi orbital of the butadiene fragment.

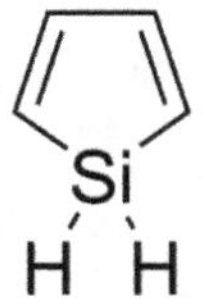

Chemical structure of silole

Hypercoordinated Silicon

Unlike carbon, silicon compounds can be coordinated to five atoms as well in a group of compounds ranging from so-called silatranes, such as phenylsilatrane, to a uniquely stable pentaorganosilicate:

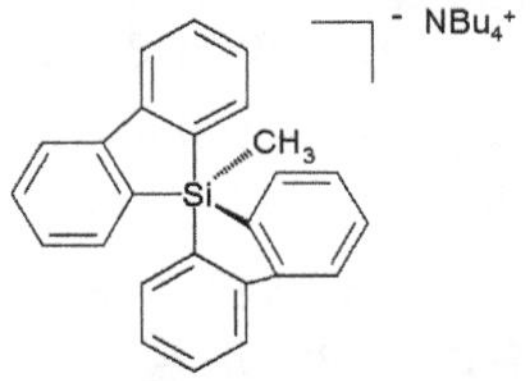

The stability of hypervalent silicon is the basis of the Hiyama coupling, a coupling reaction used in certain specialized organic synthetic applications. The reaction begins with the activation of Si-C bond by fluoride:

$$R\text{-}SiR'_3 + R''\text{-}X + F^- \rightarrow R\text{-}R'' + R'_3SiF + X^-$$

Various Reactions

Certain allyl silanes can be prepared from allylic ester such as 1 and monosilylcopper compounds such as 2 in.

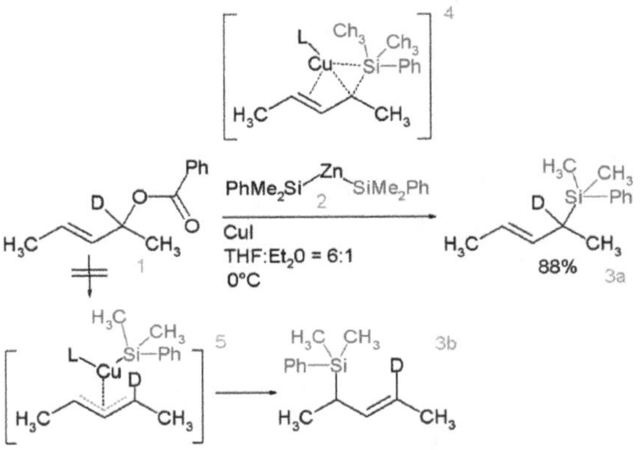

In this reaction type silicon polarity is reversed in a chemical bond with zinc and a formal allylic substitution on the benzoyloxy group takes place.

References

- García Ruano, J. L.; Cid, M. B.; Martín Castro, A. M.; Alemán, J. (2008). "Acyclic S,S-Dialkylsulfimides". In Kambe, N. Science of Synthesis. 39. Thieme. pp. 352–375. ISBN 978-1-58890-530-7.

- Drabowicz, J.; Lewkowski, J.; Kudelska, W.; Girek, T. (2008). "S,S-Dialkylsulfoximides". In Kambe, N. Science of Synthesis. 39. Thieme. pp. 154–173. ISBN 978-1-58890-530-7.

- Osawa, Eiji (2002). Perspectives of Fullerene Nanotechnology. Springer Science & Business Media. pp. 29–. ISBN 978-0-7923-7174-8.

- Fetzer, J. C. (2000). "The Chemistry and Analysis of the Large Polycyclic Aromatic Hydrocarbons". Polycyclic Aromatic Compounds. New York: Wiley. 27 (2): 143. doi:10.1080/10406630701268255. ISBN 0-471-36354-5.

- March, Jerry (1985), Advanced Organic Chemistry: Reactions, Mechanisms, and Structure (3rd ed.), New York: Wiley, ISBN 0-471-85472-7

- March, Jerry (1985), Advanced Organic Chemistry: Reactions, Mechanisms, and Structure (3rd ed.), New York: Wiley, ISBN 0-471-85472-7

- Brown, Theodore (2002). Chemistry : the central science. Upper Saddle River, NJ: Prentice Hall. p. 1001. ISBN 0130669970.

- Braverman, S.; Cherkinsky, M.; Levinger, S. (2008). "Alkylsulfur Trihalides". In Kambe, N. Science of Synthesis. 39. Thieme. pp. 187–188. ISBN 978-1-58890-530-7.

- Drabowicz, J.; Lewkowski, J.; Kudelska, W.; Girek, T. (2008). "Dialkylsulfur Tetrahalides". In Kambe, N. Science of Synthesis. 39. Thieme. pp. 123–124. ISBN 978-1-58890-530-7.

- Braverman, S.; Cherkinsky, M.; Levinger, S. (2008). "Alkanesulfinic Acids and Salts". In Kambe, N. Science of Synthesis. 39. Thieme. pp. 196–211. ISBN 978-1-58890-530-7.

- Drabowicz, J.; Kiełbasiński, P.; Łyżwa, P.; Zając, A.; Mikołajczyk, M. (2008). "Alkanesulfenic Acids". In Kambe, N. Science of Synthesis. 39. Thieme. pp. 550–557. ISBN 978-1-58890-530-7.

4

Understanding Inorganic Chemistry

Compounds that are not organic are known as inorganic compounds. The study of these inorganic compounds is termed as inorganic chemistry. Organic compounds can be made from inorganic ones. This chapter educates the reader with the basics of inorganic chemistry and its progress in contemporary times.

Inorganic Chemistry

Inorganic chemistry deals with the synthesis and behavior of inorganic and organometallic compounds. This field covers all chemical compounds except the myriad organic compounds (carbon based compounds, usually containing C-H bonds), which are the subjects of organic chemistry. The distinction between the two disciplines is far from absolute, as there is much overlap in the subdiscipline of organometallic chemistry. It has applications in every aspect of the chemical industry, including catalysis, materials science, pigments, surfactants, coatings, medications, fuels, and agriculture.

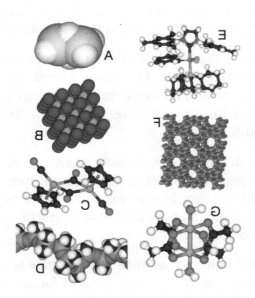

Inorganic compounds show rich variety:
A: Diborane features unusual bonding
B: Caesium chloride has an archetypal crystal structure
C: Fp_2 is an organometallic complex
D: Silicone's uses range from breast implants to Silly Putty
E: Grubbs' catalyst won the 2005 Nobel Prize for its discoverer
F: Zeolites find extensive use as molecular sieves
G: Copper(II) acetate surprised theoreticians with its diamagnetism

Key Concepts

Many inorganic compounds are ionic compounds, consisting of cations and anions joined by ionic bonding. Examples of salts (which are ionic compounds) are magnesium chloride $MgCl_2$, which consists of magnesium cations Mg^{2+} and chloride anions Cl^-; or sodium oxide Na_2O, which consists of sodium cations Na^+ and oxide anions O^{2-}. In any salt, the proportions of the ions are such that the electric charges cancel out, so that the bulk compound is electrically neutral. The ions are described by their oxidation state and their ease of formation can be inferred from the ionization potential (for cations) or from the electron affinity (anions) of the parent elements.

The structure of the ionic framework in potassium oxide, K_2O

Important classes of inorganic compounds are the oxides, the carbonates, the sulfates, and the halides. Many inorganic compounds are characterized by high melting points. Inorganic salts typically are poor conductors in the solid state. Other important features include their high meilting point and ease of crystallization. Where some salts (e.g., NaCl) are very soluble in water, others (e.g., SiO_2) are not.

The simplest inorganic reaction is double displacement when in mixing of two salts the ions are swapped without a change in oxidation state. In redox reactions one reactant, the *oxidant*, lowers its oxidation state and another reactant, the *reductant*, has its oxidation state increased. The net result is an exchange of electrons. Electron exchange can occur indirectly as well, e.g., in batteries, a key concept in electrochemistry.

When one reactant contains hydrogen atoms, a reaction can take place by exchanging protons in acid-base chemistry. In a more general definition, any chemical species capable of binding to electron pairs is called a Lewis acid; conversely any molecule that tends to donate an electron pair is referred to as a Lewis base. As a refinement of acid-base interactions, the HSAB theory takes into account polarizability and size of ions.

Inorganic compounds are found in nature as minerals. Soil may contain iron sulfide as pyrite or calcium sulfate as gypsum. Inorganic compounds are also found multitasking as biomolecules: as electrolytes (sodium chloride), in energy storage (ATP) or in construction (the polyphosphate backbone in DNA).

The first important man-made inorganic compound was ammonium nitrate for soil fertilization through the Haber process. Inorganic compounds are synthesized for use as catalysts such as vanadium(V) oxide and titanium(III) chloride, or as reagents in organic chemistry such as lithium aluminium hydride.

Subdivisions of inorganic chemistry are organometallic chemistry, cluster chemistry and bioinorganic chemistry. These fields are active areas of research in inorganic chemistry, aimed toward new catalysts, superconductors, and therapies.

Industrial Inorganic Chemistry

Inorganic chemistry is a highly practical area of science. Traditionally, the scale of a nation's economy could be evaluated by their productivity of sulfuric acid. The top 20 inorganic chemicals manufactured in Canada, China, Europe, India, Japan, and the US (2005 data): aluminium sulfate, ammonia, ammonium nitrate, ammonium sulfate, carbon black, chlorine, hydrochloric acid, hydrogen, hydrogen peroxide, nitric acid, nitrogen, oxygen, phosphoric acid, sodium carbonate, sodium chlorate, sodium hydroxide, sodium silicate, sodium sulfate, sulfuric acid, and titanium dioxide.

The manufacturing of fertilizers is another practical application of industrial inorganic chemistry.

Descriptive Inorganic Chemistry

Descriptive inorganic chemistry focuses on the classification of compounds based on their properties. Partly the classification focuses on the position in the periodic table of the heaviest element (the element with the highest atomic weight) in the compound, partly by grouping compounds by their structural similarities. When studying inorganic compounds, one often encounters parts of the different classes of inorganic chemistry (an organometallic compound is characterized by its coordination chemistry, and may show interesting solid state properties).

Different Classifications Are:

Coordination Compounds

Classical coordination compounds feature metals bound to "lone pairs" of electrons residing on the main group atoms of ligands such as H_2O, NH_3, Cl^-, and CN^-. In modern coordination compounds almost all organic and inorganic compounds can be used as ligands. The "metal" usually is a metal from the groups 3-13, as well as the *trans*-lanthanides and *trans*-actinides, but from a certain perspective, all chemical compounds can be described as coordination complexes.

EDTA chelates an octahedrally coordinated Co^{3+} ion in $[Co(EDTA)]^-$

The stereochemistry of coordination complexes can be quite rich, as hinted at by Werner's separation of two enantiomers of $[Co((OH)_2Co(NH_3)_4)_3]^{6+}$, an early demonstration that chirality is not inherent to organic compounds. A topical theme within this specialization is supramolecular coordination chemistry.

- Examples: $[Co(EDTA)]^-$, $[Co(NH_3)_6]^{3+}$, $TiCl_4(THF)_2$.

Main Group Compounds

These species feature elements from groups 1, 2 and 13-18 (excluding hydrogen) of the periodic table. Due to their often similar reactivity, the elements in group 3 (Sc, Y, and La) and group 12 (Zn, Cd, and Hg) are also generally included.

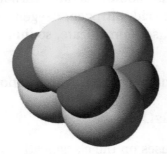

Tetrasulfur tetranitride, S_4N_4, is a main group compound that continues to intrigue chemists

Main group compounds have been known since the beginnings of chemistry, e.g., elemental sulfur and the distillable white phosphorus. Experiments on oxygen, O_2, by Lavoisier and Priestley not only identified an important diatomic gas, but opened the way for describing compounds and reactions according to stoichiometric ratios. The discovery of a practical synthesis of ammonia using iron catalysts by Carl Bosch and Fritz Haber in the early 1900s deeply impacted mankind, demonstrating the significance of inorganic chemical synthesis. Typical main group compounds are SiO_2, $SnCl_4$, and N_2O. Many main group compounds can also be classed as "organometallic", as they contain organic groups, e.g., $B(CH_3)_3$). Main group compounds also occur in nature, e.g., phosphate in DNA, and therefore may be classed as bioinorganic. Conversely, organic compounds lacking (many) hydrogen ligands can be classed as "inorganic", such as the fullerenes, buckytubes and binary carbon oxides.

- Examples: tetrasulfur tetranitride S_4N_4, diborane B_2H_6, silicones, buckminsterfullerene C_{60}.

Transition Metal Compounds

Compounds containing metals from group 4 to 11 are considered transition metal compounds. Compounds with a metal from group 3 or 12 are sometimes also incorporated into this group, but also often classified as main group compounds.

Transition metal compounds show a rich coordination chemistry, varying from tetrahedral for titanium (e.g., $TiCl_4$) to square planar for some nickel complexes to octahedral for coordination complexes of cobalt. A range of transition metals can be found in biologically important compounds, such as iron in hemoglobin.

- Examples: iron pentacarbonyl, titanium tetrachloride, cisplatin

Organometallic Compounds

Usually, organometallic compounds are considered to contain the M-C-H group. The metal (M) in these species can either be a main group element or a transition metal. Operationally, the definition of an organometallic compound is more relaxed to include also highly lipophilic complexes such as metal carbonyls and even metal alkoxides.

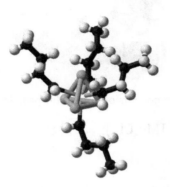

Organolithium reagents are most often found in polymeric form, such as *n*-butyllithium shown here

Organometallic compounds are mainly considered a special category because organic ligands are often sensitive to hydrolysis or oxidation, necessitating that organometallic chemistry employs more specialized preparative methods than was traditional in Werner-type complexes. Synthetic methodology, especially the ability to manipulate complexes in solvents of low coordinating power, enabled the exploration of very weakly coordinating ligands such as hydrocarbons, H_2, and N_2. Because the ligands are petrochemicals in some sense, the area of organometallic chemistry has greatly benefited from its relevance to industry.

- Examples: Cyclopentadienyliron dicarbonyl dimer $(C_5H_5)Fe(CO)_2CH_3$, Ferrocene $Fe(C_5H_5)_2$, Molybdenum hexacarbonyl $Mo(CO)_6$, Diborane B_2H_6, Tetrakis(triphenylphosphine)palladium(0) $Pd[P(C_6H_5)_3]_4$

Cluster Compounds

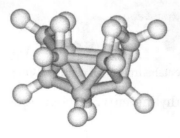

Decaborane is a powerfully toxic cluster compound of boron

Clusters can be found in all classes of chemical compounds. According to the commonly accepted definition, a cluster consists minimally of a triangular set of atoms that are directly bonded to each other. But metal-metal bonded dimetallic complexes are highly relevant to the area. Clusters occur in "pure" inorganic systems, organometallic chemistry, main group chemistry, and bioinorganic chemistry. The distinction between very large clusters and bulk solids is increasingly blurred. This interface is the chemical basis of nanoscience or nanotechnology and specifically arise from the study of quantum size effects in cadmium selenide clusters. Thus, large clusters can be described as an array of bound atoms intermediate in character between a molecule and a solid.

Iron-sulfur clusters are central components of iron-sulfur proteins, essential for human metabolism

- Examples: $Fe_3(CO)_{12}$, $B_{10}H_{14}$, $[Mo_6Cl_{14}]^{2-}$, 4Fe-4S

Bioinorganic Compounds

By definition, these compounds occur in nature, but the subfield includes anthropogenic species, such as pollutants (e.g., methylmercury) and drugs (e.g., Cisplatin). The field, which incorporates many aspects of biochemistry, includes many kinds of compounds, e.g., the phosphates in DNA, and also metal complexes containing ligands that range from biological macromolecules, commonly peptides, to ill-defined species such as humic acid, and to water (e.g., coordinated to gadolinium complexes employed for MRI). Traditionally bioinorganic chemistry focuses on electron- and energy-transfer in proteins relevant to respiration. Medicinal inorganic chemistry includes the study of both non-essential and essential elements with applications to diagnosis and therapies.

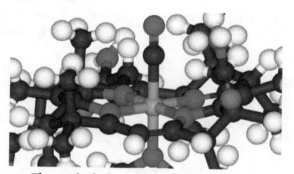

The octahedral cobalt centre of Vitamin B_{12}

- Examples: hemoglobin, methylmercury, carboxypeptidase

Solid State Compounds

This important area focuses on structure, bonding, and the physical properties of materials. In practice, solid state inorganic chemistry uses techniques such as crystallography to gain an understanding of the properties that result from collective interactions between the subunits of the solid. Included in solid state chemistry are metals and their alloys or intermetallic derivatives. Related fields are condensed matter physics, mineralogy, and materials science.

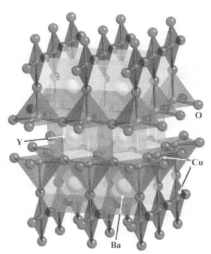

YBa$_2$Cu$_3$O$_7$, or YBCO, is a high temperature superconductor able to levitate above a magnet when colder than its critical temperature of about 90 K (−183 °C)

- Examples: silicon chips, zeolites, YBa$_2$Cu$_3$O$_7$

Theoretical Inorganic Chemistry

An alternative perspective on the area of inorganic chemistry begins with the Bohr model of the atom and, using the tools and models of theoretical chemistry and computational chemistry, expands into bonding in simple and then more complex molecules. Precise quantum mechanical descriptions for multielectron species, the province of inorganic chemistry, is difficult. This challenge has spawned many semi-quantitative or semi-empirical approaches including molecular orbital theory and ligand field theory, In parallel with these theoretical descriptions, approximate methodologies are employed, including density functional theory.

Exceptions to theories, qualitative and quantitative, are extremely important in the development of the field. For example, Cu$^{II}_2$(OAc)$_4$(H$_2$O)$_2$ is almost diamagnetic below room temperature whereas Crystal Field Theory predicts that the molecule would have two unpaired electrons. The disagreement between qualitative theory (paramagnetic) and observation (diamagnetic) led to the development of models for "magnetic coupling." These improved models led to the development of new magnetic materials and new technologies.

Qualitative Theories

Inorganic chemistry has greatly benefited from qualitative theories. Such theories are easier to learn as they require little background in quantum theory. Within main group compounds, VSEPR theory powerfully predicts, or at least rationalizes, the structures of main group compounds, such as an explanation for why NH$_3$ is pyramidal whereas ClF$_3$ is T-shaped. For the transition metals, crystal field theory allows one to understand the magnetism of many simple complexes, such as why [Fe^{III}(CN)$_6$]$^{3-}$ has only one unpaired electron, whereas [Fe^{III}(H$_2$O)$_6$]$^{3+}$ has five. A particularly powerful qualitative approach to assessing the structure and reactivity begins with classifying molecules according to electron counting, focusing on the numbers of valence electrons, usually at the central atom in a molecule.

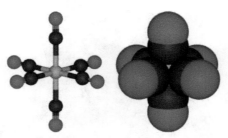

Crystal field theory explains why $[Fe^{III}(CN)_6]^{3-}$ has only one unpaired electron

Molecular Symmetry Group Theory

A central construct in inorganic chemistry is the theory of molecular symmetry. Mathematical group theory provides the language to describe the shapes of molecules according to their point group symmetry. Group theory also enables factoring and simplification of theoretical calculations.

Nitrogen dioxide, NO_2, exhibits C_{2v} symmetry

Spectroscopic features are analyzed and described with respect to the symmetry properties of the, *inter alia*, vibrational or electronic states. Knowledge of the symmetry properties of the ground and excited states allows one to predict the numbers and intensities of absorptions in vibrational and electronic spectra. A classic application of group theory is the prediction of the number of C-O vibrations in substituted metal carbonyl complexes. The most common applications of symmetry to spectroscopy involve vibrational and electronic spectra.

As an instructional tool, group theory highlights commonalities and differences in the bonding of otherwise disparate species, such as WF_6 and $Mo(CO)_6$ or CO_2 and NO_2.

Thermodynamics and Inorganic Chemistry

An alternative quantitative approach to inorganic chemistry focuses on energies of reactions. This approach is highly traditional and empirical, but it is also useful. Broad concepts that are couched in thermodynamic terms include redox potential, acidity, phase changes. A classic concept in inorganic thermodynamics is the Born-Haber cycle, which is used for assessing the energies of elementary processes such as electron affinity, some of which cannot be observed directly.

Mechanistic Inorganic Chemistry

An important and increasingly popular aspect of inorganic chemistry focuses on reaction pathways. The mechanisms of reactions are discussed differently for different classes of compounds.

Main Group Elements and Lanthanides

The mechanisms of main group compounds of groups 13-18 are usually discussed in the context of organic chemistry (organic compounds are main group compounds, after all). Elements heavier than C, N, O, and F often form compounds with more electrons than predicted by the octet rule, as explained in the article on hypervalent molecules. The mechanisms of their reactions differ from organic compounds for this reason. Elements lighter than carbon (B, Be, Li) as well as Al and Mg often form electron-deficient structures that are electronically akin to carbocations. Such electron-deficient species tend to react via associative pathways. The chemistry of the lanthanides mirrors many aspects of chemistry seen for aluminium.

Transition Metal Complexes

Mechanisms for the reactions of transition metals are discussed differently from main group compounds. The important role of d-orbitals in bonding strongly influences the pathways and rates of ligand substitution and dissociation. These themes are covered in articles on coordination chemistry and ligand. Both associative and dissociative pathways are observed.

An overarching aspect of mechanistic transition metal chemistry is the kinetic lability of the complex illustrated by the exchange of free and bound water in the prototypical complexes $[M(H_2O)_6]^{n+}$:

$$[M(H_2O)_6]^{n+} + 6\ H_2O^* \rightarrow [M(H_2O^*)_6]^{n+} + 6\ H_2O$$

where H_2O^* denotes isotopically enriched water, e.g., $H_2{}^{17}O$

The rates of water exchange varies by 20 orders of magnitude across the periodic table, with lanthanide complexes at one extreme and Ir(III) species being the slowest.

Redox Reactions

Redox reactions are prevalent for the transition elements. Two classes of redox reaction are considered: atom-transfer reactions, such as oxidative addition/reductive elimination, and electron-transfer. A fundamental redox reaction is "self-exchange", which involves the degenerate reaction between an oxidant and a reductant. For example, permanganate and its one-electron reduced relative manganate exchange one electron:

$$[MnO_4]^- + [Mn^*O_4]^{2-} \rightarrow [MnO_4]^{2-} + [Mn^*O_4]^-$$

Reactions at Ligands

Coordinated ligands display reactivity distinct from the free ligands. For example, the acidity of the ammonia ligands in $[Co(NH_3)_6]^{3+}$ is elevated relative to NH_3 itself. Alkenes bound to metal cations are reactive toward nucleophiles whereas alkenes normally are not. The large and industrially important area of catalysis hinges on the ability of metals to modify the reactivity of organic ligands. Homogeneous catalysis occurs in solution and heterogeneous catalysis occurs when gaseous or dissolved substrates interact with surfaces of solids. Traditionally homogeneous catalysis is considered part of organometallic chemistry and heterogeneous catalysis is discussed in the context of surface science, a subfield of solid state chemistry. But the basic inorganic chemical principles are the same. Transition metals, almost uniquely, react with small molecules such as CO, H_2, O_2, and C_2H_4. The industrial significance of these feedstocks drives the active area of catalysis. Ligands can also undergo ligand transfer reactions such as transmetalation.

Characterization of Inorganic Compounds

Because of the diverse range of elements and the correspondingly diverse properties of the resulting derivatives, inorganic chemistry is closely associated with many methods of analysis. Older methods tended to examine bulk properties such as the electrical conductivity of solutions, melting points, solubility, and acidity. With the advent of quantum theory and the corresponding expansion of electronic apparatus, new tools have been introduced to probe the electronic properties of inorganic molecules and solids. Often these measurements provide insights relevant to theoretical models. For example, measurements on the photoelectron spectrum of methane demonstrated that describing the bonding by the two-center, two-electron bonds predicted between the carbon and hydrogen using Valence Bond Theory is not appropriate for describing ionisation processes in a simple way. Such insights led to the popularization of molecular orbital theory as fully delocalised orbitals are a more appropriate simple description of electron removal and electron excitation.

Commonly encountered techniques are:

- X-ray crystallography: This technique allows for the 3D determination of molecular structures.

- Dual polarisation interferometer: This technique measures the conformation and conformational change of molecules.

- Various forms of spectroscopy

    o Ultraviolet-visible spectroscopy: Historically, this has been an important tool, since many inorganic compounds are strongly colored

    o NMR spectroscopy: Besides ^{1}H and ^{13}C many other "good" NMR nuclei (e.g., ^{11}B, ^{19}F, ^{31}P, and ^{195}Pt) give important information on compound properties and structure. Also the NMR of paramagnetic species can result in important structural information. Proton NMR is also important because the light hydrogen nucleus is not easily detected by X-ray crystallography.

    o Infrared spectroscopy: Mostly for absorptions from carbonyl ligands

    o Electron nuclear double resonance (ENDOR) spectroscopy

    o Mössbauer spectroscopy

    o Electron-spin resonance: ESR (or EPR) allows for the measurement of the environment of paramagnetic metal centres.

- Electrochemistry: Cyclic voltammetry and related techniques probe the redox characteristics of compounds.

Synthetic Inorganic Chemistry

Although some inorganic species can be obtained in pure form from nature, most are synthesized in chemical plants and in the laboratory.

Inorganic synthetic methods can be classified roughly according to the volatility or solubility of the component reactants. Soluble inorganic compounds are prepared using methods of organic synthesis. For metal-containing compounds that are reactive toward air, Schlenk line and glove box techniques are followed. Volatile compounds and gases are manipulated in "vacuum manifolds" consisting of glass piping interconnected through valves, the entirety of which can be evacuated to 0.001 mm Hg or less. Compounds are condensed using liquid nitrogen (b.p. 78K) or other cryogens. Solids are typically prepared using tube furnaces, the reactants and products being sealed in containers, often made of fused silica (amorphous SiO_2) but sometimes more specialized materials such as welded Ta tubes or Pt "boats". Products and reactants are transported between temperature zones to drive reactions.

Inorganic Compound

An inorganic compound is a compound that is not organic. The term is not well defined, but in its simplest definition refers to compounds that do not contain carbon, and not consisting of or deriving from living matter. Inorganic compounds are traditionally viewed as being synthesized by the agency of geological systems. In contrast, organic compounds are found in biological systems. The distinction between inorganic and organic compounds is not always clear. Organic chemists traditionally refer to any molecule containing carbon as an organic compound and by default this means that inorganic chemistry deals with molecules lacking carbon. As many minerals are of biological origin, biologists may distinguish organic from inorganic compounds in a different way that does not hinge on the presence of a carbon atom. Pools of organic matter, for example, that have been metabolically incorporated into living tissues persist in decomposing tissues, but as molecules become oxidized into the open environment, such as atmospheric CO_2, this creates a separate pool of inorganic compounds. The International Union of Pure and Applied Chemistry, an agency widely recognized for defining chemical terms, does not offer definitions of inorganic or organic. Hence, the definition for an inorganic versus an organic compound in a multidisciplinary context spans the division between organic life living (or animate) and inorganic non-living (or inanimate) matter.

Traditional Usage

The Wöhler synthesis is the conversion of ammonium cyanate into urea. This chemical reaction was discovered in 1828 by Friedrich Wöhler and is considered the starting point of modern organic chemistry.

The Wöhler synthesis is of great historical significance because for the first time an organic compound was produced from inorganic reactants. This finding went against the mainstream theory of that time called vitalism, which stated that organic matter possessed a special force or *vital force* inherent to all things living. For this reason a sharp boundary existed between organic and inorganic compounds. Urea was discovered in 1799 and could until 1828 only be obtained from biological sources such as urine. Wöhler reported to his mentor Berzelius:

"I cannot, so to say, hold my chemical water and must tell you that I can make urea without thereby needing to have kidneys, or anyhow, an animal, be it human or dog".

Modern Usage

Inorganic compounds can be defined as any compound that is not organic compound. Some simple compounds which contain carbon are usually considered inorganic. These include carbon monoxide, carbon dioxide, carbonates, cyanides, cyanates, carbides, and thiocyanates. In contrast, methane and formic acid are generally considered to be simple examples of organic compounds, although the Inorganic Crystal Structure Database (ICSD), in its definition of "inorganic" carbon compounds, states that such compounds may contain *either* C-H or C-C bonds, but not both.

Coordination Chemistry

A large class of compounds discussed in inorganic chemistry textbooks are coordination compounds. Examples range from substances that are strictly inorganic, such as $[Co(NH_3)_6]Cl_3$, to organometallic compounds, such as $Fe(C_5H_5)_2$, and extending to bioinorganic compounds, such as the hydrogenase enzymes.

Mineralogy

Minerals are mainly oxides and sulfides, which are strictly inorganic, although they may be of biological origin. In fact, most of the Earth is inorganic. Although the components of Earth's crust are well-elucidated, the processes of mineralization and the composition of the deep mantle remain active areas of investigation, which are covered mainly in geology-oriented venues.

Coordination Complex

In chemistry, a coordination complex consists of a central atom or ion, which is usually metal-lic and is called the *coordination centre*, and a surrounding array of bound molecules or ions, that are in turn known as *ligands* or complexing agents. Many metal-containing compounds, especially those of transition metals, are coordination complexes. A coordination complex whose centre is a metal atom is called a metal complex.

Cisplatin, $PtCl_2(NH_3)_2$
A platinum atom with four ligands

Nomenclature and Terminology

Coordination complexes are so pervasive that their structures and reactions are described in many ways, sometimes confusingly. The atom within a ligand that is bonded to the central metal atom or ion is called the donor atom. In a typical complex, a metal ion is bonded to several donor atoms, which can be the same or different. A polydentate (multiple bonded) ligand is a molecule or ion that bonds to the central atom through several of the ligand's atoms; ligands with 2, 3, 4 or even 6 bonds to the central atom are common. These complexes are called chelate complexes, the formation of such complexes is called chelation, complexation, and coordination.

The central atom or ion, together with all ligands comprise the coordination sphere. The central atoms or ion and the donor atoms comprise the first coordination sphere.

Coordination refers to the "coordinate covalent bonds" (dipolar bonds) between the ligands and the central atom. Originally, a complex implied a reversible association of molecules, atoms, or ions through such weak chemical bonds. As applied to coordination chemistry, this meaning has evolved. Some metal complexes are formed virtually irreversibly and many are bound together by bonds that are quite strong.

The number of donor atoms attached to the central atom or ion is called the coordination number. The most common coordination numbers are 2, 4 and especially 6. A hydrated ion is one kind of a complex ion (or simply a complex), a species formed between a central metal ion and one or more surrounding ligands, molecules or ions that contain at least one lone pair of electrons,

If all the ligands are monodentate, then the number of donor atoms equals the number of ligands. For example, the cobalt(II) hexahydrate ion or the hexaaquacobalt(II) ion $[Co(H_2O)_6]^{2+}$, is a hydrated-complex ion that consists of six water molecules attached to a metal ion Co. The oxidation state and the coordination number reflect the number of bonds formed between the metal ion and the ligands in the complex ion. However the coordination number of Pt(en)2+ 2is 4 (rather than 2) since it has two bidentate ligands, which contain four donor atoms in total.

History

Coordination complexes have been known since the beginning of modern chemistry. Early well-known coordination complexes include dyes such as Prussian blue. Their properties were first well understood in the late 1800s, following the 1869 work of Christian Wilhelm Blomstrand. Blomstrand developed what has come to be known as the complex ion chain theory. The theory claimed that the reason coordination complexes form is because in solution, ions would be bound via ammonia chains. He compared this effect to the way that various carbohydrate chains form.

Following this theory, Danish scientist Sophus Mads Jorgensen made improvements to it. In his version of the theory, Jorgensen claimed that when a molecule dissociates in a solution there were two possible outcomes: the ions would bind via the ammonia chains Blomstrand had described or the ions would bind directly to the metal.

Alfred Werner

It was not until 1893 that the most widely accepted version of the theory today was published by Alfred Werner. Werner's work included two important changes to the Blomstrand theory. The first was that Werner described the two different ion possibilities in terms of location in the coordination sphere. He claimed that if the ions were to form a chain this would occur outside of the coordination sphere while the ions that bound directly to the metal would do so within the coordination sphere. In one of Werner's most important discoveries however he disproved the majority of the chain theory. Werner was able to discover the spatial arrangements of the ligands that were involved in the formation of the complex hexacoordinate cobalt. His theory allows one to understand the difference between a coordinated ligand and a charge balancing ion in a compound, for example the chloride ion in the cobaltammine chlorides and to explain many of the previously inexplicable isomers.

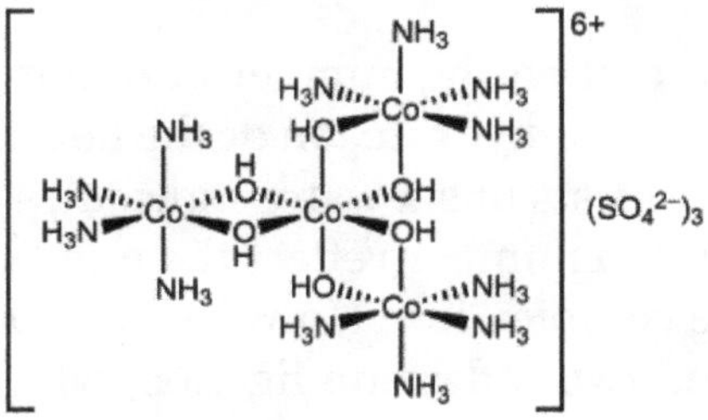

Structure of hexol

In 1914, Werner first resolved the coordination complex, called hexol, into optical isomers, overthrowing the theory that only carbon compounds could possess chirality.

Structures

The ions or molecules surrounding the central atom are called ligands. Ligands are generally bound to the central atom by a coordinate covalent bond (donating electrons from a lone electron pair into an empty metal orbital), and are said to be coordinated to the atom. There are also organic ligands such as alkenes whose pi bonds can coordinate to empty metal orbitals. An example is ethene in the complex known as Zeise's salt, $K^+[PtCl_3(C_2H_4)]^-$.

Geometry

In coordination chemistry, a structure is first described by its coordination number, the number of ligands attached to the metal (more specifically, the number of donor atoms). Usually one can count the ligands attached, but sometimes even the counting can become ambiguous. Coordination numbers are normally between two and nine, but large numbers of ligands are not uncommon for the lanthanides and actinides. The number of bonds depends on the size, charge, and electron configuration of the metal ion and the ligands. Metal ions may have more than one coordination number.

Typically the chemistry of transition metal complexes is dominated by interactions between s and p molecular orbitals of the ligands and the d orbitals of the metal ions. The s, p, and d orbitals of the metal can accommodate 18 electrons. The maximum coordination number for a certain metal is thus related to the electronic configuration of the metal ion (to be more specific, the number of empty orbitals) and to the ratio of the size of the ligands and the metal ion. Large metals and small ligands lead to high coordination numbers, e.g. $[Mo(CN)_8]^{4-}$. Small metals with large ligands lead to low coordination numbers, e.g. $Pt[P(CMe_3)]_2$. Due to their large size, lanthanides, actinides, and early transition metals tend to have high coordination numbers.

Different ligand structural arrangements result from the coordination number. Most structures follow the points-on-a-sphere pattern (or, as if the central atom were in the middle of a polyhedron where the corners of that shape are the locations of the ligands), where orbital overlap (between ligand and metal orbitals) and ligand-ligand repulsions tend to lead to certain regular geometries. The most observed geometries are listed below. There are cases that deviate from a regular geometry due to the use of ligands of different types (which results in irregular bond lengths) or due to the size of ligands.

- Linear for two-coordination

- Trigonal planar for three-coordination

- Tetrahedral or square planar for four-coordination

- Trigonal bipyramidal or square pyramidal for five-coordination

- Octahedral (orthogonal) for six-coordination

- Pentagonal bipyramidal, capped octahedral or capped trigonal prismatic for seven-coordination

- Square antiprismatic or dodecahedral for eight-coordination

- Tri-capped trigonal prismatic (triaugmented triangular prism) or capped square antiprismatic for nine-coordination.

Due to special electronic effects such as (second-order) Jahn–Teller stabilization, certain geometries (in which the coordination atoms do not follow a points-on-a-sphere pattern) are stabilized relative to the other possibilities , e.g. for some compounds the trigonal prismatic geometry is stabilized relative to octahedral structures for six-coordination.

- Bent for two-coordination

- Trigonal pyramidal for three-coordination

- Trigonal prismatic for six-coordination

Isomerism

The arrangement of the ligands is fixed for a given complex, but in some cases it is mutable by a reaction that forms another stable isomer.

There exist many kinds of isomerism in coordination complexes, just as in many other compounds.

Stereoisomerism

Stereoisomerism occurs with the same bonds in different orientations relative to one another. Stereoisomerism can be further classified into:

Cis–trans Isomerism and Facial–meridional Isomerism

Cis–trans isomerism occurs in octahedral and square planar complexes (but not tetrahedral). When two ligands are adjacent they are said to be cis, when opposite each other, trans. When three identical ligands occupy one face of an octahedron, the isomer is said to be facial, or fac. In a *fac* isomer, any two identical ligands are adjacent or *cis* to each other. If these three ligands and the metal ion are in one plane, the isomer is said to be meridional, or mer. A *mer* isomer can be considered as a combination of a *trans* and a *cis*, since it contains both trans and cis pairs of identical ligands.

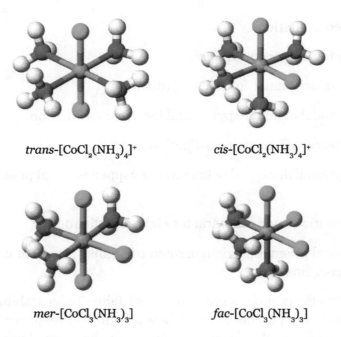

trans-[CoCl$_2$(NH$_3$)$_4$]$^+$ cis-[CoCl$_2$(NH$_3$)$_4$]$^+$

mer-[CoCl$_3$(NH$_3$)$_3$]   fac-[CoCl$_3$(NH$_3$)$_3$]

Optical Isomerism

Optical isomerism occurs when a molecule is not superimposable with its mirror image. It is so called because the two isomers are each optically active, that is, they rotate the plane of polarized light in opposite directions. The symbol Λ (*lambda*) is used as a prefix to describe the left-handed propeller twist formed by three bidentate ligands, as shown. Likewise, the symbol Δ (*delta*) is used as a prefix for the right-handed propeller twist.

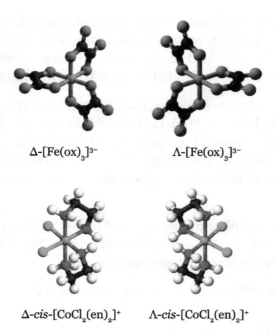

Δ-[Fe(ox)$_3$]$^{3-}$ Λ-[Fe(ox)$_3$]$^{3-}$

Δ-*cis*-[CoCl$_2$(en)$_2$]$^+$ Λ-*cis*-[CoCl$_2$(en)$_2$]$^+$

Structural Isomerism

Structural isomerism occurs when the bonds are themselves different. There are four types of structural isomerism: ionisation isomerism, solvate or hydrate isomerism, linkage isomerism and coordination isomerism.

1. **Ionisation isomerism** – the isomers give different ions in solution although they have the same composition. This type of isomerism occurs when the counter ion of the complex is also a potential ligand. For example pentaamminebromocobalt(III)sulfate [Co(NH$_3$)$_5$Br]SO$_4$ is red violet and in solution gives a precipitate with barium chloride, confirming the presence of sulfate ion, while pentaamminesulfatecobalt(III)bromide [Co(NH$_3$)$_5$SO$_4$]Br is red and tests negative for sulfate ion in solution, but instead gives a precipitate of AgBr with silver nitrate.

2. **Solvate or hydrate isomerism** – the isomers have the same composition but differ with respect to the number of solvent ligand molecules as well as the counter ion in the crystal lattice. For example [Cr(H$_2$O)$_6$]Cl$_3$ is violet colored, [CrCl(H$_2$O)$_5$]Cl$_2$·H$_2$O is blue-green, and [CrCl$_2$(H$_2$O)$_4$]Cl·2H$_2$O is dark green

3. **Linkage isomerism** occurs with ambidentate ligands that can bind in more than one place. For example, NO$_2$ is an ambidentate ligand: It can bind to a metal at either the N atom or an O atom.

4. **Coordination isomerism** – this occurs when both positive and negative ions of a salt are complex ions and the two isomers differ in the distribution of ligands between the cation and the anion. For example [Co(NH$_3$)$_6$][Cr(CN)$_6$] and [Cr(NH$_3$)$_6$][Co(CN)$_6$]

Electronic Properties

Many of the properties of transition metal complexes are dictated by their electronic structures. The electronic structure can be described by a relatively ionic model that ascribes formal charges to the metals and ligands. This approach is the essence of crystal field theory (CFT). Crystal field theory, introduced by Hans Bethe in 1929, gives a quantum mechanically based attempt at understanding complexes. But crystal field theory treats all interactions in a complex as ionic and assumes that the ligands can be approximated by negative point charges.

More sophisticated models embrace covalency, and this approach is described by ligand field theory (LFT) and Molecular orbital theory (MO). Ligand field theory, introduced in 1935 and built from molecular orbital theory, can handle a broader range of complexes and can explain complexes in which the interactions are covalent. The chemical applications of group theory can aid in the understanding of crystal or ligand field theory, by allowing simple, symmetry based solutions to the formal equations.

Chemists tend to employ the simplest model required to predict the properties of interest; for this reason, CFT has been a favorite for the discussions when possible. MO and LF theories are more complicated, but provide a more realistic perspective.

The electronic configuration of the complexes gives them some important properties:

Color of Transition Metal Complexes

Transition metal complexes often have spectacular colors caused by electronic transitions by the absorption of light. For this reason they are often applied as pigments. Most transitions that are related to colored metal complexes are either d–d transitions or charge transfer bands. In a d–d transition, an electron in a d orbital on the metal is excited by a photon to another d orbital of higher energy. A charge transfer band entails promotion of an electron from a metal-based orbital into an empty ligand-based orbital (Metal-to-Ligand Charge Transfer or MLCT). The converse also occurs: excitation of an electron in a ligand-based orbital into an empty metal-based orbital (Ligand to Metal Charge Transfer or LMCT). These phenomena can be observed with the aid of electronic spectroscopy; also known as UV-Vis. For simple compounds with high symmetry, the d–d transitions can be assigned using Tanabe–Sugano diagrams. These assignments are gaining increased support with computational chemistry.

Synthesis of copper(II)-tetraphenylporphyrin, a metal complex, from tetraphenylporphyrin and copper(II) acetate monohydrate.

Colours of Various Example Coordination Complexes						
	Fe^{2+}	Fe^{3+}	Co^{2+}	Cu^{2+}	Al^{3+}	Cr^{3+}
Hydrated Ion	$[Fe(H_2O)_6]^{2+}$ Pale green Solution	$[Fe(H_2O)_6]^{3+}$ Yellow/brown Solution	$[Co(H_2O)_6]^{2+}$ Pink Solution	$[Cu(H_2O)_6]^{2+}$ Blue Solution	$[Al(H_2O)_6]^{3+}$ Colourless Solution	$[Cr(H_2O)_6]^{3+}$ Green Solution
OH⁻, dilute	$[Fe(H_2O)_4(OH)_2]$ Dark green Precipitate	$[Fe(H_2O)_3(OH)_3]$ Brown Precipitate	$[Co(H_2O)_4(OH)_2]$ Blue/green Precipitate	$[Cu(H_2O)_4(OH)_2]$ Blue Precipitate	$[Al(H_2O)_3(OH)_3]$ White Precipitate	$[Cr(H_2O)_3(OH)_3]$ Green Precipitate
OH⁻, concentrated	$[Fe(H_2O)_4(OH)_2]$ Dark green Precipitate	$[Fe(H_2O)_3(OH)_3]$ Brown Precipitate	$[Co(H_2O)_4(OH)_2]$ Blue/green Precipitate	$[Cu(H_2O)_4(OH)_2]$ Blue Precipitate	$[Al(OH)_4]^-$ Colourless Solution	$[Cr(OH)_6]^{3-}$ Green Solution
NH₃, dilute	$[Fe(H_2O)_4(OH)_2]$ Dark green Precipitate	$[Fe(H_2O)_3(OH)_3]$ Brown Precipitate	$[Co(H_2O)_4(OH)_2]$ Blue/green Precipitate	$[Cu(H_2O)_4(OH)_2]$ Blue Precipitate	$[Al(H_2O)_3(OH)_3]$ White Precipitate	$[Cr(H_2O)_3(OH)_3]$ Green Precipitate
NH₃, concentrated	$[Fe(H_2O)_4(OH)_2]$ Dark green Precipitate	$[Fe(H_2O)_3(OH)_3]$ Brown Precipitate	$[Co(NH_3)_6]^{2+}$ Straw coloured Solution	$[Cu(NH_3)_4(H_2O)_2]^{2+}$ Deep blue Solution	$[Al(H_2O)_3(OH)_3]$ White Precipitate	$[Cr(NH_3)_6]^{3+}$ Purple Solution
CO2□ 3	$FeCO_3$ Dark green Precipitate	$[Fe(H_2O)_3(OH)_3]$ Brown Precipitate + bubbles	$CoCO_3$ Pink Precipitate	$CuCO_3$ Blue/green Precipitate		

Colors of Lanthanide Complexes

Superficially lanthanide complexes are similar to those of the transition metals in that some are coloured. However for the common Ln^{3+} ions (Ln = lanthanide) the colors are all pale, and hardly influenced by the nature of the ligand. The colors are due to 4f electron transitions. As the 4f orbitals in lanthanides are "buried" in the xenon core and shielded from the ligand by the 5s and 5p orbitals they are therefore not influenced by the ligands to any great extent leading to a much smaller crystal field splitting than in the transition metals. The absorption spectra of an Ln^{3+} ion approximates to that of the free ion where the electronic states are described by spin-orbit coupling (also called L-S coupling or Russell-Saunders coupling). This contrasts to the transition metals where the ground state is split by the crystal field. Absorptions for Ln^{3+} are weak as electric dipole transitions are parity forbidden (Laporte Rule forbidden) but can gain intensity due to the effect of a low-symmetry ligand field or mixing with higher electronic states (*e.g.* d orbitals). Also absorption bands are extremely sharp which contrasts with those observed for transition metals which generally have broad bands. This can lead to extremely unusual effects, such as significant color changes under different forms of lighting.

Magnetism

Metal complexes that have unpaired electrons are magnetic. Considering only monometallic complexes, unpaired electrons arise because the complex has an odd number of electrons or because electron pairing is destabilized. Thus, monomeric Ti(III) species have one "d-electron" and must be (para)magnetic, regardless of the geometry or the nature of the ligands. Ti(II), with two d-electrons, forms some complexes that have two unpaired electrons and others with none. This effect is illustrated by the compounds $TiX_2[(CH_3)_2PCH_2CH_2P(CH_3)_2]_2$: when X = Cl, the complex is paramagnetic (high-spin configuration), whereas when X = CH_3, it is diamagnetic (low-spin configuration). It is important to realize that ligands provide an important means of adjusting the ground state properties.

In bi- and polymetallic complexes, in which the individual centres have an odd number of electrons or that are high-spin, the situation is more complicated. If there is interaction (either direct or through ligand) between the two (or more) metal centres, the electrons may couple (antiferromagnetic coupling, resulting in a diamagnetic compound), or they may enhance each other (ferromagnetic coupling). When there is no interaction, the two (or more) individual metal centers behave as if in two separate molecules.

Reactivity

Complexes show a variety of possible reactivities:

- Electron transfers

    A common reaction between coordination complexes involving ligands are inner and outer sphere electron transfers. They are two different mechanisms of electron transfer redox reactions, largely defined by the late Henry Taube. In an inner sphere reaction, a ligand with two lone electron pairs acts as a *bridging ligand*, a ligand to which both coordination centres can bond. Through this, electrons are transferred from one centre to another.

- (Degenerate) ligand exchange

    One important indicator of reactivity is the rate of degenerate exchange of ligands. For example, the rate of interchange of coordinate water in $[M(H_2O)_6]^{n+}$ complexes varies over 20 orders of magnitude. Complexes where the ligands are released and rebound rapidly are classified as labile. Such labile complexes can be quite stable thermodynamically. Typical labile metal complexes either have low-charge (Na^+), electrons in d-orbitals that are antibonding with respect to the ligands (Zn^{2+}), or lack covalency (Ln^{3+}, where Ln is any lanthanide). The lability of a metal complex also depends on the high-spin vs. low-spin configurations when such is possible. Thus, high-spin Fe(II) and Co(III) form labile complexes, whereas low-spin analogues are inert. Cr(III) can exist only in the low-spin state (quartet), which is inert because of its high formal oxidation state, absence of electrons in orbitals that are M–L antibonding, plus some "ligand field stabilization" associated with the d^3 configuration.

- Associative processes

    Complexes that have unfilled or half-filled orbitals often show the capability to react with substrates. Most substrates have a singlet ground-state; that is, they have lone electron pairs (e.g., water, amines, ethers), so these substrates need an empty orbital to be able to react with a metal centre. Some substrates (e.g., molecular oxygen) have a triplet ground state, which results that metals with half-filled orbitals have a tendency to react with such substrates (it must be said that the dioxygen molecule also has lone pairs, so it is also capable to react as a 'normal' Lewis base).

If the ligands around the metal are carefully chosen, the metal can aid in (stoichiometric or catalytic) transformations of molecules or be used as a sensor.

Classification

Metal complexes, also known as coordination compounds, include all metal compounds, aside from metal vapors, plasmas, and alloys. The study of "coordination chemistry" is the study of "inorganic chemistry" of all alkali and alkaline earth metals, transition metals, lanthanides, actinides, and metalloids. Thus, coordination chemistry is the chemistry of the majority of the periodic table. Metals and metal ions exist, in the condensed phases at least, only surrounded by ligands.

The areas of coordination chemistry can be classified according to the nature of the ligands, in broad terms:

- Classical (or "Werner Complexes"): Ligands in classical coordination chemistry bind to metals, almost exclusively, via their "lone pairs" of electrons residing on the main group atoms of the ligand. Typical ligands are H_2O, NH_3, Cl^-, CN^-, en. Some of the simplest members of such complexes are described in metal aquo complexes, metal ammine complexes,

  Examples: $[Co(EDTA)]^-$, $[Co(NH_3)_6]Cl_3$, $[Fe(C_2O_4)_3]K_3$

- Organometallic Chemistry: Ligands are organic (alkenes, alkynes, alkyls) as well as "organic-like" ligands such as phosphines, hydride, and CO.

  Example: $(C_5H_5)Fe(CO)_2CH_3$

- Bioinorganic Chemistry: Ligands are those provided by nature, especially including the side chains of amino acids, and many cofactors such as porphyrins.

  Example: hemoglobin contains heme, a porphyrin complex of iron

  Example: chlorophyll contains a porphyrin complex of magnesium

  Many natural ligands are "classical" especially including water.

- Cluster Chemistry: Ligands are all of the above also include other metals as ligands.

  Example $Ru_3(CO)_{12}$

- In some cases there are combinations of different fields:

  Example: $[Fe_4S_4(Scysteinyl)_4]^{2-}$, in which a cluster is embedded in a biologically active species.

Mineralogy, materials science, and solid state chemistry – as they apply to metal ions – are subsets of coordination chemistry in the sense that the metals are surrounded by ligands. In many cases these ligands are oxides or sulfides, but the metals are coordinated nonetheless, and the principles and guidelines discussed below apply. In hydrates, at least some of the ligands are water molecules. It is true that the focus of mineralogy, materials science, and solid state chemistry differs from the usual focus of coordination or inorganic chemistry. The former are concerned primarily with polymeric structures, properties arising from a collective effects of many highly interconnected metals. In contrast, coordination chemistry focuses on reactivity and properties of complexes containing individual metal atoms or small ensembles of metal atoms.

Older Classifications of Isomerism

Traditional classifications of the kinds of isomer have become archaic with the advent of modern structural chemistry. In the older literature, one encounters:

- Ionisation isomerism describes the possible isomers arising from the exchange between the outer sphere and inner sphere. This classification relies on an archaic classification of the inner and outer sphere. In this classification, the "outer sphere ligands," when ions in solution, may be switched with "inner sphere ligands" to produce an isomer.

- Solvation isomerism occurs when an inner sphere ligand is replaced by a solvent molecule. This classification is obsolete because it considers solvents as being distinct from other ligands. Some of the problems are discussed under water of crystallization.

Naming Complexes

The basic procedure for naming a complex:

1. When naming a complex ion, the ligands are named before the metal ion.

2. Write the names of the ligands in alphabetical order. (Numerical prefixes do not affect the order.)

   o Multiple occurring monodentate ligands receive a prefix according to the number of occurrences: *di-*, *tri-*, *tetra-*, *penta-*, or *hexa*. Polydentate ligands (e.g., ethylene-diamine, oxalate) receive *bis-*, *tris-*, *tetrakis-*, etc.

   o Anions end in *o*. This replaces the final 'e' when the anion ends with '-ide', '-ate' or '-ite', e.g. *chloride* becomes *chlorido* and *sulfate* becomes *sulfato*. Formerly, '-ide' was changed to '-o' (e.g. *chloro* and *cyano*), but this rule has been modified in the 2005 IUPAC recommendations and the correct forms for these ligands are now *chloro* and *cyanido*.

   o Neutral ligands are given their usual name, with some exceptions: NH_3 becomes *ammine*; H_2O becomes *aqua* or *aquo*; CO becomes *carbonyl*; NO becomes *nitrosyl*.

3. Write the name of the central atom/ion. If the complex is an anion, the central atom's name will end in *-ate*, and its Latin name will be used if available (except for mercury).

1. If the central atom's oxidation state needs to be specified (when it is one of several possible, or zero), write it as a Roman numeral (or 0) in parentheses.

2. Name cation then anion as separate words (if applicable, as in last example)

Examples:

metal	changed to
cobalt	cobaltate
aluminium	aluminate
chromium	chromate
vanadium	vanadate
copper	cuprate
iron	ferrate

$[NiCl_4]^{2-} \rightarrow$ tetrachloronickelate(II) ion

$[CuCl_5NH_3]^{3-} \rightarrow$ amminepentachlorocuprate(II) ion

$[Cd(CN)_2(en)_2] \rightarrow$ dicyanobis(ethylenediamine)cadmium(II)

$[CoCl(NH_3)_5]SO_4 \rightarrow$ pentaamminechlorocobalt(III) sulfate

The coordination number of ligands attached to more than one metal (bridging ligands) is indicated by a subscript to the Greek symbol μ placed before the ligand name. Thus the dimer of aluminium trichloride is described by $Al_2Cl_4(\mu_2\text{-}Cl)_2$.

Stability Constant

The affinity of metal ions for ligands is described by stability constant. This constant, also referred to as the formation constant, is given the notation of K_f and can be calculated through the following method for simple cases:

$$(X)\text{Metal}_{(aq)} + (Y)\text{Lewis Base}_{(aq)} = (Z)\text{Complex Ion}_{(aq)}$$

$$K_f = \frac{[\text{Complex ion}]^Z}{[\text{Metal ion}]^X [\text{Lewis base}]^Y}$$

Formation constants vary widely. Large values indicate that the metal has high affinity for the ligand, provided the system is at equilibrium.

Sometimes the stability constant will be in a different form known as the constant of destability. This constant is expressed as the inverse of the constant of formation and is denoted as $K_d = 1/K_f$. This constant represents the reverse reaction for the decomposition of a complex ion into its individual metal and ligand components. When comparing the values for K_d, the larger the value is the more unstable the complex ion is.

As a result of these complex ions forming in solutions they also can play a key role in solubility of other compounds. When a complex ion is formed it can alter the concentrations of its components in the solution. For example:

$$Ag^+_{(aq)} + 2NH_4OH_{(aq)} = Ag(NH_3)^+_2 + H_2O$$

$$AgCl_{(s)} + H_2O_{(l)} = Ag^+_{(aq)} + Cl^-_{(aq)}$$

In these reactions which both occurred in the same reaction vessel, the solubility of the silver chloride would be increased as a result of the formation of the complex ion. The complex ion formation is favorable takes away a significant portion of the silver ions in solution, as a result the equilibrium for the formation of silver ions from silver chloride will shift to the right to make up for the deficit.

This new solubility can be calculated given the values of K_f and K_{sp} for the original reactions. The solubility is found essentially by combining the two separate equilibria into one combined equilibrium reaction and this combined reaction is the one that determines the new solubility. So K_c, the new solubility constant, is denoted by $K_c = K_{sp} * K_f$.

Application of Coordination Compounds

Metals only exist in solution as coordination complexes, it follows then that this class of compounds are useful. Coordination compounds are found both in the natural world and artificially in industry. Some common complex ions include such substances as vitamin B12 , hemoglobin , chlorophyll, and some dyes and pigments. One major use of coordination compounds is in homogeneous catalysis for the production of organic substances.

Coordination compounds have uses in both nature and in industry. Coordination compounds are vital to many living organisms. For example many enzymes are metal complexes, like carboxypeptidase, a hydrolytic enzyme important in digestion. This enzyme consists of a zinc ion surrounded by many amino acid residues. Another complex ion enzyme is catalase, which decomposes the cell waste hydrogen peroxide. This enzyme contains iron-porphyrin complexes, similar to that in hemoglobin. Chlorophyll contains a magnesium-porphyrin complexes, and vitamin B12 is a complex with cobalt and corrin.

Coordination compounds are also widely used in industry. The intense colors of many compounds render them of great use as dyes and pigments. Specifically Phthalocyanine complexes are an important class of dyes for fabrics. Nickel, cobalt, and copper can be extracted using hydrometallurgical processes involving complex ions. They are extracted from their ores as ammine complexes with aqueous ammonia. Metals can also be separated using the selective precipitation and solubility of complex ions, as explained in later sections. Cyanide complexes are often used in electroplating.

Coordination compounds can also be used to identify unknown substances in a solution. This analysis can be done by utilizing the selective precipitation of the complex ions, the formation of color complexes which can be measured spectrophotometrically, or the preparation of complexes, such as metal acetylacetonates, which can be separated with organic solvents.

A combination of titanium trichloride and triethylaluminum brings about the polymerization of organic compounds with carbon-carbon double bonds to form polymers of high molecular weight and ordered structures. Many of these polymers are of great commercial importance because they are used in common fibers, films, and plastics.

Other common uses of coordination compounds in industry include the following:

- They are used in photography, i.e., AgBr forms a soluble complex with sodium thiosulfate in photography.
- $K[Ag(CN)_2]$ is used forelectroplating of silver, and $K[Au(CN)_2]$ is used for gold plating.
- Some ligands oxidise Co^{2+} to Co^{3+} ion.
- Ethylenediaminetetraacetic acid (EDTA) is used for estimation of Ca^{2+} and Mg^{2+} in hard water.
- Silver and gold are extracted by treating zinc with their cyanide complexes

References

- Cotton, Frank Albert; Geoffrey Wilkinson; Carlos A. Murillo (1999). Advanced Inorganic Chemistry. p. 1355. ISBN 978-0-471-19957-1.
- Cotton, F. Albert; Wilkinson, Geoffrey; Murillo, Carlos A.; Bochmann, Manfred (1999), Advanced Inorganic Chemistry (6th ed.), New York: Wiley-Interscience, ISBN 0-471-19957-5

- Greenwood, Norman N.; Earnshaw, Alan (1997). Chemistry of the Elements (2nd ed.). Butterworth-Heinemann. ISBN 0-08-037941-9.

- Elschenbroich, C.; Salzer, A. (1992). Organometallics: A Concise Introduction (2nd ed.). Weinheim: Wiley-VCH. ISBN 3527281649.

- S. J. Lippard; J. M. Berg (1994). Principles of Bioinorganic Chemistry. Mill Valley, CA: University Science Books. ISBN 0-935702-73-3.

- R. G. Wilkins (1991). Kinetics and Mechanism of Reactions of Transition Metal Complexes (2nd ed.). Wiley-VCH. ISBN 3-527-28389-7.

- Girolami, G. S.; Rauchfuss, T. B.; Angelici, R. J. (1999). Synthesis and Technique in Inorganic Chemistry (3rd ed.). Mill Valley, CA: University Science Books. ISBN 978-0935702484

5

IUPAC Nomenclature of Organic and Inorganic Chemistry

The nomenclature of organic chemistry is commonly referred to as the blue book and is explained in this chapter along with IUPAC nomenclature of organic chemistry, international chemical identifier and simplified molecular- input line- entry system. The topics discussed in the chapter are of great importance to broaden the existing knowledge on organic chemistry and inorganic chemistry.

IUPAC Nomenclature of Organic Chemistry

In chemical nomenclature, the IUPAC nomenclature of organic chemistry is a systematic method of naming organic chemical compounds as recommended by the International Union of Pure and Applied Chemistry (IUPAC). It is published in the *Nomenclature of Organic Chemistry* (informally called the Blue Book). Ideally, every possible organic compound should have a name from which an unambiguous structural formula can be created. There is also an IUPAC nomenclature of inorganic chemistry.

For ordinary communication, to spare a tedious description, the official IUPAC naming recommendations are not always followed in practice, except when it is necessary to give an unambiguous and absolute definition to a compound, or when the IUPAC name is simpler (e.g. ethanol instead of ethyl alcohol). Otherwise the common or trivial name may be used, often derived from the source of the compound. In addition, very long names may be less concise than structural formulae.

Basic Principles

In chemistry, a number of prefixes, suffixes and infixes are used to describe the type and position of functional groups in the compound.

The steps for naming an organic compound are:

1. Identification of the parent hydrocarbon chain. This chain must obey the following rules, in order of precedence:

    1. It should have the maximum number of substituents of the suffix functional group. By suffix, it is meant that the parent functional group should have a suffix, unlike halogen substituents. If more than one functional group is present, the one with highest precedence should be used.

    2. It should have the maximum number of multiple bonds

3. It should have the maximum number of single bonds.

4. It should have the maximum length.

2. Identification of the parent functional group, if any, with the highest order of precedence.

3. Identification of the side-chains. *Side chains are the carbon chains that are not in the parent chain, but are branched off from it.*

4. Identification of the remaining functional groups, if any, and naming them by their ionic prefixes (such as hydroxy for -OH, oxy for =O, oxyalkane for O-R, etc.). Different side-chains and functional groups will be grouped together in alphabetical order. (The prefixes di-, tri-, etc. are not taken into consideration for grouping alphabetically. For example, ethyl comes before dihydroxy or dimethyl, as the "e" in "ethyl" precedes the "h" in "dihydroxy" and the "m" in "dimethyl" alphabetically. The "di" is not considered in either case). When both side chains and secondary functional groups are present, they should be written mixed together in one group rather than in two separate groups.

5. Identification of double/triple bonds.

6. Numbering of the chain. This is done by first numbering the chain in both directions (left to right and right to left), and then choosing the numbering which follows these rules, in order of precedence

    1. Has the lowest-numbered locant (or locants) for the suffix functional group. Locants are the numbers on the carbons to which the substituent is directly attached.

    2. Has the lowest-numbered locants for multiple bonds (The locant of a multiple bond is the number of the adjacent carbon with a lower number).

    3. Has the lowest-numbered locants for prefixes.

7. Numbering of the various substituents and bonds with their locants. If there is more than one of the same type of substituent/double bond, a prefix is added showing how many there are (di – 2 tri – 3 tetra – 4 then as for the number of carbons below with 'a' added)

The numbers for that type of side chain will be grouped in ascending order and written before the name of the side-chain. If there are two side-chains with the same alpha carbon, the number will be written twice. Example: 2,2,3-trimethyl- . If there are both double bonds and triple bonds, "en" (double bond) is written before "yne" (triple bond). When the main functional group is a terminal functional group (A group which can only exist at the end of a chain, like formyl and carboxyl groups), there is no need to number it.

1. Arrangement in this form: Group of side chains and secondary functional groups with numbers made in step 3 + prefix of parent hydrocarbon chain (eth, meth) + double/triple bonds with numbers (or "ane") + primary functional group suffix with numbers. Wherever it says "with numbers", it is understood that between the word and the numbers, the prefix(di-, tri-) is used.

2. Adding of punctuation:

1. Commas are put between numbers (2 5 5 becomes 2,5,5)

2. Hyphens are put between a number and a letter (2 5 5 trimethylheptane becomes 2,5,5-trimethylheptane)

3. Successive words are merged into one word (trimethyl heptane becomes trimethylheptane)

Note: IUPAC uses one-word names throughout. This is why all parts are connected.

The finalized name should look like this:

 #,#-di<side chain>-#-<secondary functional group>-#-<side chain>-#,#,#-tri<secondary functional group><parent chain prefix><If all bonds are single bonds, use "ane">-#,#-di<double bonds>-#-<triple bonds>-#-<primary functional group>

Note: # is used for a number. The group secondary functional groups and side chains may not look the same as shown here, as the side chains and secondary functional groups are arranged alphabetically. The di- and tri- have been used just to show their usage. (di- after #,#, tri- after #,#,#, etc.)

Example

Here is a sample molecule with the parent carbons numbered:

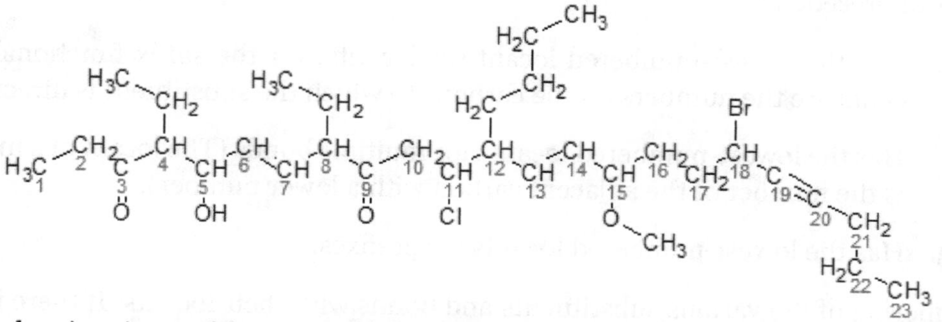

For simplicity, here is an image of the same molecule, where the hydrogens in the parent chain are removed and the carbons are shown by their numbers:

Now, following the above steps:

1. The parent hydrocarbon chain has 23 carbons. It is called tricosa-.

2. The functional groups with the highest precedence are the two ketone groups.

    1. The groups are on carbon atoms 3 and 9. As there are two, we write 3,9-dione.

    2. The numbering of the molecule is based on the ketone groups. When numbering from left to right, the ketone groups are numbered 3 and 9. When numbering from right to left, the ketone groups are numbered 15 and 21. 3 is less than 15, therefore the ketones are numbered 3 and 9. The smaller number is always used, not the sum of the constituents numbers.

3. The side chains are: an ethyl- at carbon 4, an ethyl- at carbon 8, and a butyl- at carbon 12.

Note:The -O-CH$_3$ at carbon atom 15 is not a side chain, but it is a methoxy functional group

- o   There are two ethyl- groups. They are combined to create, 4,8-diethyl.

- o   The side chains are grouped like this: 12-butyl-4,8-diethyl. (But this is not the final grouping, as functional groups may be added in between.)

1. The secondary functional groups are: a hydroxy- at carbon 5, a chloro- at carbon 11, a methoxy- at carbon 15, and a bromo- at carbon 18. Grouped with the side chains, this gives 18-bromo-12-butyl-11-chloro-4,8-diethyl-5-hydroxy-15-methoxy

2. There are two double bonds: one between carbons 6 and 7, and one between carbons 13 and 14. They would be called "6,13-diene", but the presence of alkynes switches it to 6,13-dien. There is one triple bond between carbon atoms 19 and 20. It will be called 19-yne

3. The arrangement (with punctuation) is: 18-bromo-12-butyl-11-chloro-4,8-diethyl-5-hydroxy-15-methoxytricosa-6,13-dien-19-yne-3,9-dione

4. Finally, due to Cis-trans isomerism, we have to specify the relative orientation of functional groups around each double bond. For this example, we have (6E,13E)

The final name is (6E,13E)-18-bromo-12-butyl-11-chloro-4,8-diethyl-5-hydroxy-15-methoxytricosa-6,13-dien-19-yne-3,9-dione.

Alkanes

Straight-chain alkanes take the suffix "-ane" and are prefixed depending on the number of carbon atoms in the chain, following standard rules. The first few are:

Number of carbons	1	2	3	4	5	6	7	8	9	10
Prefix	Meth	Eth	Prop	But	Pent	Hex	Hept	Oct	Non	Dec

11	12	13	14	15	20	30	40	50
Undec	Dodec	Tridec	Tetradec	Pentadec	Eicos	Triacont	Tetracont	Pentacont

For example, the simplest alkane is CH$_4$ methane, and the nine-carbon alkane CH$_3$(CH$_2$)$_7$CH$_3$ is named nonane. The names of the first four alkanes were derived from methanol, ether, propionic acid and butyric acid, respectively. The rest are named with a Greek numeric prefix, with the exceptions of nonane which has a Latin prefix, and undecane and tridecane which have mixed-language prefixes.

Cyclic alkanes are simply prefixed with "cyclo-": for example, C$_4$H$_8$ is cyclobutane and C$_6$H$_{12}$ is cyclohexane.

Branched alkanes are named as a straight-chain alkane with attached alkyl groups. They are prefixed with a number indicating the carbon the group is attached to, counting from the end of the alkane chain. For example, (CH$_3$)$_2$CHCH$_3$, commonly known as isobutane, is treated as a propane chain with a methyl group bonded to the middle (2) carbon, and given the systematic name 2-methylpropane. However, although the name 2-methylpropane *could* be used, it is easier and more logical to call it simply methylpropane – the methyl group could not possibly occur on any of the other carbon atoms (that would lengthen the chain and result in butane, not propane) and therefore the use of the number "2" is unnecessary.

2-methylpropane   2-methylbutane   3-methylbutane
(isobutane)   (correct numbering)   (incorrect numbering)

If there is ambiguity in the position of the substituent, depending on which end of the alkane chain is counted as "1", then numbering is chosen so that the smaller number is used. For example, $(CH_3)_2CHCH_2CH_3$ (isopentane) is named 2-methylbutane, not 3-methylbutane.

2,2-dimethylpropane
(neopentane)

If there are multiple side-branches of the same size alkyl group, their positions are separated by commas and the group prefixed with di-, tri-, tetra-, etc., depending on the number of branches. For example, $C(CH_3)_4$ (neopentane) is named 2,2-dimethylpropane. If there are different groups, they are added in alphabetical order, separated by commas or hyphens: . The longest possible main alkane chain is used; therefore 3-ethyl-4-methylhexane instead of 2,3-diethylpentane, even though these describe equivalent structures. The di-, tri- etc. prefixes are ignored for the purpose of alphabetical ordering of side chains (e.g. 3-ethyl-2,4-dimethylpentane, not 2,4-dimethyl-3-ethylpentane).

3-ethyl-4-methylhexane   2,3-diethylpentane   3-methyl-4-propyloctane
                                (incorrect)

Alkenes and Alkynes

but-1-ene   buta-1,3-diene

trans-but-2-ene   cis-but-2-ene
(E)-but-2-ene   (Z)-but-2-ene

Alkenes are named for their parent alkane chain with the suffix "-ene" and an infixed number indicating the position of the carbon with the lower number for each double bond in the chain: $CH_2=CHCH_2CH_3$ is but-1-ene. Multiple double bonds take the form -diene, -triene, etc., with the size prefix of the chain taking an extra "a": $CH_2=CHCH=CH_2$ is buta-1,3-diene. Simple cis and trans isomers may be indicated with a prefixed *cis-* or *trans-*: *cis*-but-2-ene, *trans*-but-2-ene. However, *cis-* and *trans-* are *relative* descriptors. It is IUPAC convention to describe all alkenes using *absolute* descriptors of *Z-* (same side) and *E-* (opposite) with the Cahn–Ingold–Prelog priority rules.

ethyne
(acetylene)

propyne
(methylacetylene)

Alkynes are named using the same system, with the suffix "-yne" indicating a triple bond: ethyne (acetylene), propyne (methylacetylene).

Functional Groups

Alcohols

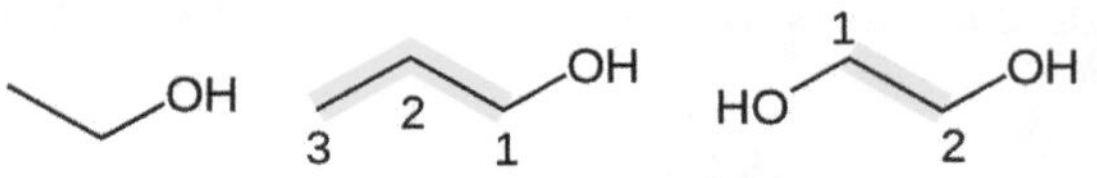

ethanol
(ethyl alcohol)

propan-1-ol
(*n*-propyl alcohol)

ethane-1,2-diol
(ethylene glycol)

Alcohols (R-OH) take the suffix "-ol" with an infix numerical bonding position: $CH_3CH_2CH_2OH$ is propan-1-ol. The suffixes -diol, -triol, -tetraol, etc., are used for multiple -OH groups: Ethylene glycol CH_2OHCH_2OH is ethane-1,2-diol.

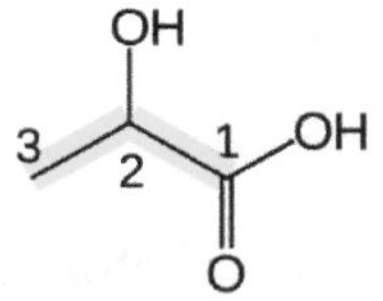

2-hydroxypropanoic acid

If higher precedence functional groups are present, the prefix "hydroxy" is used with the bonding position: $CH_3CHOHCOOH$ is 2-hydroxypropanoic acid.

Halogens (Alkyl Halides)

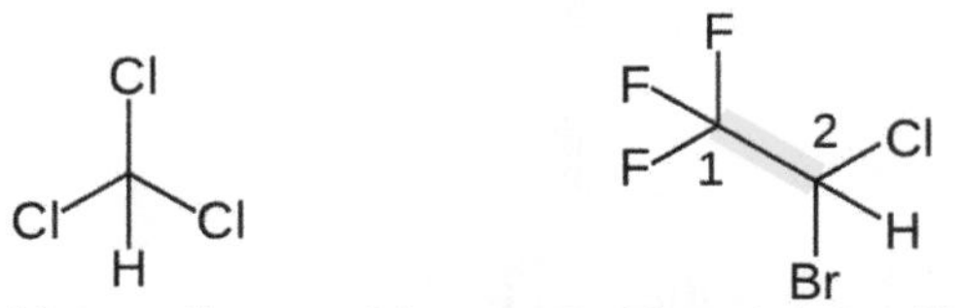

trichloromethane
(chloroform)

2-bromo-2-chloro-1,1,1,-trifluoroethane
(halothane)

Halogen functional groups are prefixed with the bonding position and take the form fluoro-, chloro-, bromo-, iodo-, etc., depending on the halogen. Multiple groups are dichloro-, trichloro-, etc., and dissimilar groups are ordered alphabetically as before. For example, $CHCl_3$ (chloroform) is trichloromethane. The anesthetic Halothane ($CF_3CHBrCl$) is 2-bromo-2-chloro-1,1,1-trifluoroethane.

Ketones

In general ketones (R-CO-R) take the suffix "-one" (pronounced *own*, not *won*) with an infix position number: $CH_3CH_2CH_2COCH_3$ is pentan-2-one. If a higher precedence suffix is in use, the prefix "oxo-" is used: $CH_3CH_2CH_2COCH_2CHO$ is 3-oxohexanal.

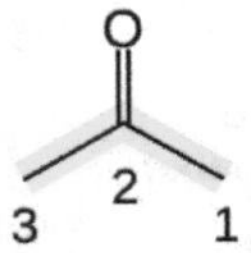

propan-2-one
(acetone)

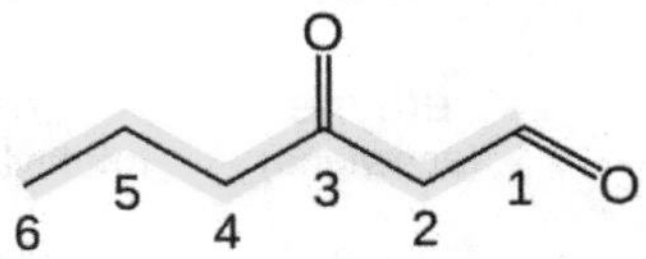

3-oxohexanal

Aldehydes

Aldehydes (R-CHO) take the suffix "-al". If other functional groups are present, the chain is numbered such that the aldehyde carbon is in the "1" position, unless functional groups of higher precedence are present.

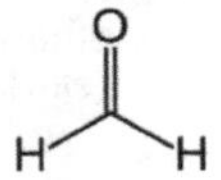

methanal
(formaldehyde)

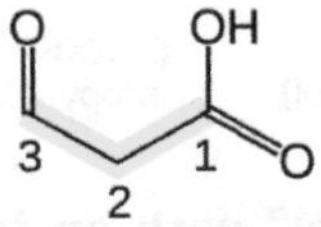

3-oxopropanoic acid

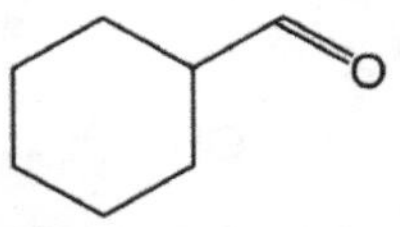

cyclohexanecarbaldehyde

If a prefix form is required, "oxo-" is used (as for ketones), with the position number indicating the end of a chain: $CHOCH_2COOH$ is 3-oxopropanoic acid. If the carbon in the carbonyl group cannot be included in the attached chain (for instance in the case of cyclic aldehydes), the prefix "formyl-" or the suffix "-carbaldehyde" is used: $C_6H_{11}CHO$ is cyclohexanecarbaldehyde. If an aldehyde is attached to a benzene and is the main functional group, the suffix becomes benzaldehyde.

Carboxylic Acids

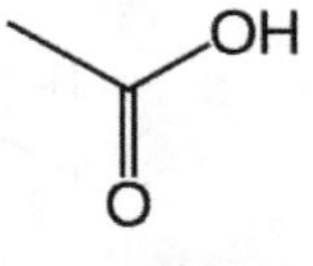

ethanoic acid
(acetic acid)

In general carboxylic acids are named with the suffix -*oic acid* (etymologically a back-formation from benzoic acid). Similar to aldehydes, they take the "1" position on the parent chain, but do not have their position number indicated. For example, $CH_3CH_2CH_2CH_2COOH$ (valeric acid) is named pentanoic acid. For common carboxylic acids some traditional names such as acetic acid are in such widespread use they are considered retained IUPAC names, although "systematic" names such as ethanoic acid are also acceptable. For carboxylic acids attached to a benzene ring such as Ph-COOH, these are named as benzoic acid or its derivatives.

If there are multiple carboxyl groups on the same parent chain, the suffix "-carboxylic acid" can be used (as -dicarboxylic acid, -tricarboxylic acid, etc.). In these cases, the carbon in the carboxyl group does *not* count as being part of the main alkane chain. The same is true for the prefix form, "carboxyl-". Citric acid is one example; it is named 2-hydroxypropane- 1,2,3-tricarboxylic acid, rather than 3-carboxy-3-hydroxypentanedioic acid.

Ethers

Ethers (R-O-R) consist of an oxygen atom between the two attached carbon chains. The shorter of the two chains becomes the first part of the name with the -ane suffix changed to -oxy, and the longer alkane chain becomes the suffix of the name of the ether. Thus, CH_3OCH_3 is methoxymethane, and $CH_3OCH_2CH_3$ is methoxyethane (*not* ethoxymethane). If the oxygen is not attached to the end of the main alkane chain, then the whole shorter alkyl-plus-ether group is treated as a side-chain and prefixed with its bonding position on the main chain. Thus $CH_3OCH(CH_3)_2$ is 2-methoxypropane.

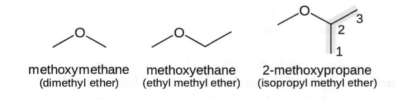

methoxymethane	methoxyethane	2-methoxypropane
(dimethyl ether)	(ethyl methyl ether)	(isopropyl methyl ether)

Esters

Esters (R-CO-O-R') are named as alkyl derivatives of carboxylic acids. The alkyl (R') group is named first. The R-CO-O part is then named as a separate word based on the carboxylic acid name, with the ending changed from -*oic acid* to -*oate*. For example, $CH_3CH_2CH_2CH_2COOCH_3$ is *methyl pentanoate*, and $(CH_3)_2CHCH_2CH_2COOCH_2CH_3$ is *ethyl 4-methylpentanoate*. For esters such as ethyl acetate ($CH_3COOCH_2CH_3$), ethyl formate ($HCOOCH_2CH_3$) or dimethyl phthalate that are based on common acids, IUPAC recommends use of these established names, called retained names. The -*oate* changes to -*ate*. Some simple examples, named both ways, are shown in the figure above.

methyl methanoate	methyl ethanoate	ethyl methanoate
(formate)	(acetate)	(ethyl formate)

If the alkyl group is not attached at the end of the chain, the bond position to the ester group is infixed before "-yl": $CH_3CH_2CH(CH_3)OOCCH_2CH_3$ may be called but-2-yl propanoate or but-2-yl propionate.

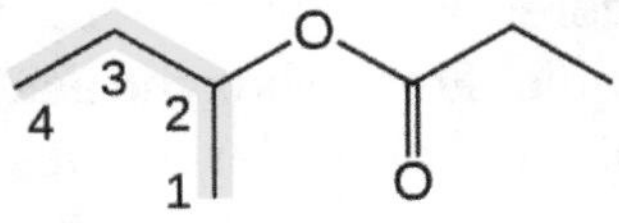

but-2-yl propanoate

Amines and Amides

Amines ($R-NH_2$) are named for the attached alkane chain with the suffix "-amine" (e.g. CH_3NH_2 methanamine). If necessary, the bonding position is infixed: $CH_3CH_2CH_2NH_2$ propan-1-amine, $CH_3CHNH_2CH_3$ propan-2-amine. The prefix form is "amino-".

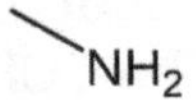

methanamine
(methylamine)

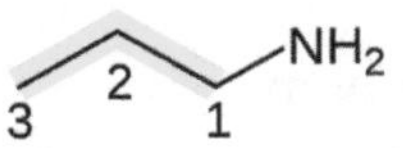

propan-1-amine
(*n*-propyl amine)

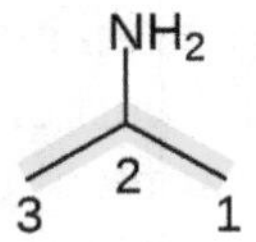

propan-2-amine
(isopropyl amine)

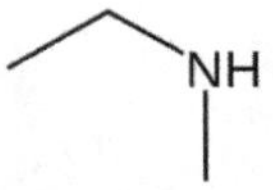

N-methylethanamine 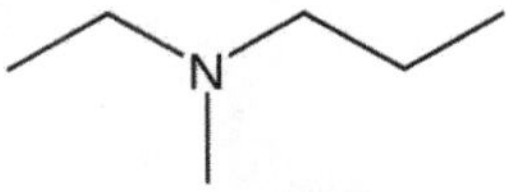　　*N*-ethyl-*N*-methylpropanamine

For secondary amines (of the form R-NH-R), the longest carbon chain attached to the nitrogen atom becomes the primary name of the amine; the other chain is prefixed as an alkyl group with location prefix given as an italic *N*: $CH_3NHCH_2CH_3$ is *N*-methylethanamine. Tertiary amines (R-NR-R) are treated similarly: $CH_3CH_2N(CH_3)CH_2CH_2CH_3$ is *N*-ethyl-*N*-methylpropanamine. Again, the substituent groups are ordered alphabetically.

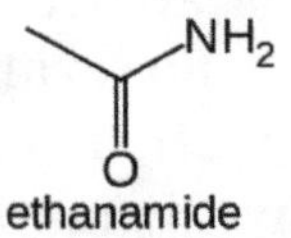

ethanamide
(acetamide)

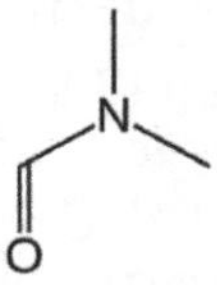

N,N-dimethylmethanamide

Amides ($R-CO-NH_2$) take the suffix "-amide", or "-carboxamide" if the carbon in the amide group cannot be included in the main chain. The prefix form is both "carbamoyl-" and "amido-".

Amides that have additional substituents on the nitrogen are treated similarly to the case of amines: they are ordered alphabetically with the location prefix *N*: $HCON(CH_3)_2$ is *N,N*-dimethyl-methanamide.

Cyclic Compounds

Cycloalkanes and aromatic compounds can be treated as the main parent chain of the compound, in which case the position of substituents are numbered around the ring structure. For example, the three isomers of xylene $CH_3C_6H_4CH_3$, commonly the *ortho-*, *meta-*, and *para-* forms, are

1,2-dimethylbenzene, 1,3-dimethylbenzene, and 1,4-dimethylbenzene. The cyclic structures can also be treated as functional groups themselves, in which case they take the prefix "cyclo*alkyl*-" (e.g. "cyclohexyl-") or for benzene, "phenyl-".

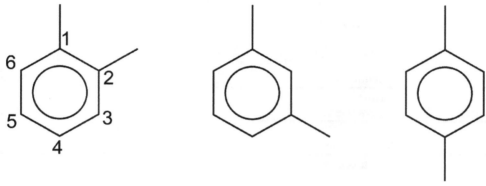

1,2-dimethylbenzene   **1,3-dimethylbenzene**   **1,4-dimethylbenzene**
(*ortho*-xylene)   (*meta*-xylene)   (*para*-xylene)

The IUPAC nomenclature scheme becomes rapidly more elaborate for more complex cyclic structures, with notation for compounds containing conjoined rings, and many common names such as phenol being accepted as base names for compounds derived from them.

Order of Precedence of Groups

When compounds contain more than one functional group, the order of precedence determines which groups are named with prefix or suffix forms. The highest-precedence group takes the suffix, with all others taking the prefix form. However, double and triple bonds only take suffix form (-en and -yn) and are used with other suffixes.

Prefixed substituents are ordered alphabetically (excluding any modifiers such as di-, tri-, etc.), e.g. chlorofluoromethane, *not* fluorochloromethane. If there are multiple functional groups of the same type, either prefixed or suffixed, the position numbers are ordered numerically (thus ethane-1,2-diol, *not* ethane-2,1-diol.) The *N* position indicator for amines and amides comes before "1", e.g. $CH_3CH(CH_3)CH_2NH(CH_3)$ is *N*,2-dimethylpropanamine.

Priority	Functional group	Formula	Prefix	Suffix
1	Cations e.g. Ammonium	NH_4^+	-onio- ammonio-	-onium -ammonium
2	Carboxylic acids	–COOH	carboxy-	-oic acid*
	Carbothioic *S*-acids	–COSH	sulfanylcarbonyl-	-thioic *S*-acid*
	Carboselenoic *Se*-acids	–COSeH	selanylcarbonyl-	-selenoic *Se*-acid*
	Sulfonic acids	–SO$_3$H	sulfo-	-sulfonic acid
	Sulfinic acids	–SO$_2$H	sulfino-	-sulfinic acid

		Carboxylic acid derivatives			
3		Esters	−COOR	R-oxycarbonyl-	-R-oate
		Acyl halides	−COX	halocarbonyl-	-oyl halide*
		Amides	−CONH$_2$	carbamoyl-	-amide*
		Imides	−CON=C<	-imido-	-imide*
		Amidines	−C(=NH)NH$_2$	amidino-	-amidine*
4		Nitriles	−CN	cyano-	-nitrile*
		Isocyanides	−NC	isocyano-	isocyanide
5		Aldehydes	−CHO	formyl-	-al*
		Thioaldehydes	−CHS	thioformyl-	-thial*
6		Ketones	=O	oxo-	-one
		Thiones	=S	sulfanylidene-	-thione
		Selones	=Se	selanylidene-	-selone
		Tellones	=Te	tellanylidene-	-tellone
7		Alcohols	−OH	hydroxy-	-ol
		Thiols	−SH	sulfanyl-	-thiol
		Selenols	−SeH	selanyl-	-selenol
		Tellurols	−TeH	tellanyl-	-tellurol
8		Hydroperoxides Peroxols	-OOH	hydroperoxy-	-peroxol
		Thioperoxols (Sulfenic acid)	-SOH	hydroxysulfanyl-	-*SO*-thioperoxol
		Dithioperoxols	-SSH	disulfanyl-	-dithioperoxol
9		Amines	−NH$_2$	amino-	-amine
		Imines	=NH	imino-	-imine
		Hydrazines	−NHNH$_2$	hydrazino-	-hydrazine

*Note: These suffixes, in which the carbon atom is counted as part of the preceding chain, are the most commonly used.

The order of remaining functional groups is only needed for substituted benzene and hence is not mentioned here.

Common Nomenclature – trivial Names

Common nomenclature is an older system of naming organic compounds. Instead of using the prefixes for the carbon skeleton above, another system is used. The pattern can be seen below.

Number of carbons	Prefix as in new system	Common name for alcohol	Common name for aldehyde	Common name for acid
1	Meth-	Methyl alcohol (wood alcohol)	Formaldehyde	Formic acid

2	Eth-	Ethyl alcohol (grain alcohol)	Acetaldehyde	Acetic acid
3	Prop-	Propyl alcohol	Propionaldehyde	Propionic acid
4	But-	Butyl alcohol	Butyraldehyde	Butyric acid
5	Pent-	Amyl alcohol	Valeraldehyde	Valeric acid
6	Hex-	Caproyl alcohol	Caproaldehyde	Caproic acid
7	Hept-	Enanthyl alcohol	Enanthaldehyde	Enanthoic acid
8	Oct-	Capryl alcohol	Caprylaldehyde	Caprylic acid
9	Non-	Pelargonic alcohol	Pelargonaldehyde	Pelargonic acid
10	Dec-	Capric alcohol	Capraldehyde	Capric acid
11	Undec-	-	-	-
12	Dodec-	Lauryl alcohol	Lauraldehyde	Lauric acid
13	Tridec-	-	-	-
14	Tetradec-	Myristyl alcohol	Myristaldehyde	Myristic acid
15	pentadec-			
16	Hexadec-	Cetyl alcohol Palmityl alcohol	Palmitaldehyde	Palmitic acid
17	Heptadec-	-	-	Margaric acid
18	Octadec-	Stearyl alcohol	Stearaldehyde	Stearic acid
20	Eicos-	Arachidyl alcohol	-	Arachidic acid
22	Docos-	Behenyl alcohol	-	Behenic acid
24	Tetracos-	Lignoceryl alcohol	-	Lignoceric acid
26	Hexacos-	Ceryl alcohol	-	Cerotic acid
28	Octacos-	Montanyl alcohol	-	Montanic acid
30	Triacont-	Melissyl alcohol	-	Melissic acid
32	Dotriacont-	Lacceryl alcohol	-	Lacceroic acid
33	Tritriacont-	Psyllic alcohol	-	Psyllic acid
34	Tetratriacont-	Geddyl alcohol	-	Geddic acid
35	Pentatriacont-	-	-	Ceroplastic acid
40	Tetracont-	-	-	-

Ketones

Common names for ketones can be derived by naming the two alkyl or aryl groups bonded to the carbonyl group as separate words followed by the word *ketone*.

- Acetone
- Acetophenone
- Benzophenone
- Ethyl isopropyl ketone
- Diethyl ketone

The first three of the names shown above are still considered to be acceptable IUPAC names.

Aldehydes

The common name for an aldehyde is derived from the common name of the corresponding carboxylic acid by dropping the word *acid* and changing the suffix from -ic or -oic to -aldehyde.

- Formaldehyde
- Acetaldehyde

Ions

The IUPAC nomenclature also provides rules for naming ions.

Hydron

Hydron is a generic term for hydrogen cation; protons, deuterons and tritons are all hydrons. The Hydrons are not found in heavier isotopes, however.

Parent Hydride Cations

Simple cations formed by adding a hydron to a hydride of a halogen, chalcogen or pnictogen are named by adding the suffix "-onium" to the element's root: H_4N^+ is ammonium, H_3O^+ is oxonium, and H_2F^+ is fluoronium. Ammonium was adopted instead of nitronium, which commonly refers to NO_2^+.

If the cationic center of the hydride is not a halogen, chalcogen or pnictogen then the suffix "-ium" is added to the name of the neutral hydride after dropping any final 'e'. H_5C^+ is methanium, $HO-(O^+)-H_2$ is dioxidanium (HO-OH is dioxidane), and $H_2N-(N^+)-H_3$ is diazanium (H_2N-NH_2 is diazane).

Cations and Substitution

The above cations except for methanium are not, strictly speaking, organic, since they do not contain carbon. However, many organic cations are obtained by substituting another element or some functional group for a hydrogen.

The name of each substitution is prefixed to the hydride cation name. If many substitutions by the same functional group occur, then the number is indicated by prefixing with "di-", "tri-" as with halogenation. $(CH_3)_3O^+$ is trimethyloxonium. $CH_3F_3N^+$ is trifluoromethylammonium.

IUPAC Nomenclature of Inorganic Chemistry

In chemical nomenclature, the IUPAC nomenclature of inorganic chemistry is a systematic method of naming inorganic chemical compounds, as recommended by the International Union of Pure and Applied Chemistry (IUPAC). It is published in *Nomenclature of Inorganic Chemistry* (which is informally called the Red Book). Ideally, every inorganic compound should have a name from which an unambiguous formula can be determined. There is also an IUPAC nomenclature of organic chemistry.

System

The names "caffeine" and "3,7-dihydro-1,3,7-trimethyl-1H-purine-2,6-dione" both signify the same chemical. The systematic name encodes the structure and composition of the caffeine molecule in some detail, and provides an unambiguous reference to this compound, whereas the name "caffeine" just names it. These advantages make the systematic name far superior to the common name when absolute clarity and precision are required. However, for the sake of brevity, even professional chemists will use the non-systematic name almost all of the time, because caffeine is a well-known common chemical with a unique structure. Similarly, H_2O is most often simply called water in English, though other chemical names do exist.

1. Single atom anions are named with an *-ide* suffix: for example, H^- is hydride.

2. Compounds with a positive ion (cation): The name of the compound is simply the cation's name (usually the same as the element's), followed by the anion. For example, NaCl is *sodium chloride*, and CaF_2 is *calcium fluoride*.

3. Cations which have taken on more than one positive charge are labeled with Roman numerals in parentheses. For example, Cu^+ is copper(I), Cu^{2+} is copper(II). An older, deprecated notation is to append *-ous* or *-ic* to the root of the Latin name to name ions with a lesser or greater charge. Under this naming convention, Cu^+ is cuprous and Cu^{2+} is cupric.

4. Oxyanions (polyatomic anions containing oxygen) are named with *-ite* or *-ate*, for a lesser or greater quantity of oxygen, respectively. For example, $NO-2$ is nitrite, while $NO-3$ is nitrate. If four oxyanions are possible, the prefixes *hypo-* and *per-* are used: hypochlorite is ClO^-3, perchlorate is $ClO-4$.

5. The prefix *bi-* is a deprecated way of indicating the presence of a single hydrogen ion, as in "sodium bicarbonate" ($NaHCO_3$). The modern method specifically names the hydrogen atom. Thus, $NaHCO_3$ would be pronounced sodium hydrogen carbonate.

Positively charged ions are called cations and negatively charged ions are called anions. The cation is always named first. Ions can be metals or polyatomic ions. Therefore the name of the metal or positive polyatomic ion is followed by the name of the non-metal or negative polyatomic ion. The positive ion retains its element name whereas for a single non-metal anion the ending is changed to *-ide*.

Example: sodium chloride, potassium oxide, or calcium carbonate.

When the metal has more than one possible ionic charge or oxidation number the name becomes ambiguous. In these cases the oxidation number (the same as the charge) of the metal ion is represented by a Roman numeral in parentheses immediately following the metal ion name. For example in uranium(VI) fluoride the oxidation number of uranium is 6. Another example is the iron oxides. FeO is iron(II) oxide and Fe_2O_3 is iron(III) oxide.

An older system used prefixes and suffixes to indicate the oxidation number, according to the following scheme:

Oxidation state	Cations and acids	Anions
Lowest	*hypo- -ous*	*hypo- -ite*
	-ous	*-ite*
	-ic	*-ate*
	per- -ic	*per- -ate*
Highest	*hyper- -ic*	*hyper- -ate*

Thus the four oxyacids of chlorine are called hypochlorous acid (HOCl), chlorous acid (HOClO), chloric acid ($HOClO_2$) and perchloric acid ($HOClO_3$), and their respective conjugate bases are the hypochlorite, chlorite, chlorate and perchlorate ions. This system has partially fallen out of use, but survives in the common names of many chemical compounds: the modern literature contains few references to "ferric chloride" (instead calling it "iron(III) chloride"), but names like "potassium permanganate" (instead of "potassium manganate(VII)") and "sulfuric acid" abound.

Traditional Naming

Naming Simple Ionic Compounds

An ionic compound is named by its cation followed by its anion.

For cations that take on multiple charges, the charge is written using Roman numerals in parentheses immediately following the element name. For example, $Cu(NO_3)_2$ is *copper(II) nitrate*, because the charge of two nitrate ions ($NO-3$) is $2 \times -1 = -2$, and since the net charge of the ionic compound must be zero, the Cu ion has a 2+ charge. This compound is therefore copper(II) nitrate. In the case of cations with a +4 oxidation state, the only acceptable format for the Roman numeral 4 is IV and not IIII.

The Roman numerals in fact show the oxidation number, but in simple ionic compounds (i.e., not metal complexes) this will always equal the ionic charge on the metal.

List of Common Ion Names

Monatomic anions:

Cl^- chloride

S^{2-} sulfide

P^{3-} phosphide

Polyatomic ions:

NH_4^+ ammonium

H_3O^+ hydronium

NO_3^- nitrate

NO_2^- nitrite

ClO^- hypochlorite

ClO_2^- chlorite

ClO_3^- chlorate

ClO_4^- perchlorate

SO_3^{2-} sulfite

SO_4^{2-} sulfate

HSO_3^- hydrogen sulfite (or bisulfite)

HCO_3^- hydrogen carbonate (or bicarbonate)

CO_3^{2-} carbonate

PO_4^{3-} phosphate

HPO_4^{2-} hydrogen phosphate

$H_2PO_4^-$ dihydrogen phosphate

CrO_4^{2-} chromate

$Cr_2O_7^{2-}$ dichromate

BO_3^{3-} borate

AsO_4^{3-} arsenate

$C_2O_4^{2-}$ oxalate

CN^- cyanide

SCN^- thiocyanate

MnO_4^- permanganate

Naming Hydrates

Hydrates are ionic compounds that have absorbed water. They are named as the ionic compound followed by a numerical prefix and -*hydrate*. The numerical prefixes used are listed below:

1. mono-

2. di-

3. tri-

4. tetra-

5. penta-

6. hexa-

7. hepta-

8. octa-

9. nona-

10. deca-

For example, $CuSO_4 \cdot 5H_2O$ is "copper(II) sulfate pentahydrate".

Naming Molecular Compounds

Inorganic molecular compounds are named with a prefix before each element. The more electronegative element is written last and with an *-ide* suffix. For example, H_2O (water) can be called *dihydrogen monoxide*. Organic molecules do not follow this rule. In addition, the prefix mono- is not used with the first element; for example, SO_2 is *sulfur dioxide*, not "monosulfur dioxide". Sometimes prefixes are shortened when the ending vowel of the prefix "conflicts" with a starting vowel in the compound. This makes the name easier to pronounce; for example, CO is "carbon monoxide" (as opposed to "monooxide").

Common Exceptions

There are a number of exceptions and special cases that violate the above rules. Sometimes the prefix is left off of the initial atom: S_2O_7 is known as *sulfur heptoxide*, but it should be called *disulfur heptoxide*. S_2O_3 is called *sulfur sesquioxide* (*sesqui-* means 1 $\frac{1}{2}$).

The main oxide of phosphorus is called *phosphorus pentoxide*. It should actually be *diphosphorus pentoxide*, but everyone knows that there are two phosphorus atoms (P_2O_5) needed in order to balance the oxidation numbers of the five oxygen atoms. However, people have known for years that the real form of the molecule is P_4O_{10}, not P_2O_5, yet it is not normally called *tetraphosphorus decaoxide*.

In writing formulas, *ammonia* is NH_3 even though nitrogen is more electronegative. Likewise, *methane* is written as CH_4 even though carbon is more electronegative.

2005 Revision of Iupac's Nomenclature for Inorganic Compounds

Nomenclature of Inorganic Chemistry

Nomenclature of Inorganic Chemistry, by chemists commonly referred to as the *Red Book*, is a collection of recommendations on IUPAC nomenclature , published at irregular intervals by the IUPAC. The last full edition was published in 2005, in both paper and electronic versions.

The front cover of the 2005 edition of the *Red Book*

Published editions			
Release year	**Title**	**Publisher**	**ISBN**
2005	Recommendations 2005 (Red Book)	RSC Publishing	0-85404-438-8
2001	Recommendations 2000 (Red Book II) (supplement)	RSC Publishing	0-85404-487-6
1990	Recommendations 1990 (Red Book I)	Blackwell	0-632-02494-1
1971	Definitive Rules 1970	Butterworth	0-408-70168-4
1959	1957 Rules	Butterworth	
1940/1941	1940 Rules	Scientific journals	

International Chemical Identifier

The IUPAC International Chemical Identifier is a textual identifier for chemical substances, designed to provide a standard and human-readable way to encode molecular information and to facilitate the search for such information in databases and on the web. Initially developed by IUPAC and NIST from 2000 to 2005, the format and algorithms are non-proprietary.

The continuing development of the standard has been supported since 2010 by the not-for-profit InChI Trust, of which IUPAC is a member. The current version is 1.04 and was released in September 2011.

Prior to 1.04, the software was freely available under the open source LGPL license, but it now uses a custom license called IUPAC-InChI Trust License.

Overview

The identifiers describe chemical substances in terms of *layers* of information — the atoms and their bond connectivity, tautomeric information, isotope information, stereochemistry, and electronic charge information. Not all layers have to be provided; for instance, the tautomer layer can be omitted if that type of information is not relevant to the particular application.

InChIs differ from the widely used CAS registry numbers in three respects:

- they are freely usable and non-proprietary;

- they can be computed from structural information and do not have to be assigned by some organization;

- most of the information in an InChI is human readable (with practice).

InChIs can thus be seen as akin to a general and extremely formalized version of IUPAC names. They can express more information than the simpler SMILES notation and differ in that every structure has a unique InChI string, which is important in database applications. Information about the 3-dimensional coordinates of atoms is not represented in InChI; for this purpose a format such as PDB can be used.

The InChI algorithm converts input structural information into a unique InChI identifier in a three-step process: normalization (to remove redundant information), canonicalization (to generate a unique number label for each atom), and serialization (to give a string of characters).

The InChIKey, sometimes referred to as a hashed InChI, is a fixed length (25 character) condensed digital representation of the InChI that is not human-understandable. The InChIKey specification was released in September 2007 in order to facilitate web searches for chemical compounds, since these were problematic with the full-length InChI. It should be noted that, unlike the InChI, the InChIKey is not unique: though collisions can be calculated to be very rare, they happen.

In January 2009 the final 1.02 version of the InChI software was released. This provided a means to generate so called standard InChI, which does not allow for user selectable options in dealing with the stereochemistry and tautomeric layers of the InChI string. The standard InChIKey is then the hashed version of the standard InChI string. The standard InChI will simplify comparison of InChI strings and keys generated by different groups, and subsequently accessed via diverse sources such as databases and web resources.

Format and Layers

Every InChI starts with the string "InChI=" followed by the version number, currently 1. This is followed by the letter S for standard InChIs. The remaining information is structured as a sequence of layers and sub-layers, with each layer providing one specific type of information. The layers and sub-layers are separated by the delimiter "/" and start with a characteristic prefix letter (except for the chemical formula sub-layer of the main layer). The six layers with important sublayers are:

1. Main layer

   o Chemical formula (no prefix). This is the only sublayer that must occur in every InChI.

   o Atom connections (prefix: "c"). The atoms in the chemical formula (except for hydrogens) are numbered in sequence; this sublayer describes which atoms are connected by bonds to which other ones.

- - Hydrogen atoms (prefix: "h"). Describes how many hydrogen atoms are connected to each of the other atoms.

2. Charge layer

   - proton sublayer (prefix: "p" for "protons")

   - charge sublayer (prefix: "q")

3. Stereochemical layer

   - double bonds and cumulenes (prefix: "b")

   - tetrahedral stereochemistry of atoms and allenes (prefixes: "t", "m")

   - type of stereochemistry information (prefix: "s")

4. Isotopic layer (prefixes: "i", "h", as well as "b", "t", "m", "s" for isotopic stereochemistry)

5. Fixed-H layer (prefix: "f"); contains some or all of the above types of layers except atom connections; may end with "o" sublayer; never included in standard InChI

6. Reconnected layer (prefix: "r"); contains the whole InChI of a structure with reconnected metal atoms; never included in standard InChI

The delimiter-prefix format has the advantage that a user can easily use a wildcard search to find identifiers that match only in certain layers.

Examples

CH_3CH_2OH ethanol	InChI=1/C2H6O/c1-2-3/h3H,2H2,1H3 InChI=1S/C2H6O/c1-2-3/h3H,2H2,1H3 (standard InChI)
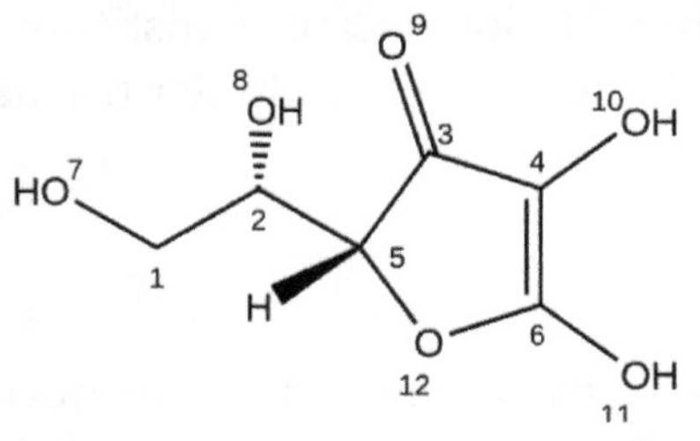 L-ascorbic acid	InChI=1/C6H8O6/c7-1-2(8)5-3(9)4(10)6(11)12-5/h2,5,7-10H,1H2/t2-,5+/m0/s1 InChI=1S/C6H8O6/c7-1-2(8)5-3(9)4(10)6(11)12-5/h2,5,7-8,10-11H,1H2/t2-,5+/m0/s1 (standard InChI)

InChIKey

The condensed, 27 character standard InChIKey is a hashed version of the full standard InChI (using the SHA-256 algorithm), designed to allow for easy web searches of chemical compounds. Most chemical structures on the Web up to 2007 have been represented as GIF files, which are not searchable for chemical content. The full InChI turned out to be too lengthy for easy searching, and therefore the InChIKey was developed. There is a very small, but nonzero chance of two different molecules having the same InChIKey, but the probability for duplication of only the first 14 characters has been estimated as only one duplication in 75 databases each containing one billion unique structures. With all databases currently having below 50 million structures, such duplication appears unlikely at present. A recent study more extensively studies the collision rate finding that the experimental collision rate is in agreement with the theoretical expectations.

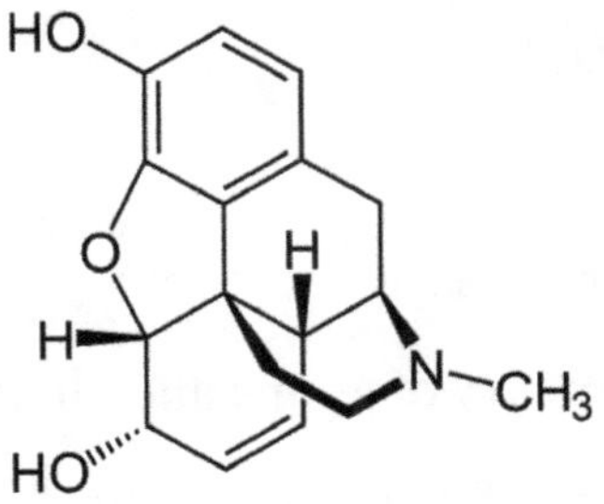

Morphine structure

InChIKeys consist of 14 characters resulting from a hash of the connectivity information of the InChI, followed by a hyphen, followed by 10 characters resulting from a hash of the remaining layers of the InChI, followed by a single character indicating the version of InChI used, another hyphen, followed by single checksum character.

Example: Morphine has the structure shown on the right. The standard InChI for morphine is InChI=1S/C17H19NO3/c1-18-7-6-17-10-3-5-13(20)16(17)21-15-12(19)4-2-9(14(15)17)8-11(10)18/h2-5,10-11,13,16,19-20H,6-8H2,1H3/t10-,11+,13-,16-,17-/m0/s1 and the standard InChIKey for morphine is BQJCRHHNABKAKU-KBQPJGBKSA-N.

InChI Resolvers

As the InChI cannot be reconstructed from the InChIKey, an InChIKey always needs to be linked to the original InChI to get back to the original structure. InChI Resolvers act as a lookup service to make these links, and prototype services are available from National Cancer Institute, the UniChem service at the European Bioinformatics Institute, and PubChem. ChemSpider has had a resolver until July 2015 when it was decommissioned.

Name

The format was originally called IChI (IUPAC Chemical Identifier), then renamed in July 2004 to INChI (IUPAC-NIST Chemical Identifier), and renamed again in November 2004 to InChI (IUPAC International Chemical Identifier), a trademark of IUPAC.

Continuing Development

Scientific direction of the InChI standard is carried out by the IUPAC Division VIII Subcommittee, and funding of subgroups investigating and defining the expansion of the standard is carried out by both IUPAC and the InChI Trust. The InChI Trust funds the development, testing and documentation of the InChI. Current extensions are being defined to handle polymers and mixtures, Markush structures, reactions and organometallics, and once accepted by the Division VIII Subcommittee will be added to the algorithm.

Adoption

The InChI has been adopted by many larger and smaller databases, including ChemSpider, ChEM-BL, Golm Metabolome Database, OpenPHACTS, and PubChem. However, the adoption is not straightforward, and many databases show a discrepancy between the chemical structures and the InChI they contain, which is a problem for linking databases.

Simplified Molecular-input Line-entry System

The simplified molecular-input line-entry system (SMILES) is a specification in form of a line notation for describing the structure of chemical species using short ASCII strings. SMILES strings can be imported by most molecule editors for conversion back into two-dimensional drawings or three-dimensional models of the molecules.

N1CCN(CC1)C(C(F)=C2)=CC(=C2C4=O)N(C3CC3)C=C4C(=O)O

Generation of SMILES: Break cycles, then write as branches off a main backbone. (Ciprofloxacin)

The original SMILES specification was initiated in the 1980s. It has since been modified and extended. In 2007, an open standard called "OpenSMILES" was developed in the open-source chemistry community. Other 'linear' notations include the Wiswesser Line Notation (WLN), ROSDAL and SLN.

History

The original SMILES specification was initiated by David Weininger at the USEPA Mid-Continent Ecology Division Laboratory in Duluth in the 1980s. Acknowledged for their parts in the early development were "Gilman Veith and Rose Russo (USEPA) and Albert Leo and Corwin Hansch (Pomona College) for supporting the work, and Arthur Weininger (Pomona; Daylight CIS) and

Jeremy Scofield (Cedar River Software, Renton, WA) for assistance in programming the system." The Environmental Protection Agency funded the initial project to develop SMILES.

It has since been modified and extended by others, most notably by Daylight Chemical Information Systems. In 2007, an open standard called "OpenSMILES" was developed by the Blue Obelisk open-source chemistry community. Other 'linear' notations include the Wiswesser Line Notation (WLN), ROSDAL and SLN (Tripos Inc).

In July 2006, the IUPAC introduced the InChI as a standard for formula representation. SMILES is generally considered to have the advantage of being slightly more human-readable than InChI; it also has a wide base of software support with extensive theoretical (e.g., graph theory) backing.

Terminology

The term SMILES refers to a line notation for encoding molecular structures and specific instances should strictly be called SMILES strings. However, the term SMILES is also commonly used to refer to both a single SMILES string and a number of SMILES strings; the exact meaning is usually apparent from the context. The terms "canonical" and "isomeric" can lead to some confusion when applied to SMILES. The terms describe different attributes of SMILES strings and are not mutually exclusive.

Typically, a number of equally valid SMILES strings can be written for a molecule. For example, CCO, OCC and C(O)C all specify the structure of ethanol. Algorithms have been developed to generate the same SMILES string for a given molecule; of the many possible strings, these algorithms choose only one of them. This SMILES is unique for each structure, although dependent on the canonicalization algorithm used to generate it, and is termed the canonical SMILES. These algorithms first convert the SMILES to an internal representation of the molecular structure; an algorithm then examines that structure and produces a unique SMILES string. Various algorithms for generating canonical SMILES have been developed and include those by Daylight Chemical Information Systems, OpenEye Scientific Software, MEDIT, Chemical Computing Group, MolSoft LLC, and the Chemistry Development Kit. A common application of canonical SMILES is indexing and ensuring uniqueness of molecules in a database.

The original paper that described the CANGEN algorithm claimed to generate unique SMILES strings for graphs representing molecules, but the algorithm fails for a number of simple cases (e.g. cuneane, 1,2-dicyclopropylethane) and cannot be considered a correct method for representing a graph canonically. There is currently no systematic comparison across commercial software to test if such flaws exist in those packages.

SMILES notation allows the specification of configuration at tetrahedral centers, and double bond geometry. These are structural features that cannot be specified by connectivity alone and SMILES which encode this information are termed isomeric SMILES. A notable feature of these rules is that they allow rigorous partial specification of chirality. The term isomeric SMILES is also applied to SMILES in which isotopes are specified.

Graph-based Definition

In terms of a graph-based computational procedure, SMILES is a string obtained by printing the symbol nodes encountered in a depth-first tree traversal of a chemical graph. The chemical graph is first trimmed to remove hydrogen atoms and cycles are broken to turn it into a spanning tree. Where cycles have been broken, numeric suffix labels are included to indicate the connected nodes. Parentheses are used to indicate points of branching on the tree.

Examples

Atoms

Atoms are represented by the standard abbreviation of the chemical elements, in square brackets, such as [Au] for gold. Brackets can be omitted for the "organic subset" of B, C, N, O, P, S, F, Cl, Br, and I. All other elements must be enclosed in brackets. If the brackets are omitted, the proper number of implicit hydrogen atoms is assumed; for instance the SMILES for water is simply O.

An atom holding one or more electrical charges is enclosed in brackets. The symbol H is added if the atom in brackets is bonded to one or more hydrogen, followed by the number of hydrogen atoms (again one is omitted, for example: NH4 for ammonium), then by the sign '+' for a positive charge or by '-' for a negative charge. The number of charges is specified after the sign (except if there is one only); however, it is also possible write the sign as many times as the ion has charges: instead of "Ti+4", one can also write "Ti++++" (Titanium IV, Ti^{4+}). Thus, the hydroxide anion is represented by [OH-], the oxonium cation is [OH3+] and the cobalt III cation (Co^{3+}) is either [Co+3] or [Co+++].

Bonds

Bonds between aliphatic atoms are assumed to be single unless specified otherwise and are implied by adjacency in the SMILES string. For example, the SMILES for ethanol can be written as CCO. Ring closure labels are used to indicate connectivity between non-adjacent atoms in the SMILES string, which for cyclohexane and dioxane can be written as C1CCCCC1 and O1CCOCC1 respectively. For a second ring, the label will be 2 (naphthalene: c1cccc2c1cccc2 (note the lower case for aromatic compounds)), and so on. After reaching 9, the label must be preceded by a '%', in order to differentiate it from two different labels bonded to the same atom (~C12~ will mean the atom of carbon holds the ring closure labels 1 and 2, whereas ~C%12~ will indicate one label only, 12). Double, triple, and quadruple bonds are represented by the symbols '=', '#', and '$' respectively as illustrated by the SMILES O=C=O (carbon dioxide), C#N (hydrogen cyanide) and [Ga-]$[As+] (gallium arsenide).

Aromaticity

Aromatic C, O, S and N atoms are shown in their lower case 'c', 'o', 's' and 'n' respectively. Benzene, pyridine and furan can be represented respectively by the SMILES c1ccccc1, n1ccccc1 and o1cccc1. Bonds between aromatic atoms are, by default, aromatic although these can be specified explicitly using the ':' symbol. Aromatic atoms can be singly bonded to each other and biphenyl can be represented by c1ccccc1-c2ccccc2. Aromatic nitrogen bonded to hydrogen, as found in pyrrole must be represented as [nH] and imidazole is written in SMILES notation as n1c[nH]cc1.

The Daylight and OpenEye algorithms for generating canonical SMILES differ in their treatment of aromaticity.

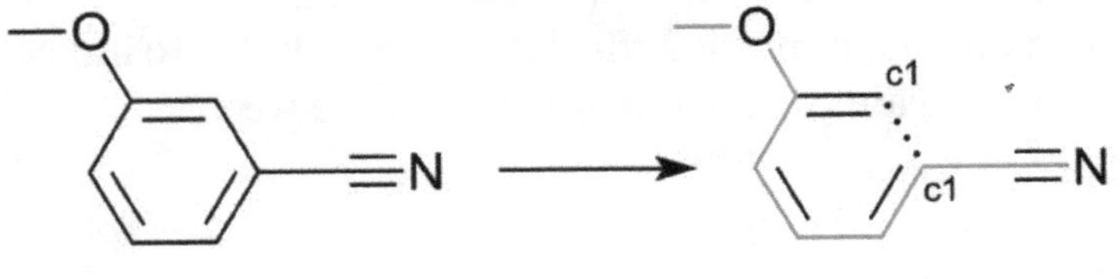

Visualization of 3-cyanoanisole as COc(c1)cccc1C#N.

Branching

Branches are described with parentheses, as in CCC(=O)O for propionic acid and C(F)(F)F for fluoroform. Substituted rings can be written with the branching point in the ring as illustrated by the SMILES COc(c1)cccc1C#N and COc(cc1)ccc1C#N which encode the 3 and 4-cyanoanisole isomers. Writing SMILES for substituted rings in this way can make them more human-readable.

Stereochemistry

Configuration around double bonds is specified using the characters "/" and "\". For example, F/C=C/F is one representation of *trans*-difluoroethene, in which the fluorine atoms are on opposite sides of the double bond, whereas F/C=C\F is one possible representation of *cis*-difluoroethene, in which the Fs are on the same side of the double bond, as shown in the figure.

Configuration at tetrahedral carbon is specified by @ or @@. L-Alanine, the more common enantiomer of the amino acid alanine can be written as N[C@@H](C)C(=O)O. The @@ specifier indicates that, when viewed from nitrogen along the bond to the chiral center, the sequence of substituents hydrogen (H), methyl (C) and carboxylate (C(=O)O) appear clockwise. D-Alanine can be written as N[C@H](C)C(=O)O. The order of the substituents in the SMILES string is very important and D-alanine can also be encoded as N[C@@H](C(=O)O)C.

Isotopes

Isotopes are specified with a number equal to the integer isotopic mass preceding the atomic symbol. Benzene in which one atom is carbon-14 is written as [14c]1ccccc1 and deuterochloroform is [2H]C(Cl)(Cl)Cl.

Examples

Molecule	Structure	SMILES Formula
Dinitrogen	N≡N	N#N
Methyl isocyanate (MIC)	$CH_3-N=C=O$	CN=C=O
Copper(II) sulfate	$Cu^{2+}\ SO_4^{2-}$	[Cu+2].[O-]S(=O)(=O)[O-]
Oenanthotoxin ($C_{17}H_{22}O_2$)	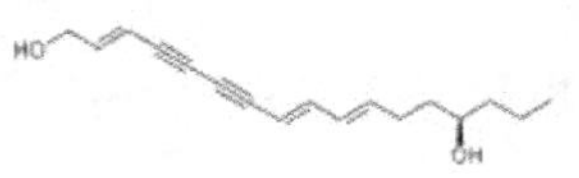	CCC[C@@H](O)CC\C=C\C=C\C#CC#C\C=C\CO

Pyrethrin II (C$_{22}$H$_{28}$O$_5$)		COC(=O)C(\C)=C\C1C(C)(C)[C@H]1C(=O)O[C@@H]2C(C)=C(C(=O)C2)CC=CC=C
Aflatoxin B1 (C$_{17}$H$_{12}$O$_6$)		O1C=C[C@H]([C@H]1O2)c3c2cc(OC)c4c3OC(=O)C5=C4CCC(=O)5
Glucose (glucopyranose) (C$_6$H$_{12}$O$_6$)		OC[C@@H](O1)[C@@H](O)[C@H](O)[C@@H](O)[C@@H](O)1
Bergenin (cuscutin) (a resin) (C$_{14}$H$_{16}$O$_9$)		OC[C@@H](O1)[C@@H](O)[C@H](O)[C@@H]2[C@@H]1c3c(O)c(OC)c(O)cc3C(=O)O2
A pheromone of the Californian scale insect		CC(=O)OCCC(/C)=C\C[C@H](C(C)=C)CCC=C
2S,5R-Chalcogran: a pheromone of the bark beetle Pityogenes chalcographus		CC[C@H](O1)CC[C@@]12CCCO2
Vanillin		O=Cc1ccc(O)c(OC)c1
Melatonin (C$_{13}$H$_{16}$N$_2$O$_2$)		CC(=O)NCCC1=CNc2c1cc(OC)cc2
Flavopereirin (C$_{17}$H$_{15}$N$_2$)		CCc(c1)ccc2[n+]1ccc3c2Nc4c3cccc4
Nicotine (C$_{10}$H$_{14}$N$_2$)		CN1CCC[C@H]1c2cccnc2

Alpha-thujone ($C_{10}H_{16}O$)	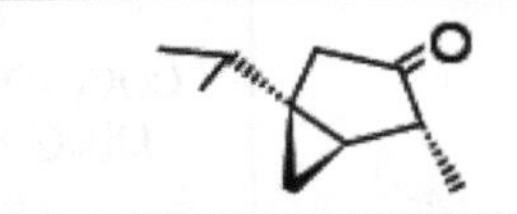	CC(C)[C@@]12C[C@@H]1[C@@H](C)C(=O)C2
Thiamin ($C_{12}H_{17}N_4OS^+$) (vitamin B1)	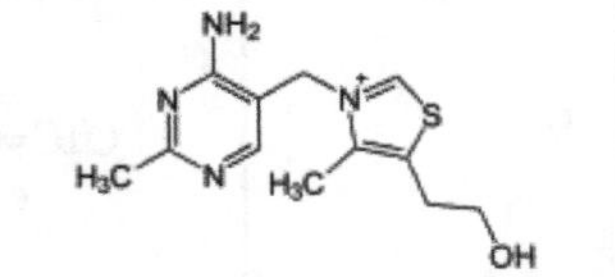	OCCc1c(C)[n+](=cs1)Cc2cnc(C)nc(N)2

Illustration with a molecule with more than 9 rings, Cephalostatin-1 (a steroidic trisdecacyclic pyrazine with the empirical formula $C_{54}H_{74}N_2O_{10}$ isolated from the Indian Ocean hemichordate Cephalodiscus gilchristi):

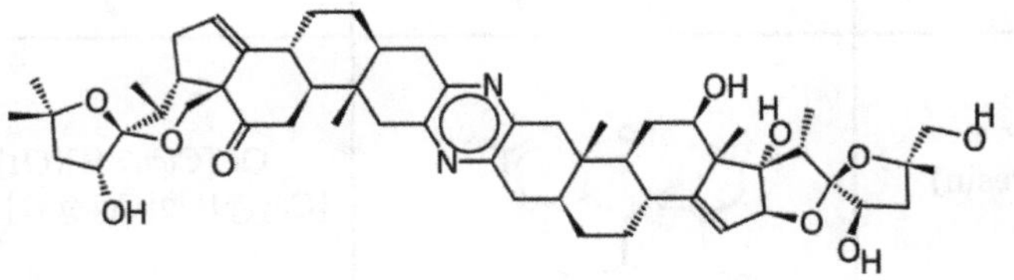

Starting with the left-most methyl group in the figure:

C[C@@](C)(O1)C[C@@H](O)[C@@]1(O2)[C@@H](C)[C@@H]3CC=C4[C@]3(C2)C(=O)C[C@H]5[C@H]4CC[C@@H](C6)[C@]5(C)Cc(n7)c6nc(C[C@@]89(C))c7C[C@@H]8C-C[C@@H]%10[C@@H]9C[C@@H](O)[C@@]%11(C)C%10=C[C@H](O%12)[C@]%11(O)[C@H](C)[C@]%12(O%13)[C@H](O)C[C@@]%13(C)CO

Note that '%' appears in front of the index of ring closure labels above 9; see section Bonds above.

Other Examples of SMILES

The SMILES notation is described extensively in the SMILES theory manual provided by Daylight Chemical Information Systems and a number of illustrative examples are presented. Daylight's depict utility provides users with the means to check their own examples of SMILES and is a valuable educational tool.

Extensions

SMARTS is a line notation for specification of substructural patterns in molecules. While it uses many of the same symbols as SMILES, it also allows specification of wildcard atoms and bonds, which can be used to define substructural queries for chemical database searching. One common misconception is that SMARTS-based substructural searching involves matching of SMILES and SMARTS strings. In fact, both SMILES and SMARTS strings are first converted to internal graph representations which are searched for subgraph isomorphism. SMIRKS is a line notation for specifying reaction transforms.

Conversion

SMILES can be converted back to 2-dimensional representations using Structure Diagram Generation algorithms (Helson, 1999). This conversion is not always unambiguous. Conversion to

3-dimensional representation is achieved by energy minimization approaches. There are many downloadable and web-based conversion utilities.

References

- Favre, Henri A.; Powell, Warren H. (2013). Nomenclature of Organic Chemistry: IUPAC Recommendations and Preferred Names 2013. Royal Society of Chemistry. ISBN 978-0-85404-182-4.

- International Union of Pure and Applied Chemistry (2005). Nomenclature of Inorganic Chemistry (IUPAC Recommendations 2005). Cambridge (UK): RSC–IUPAC. ISBN 0-85404-438-8.

- Heller, S.R.; McNaught, A.; Pletnev, I.; Stein, S.; Tchekhovskoi, D. (2015). "InChI, the IUPAC International Chemical Identifier". Journal of Cheminformatics. 7. doi:10.1186/s13321-015-0068-4.

- Warr, W.A. (2015). "Many InChIs and quite some feat". Journal of Computer-Aided Molecular Design. doi:10.1007/s10822-015-9854-3.

- Heller, S.; McNaught, A.; Stein, S.; Tchekhovskoi, D.; Pletnev, I. (2013). "InChI - the worldwide chemical structure identifier standard". Journal of Cheminformatics. 5 (1): 7. doi:10.1186/1758-2946-5-7. PMC 3599061. PMID 23343401.

- Pletnev, I.; Erin, A.; McNaught, A.; Blinov, K.; Tchekhovskoi, D.; Heller, S. (2012). "InChIKey collision resistance: An experimental testing". Journal of Cheminformatics. 4 (1): 39. doi:10.1186/1758-2946-4-39. PMC 3558395. PMID 23256896.

Applications of Inorganic Chemistry

Inorganic chemistry has a number of applications. Some of these are catalysis, materials science and pigment. Chemical reactions increase with the participation of added substances. This process is known as the catalysis. The reader is provided with an in-depth understanding of the applications of inorganic chemistry.

Catalysis

Catalysis is the increase in the rate of a chemical reaction due to the participation of an additional substance called a catalyst. With a catalyst, reactions occur faster because they require less activation energy. Furthermore since they are not consumed in the catalyzed reaction, catalysts can continue to act repeatedly. Often only tiny amounts are required in principle.

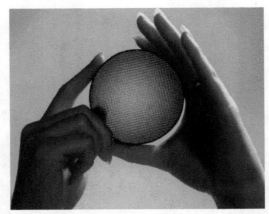

An air filter that utilizes low-temperature oxidation catalyst used to convert carbon monoxide to less toxic carbon dioxide at room temperature. It can also remove formaldehyde from the air.

Technical Perspective

In the presence of a catalyst, less free energy is required to reach the transition state, but the total free energy from reactants to products does not change. A catalyst may participate in multiple chemical transformations. The effect of a catalyst may vary due to the presence of other substances known as inhibitors or poisons (which reduce the catalytic activity) or promoters (which increase the activity). The opposite of a catalyst, a substance that reduces the rate of a reaction, is an inhibitor.

Catalyzed reactions have a lower activation energy (rate-limiting free energy of activation) than the corresponding uncatalyzed reaction, resulting in a higher reaction rate at the same temperature and for the same reactant concentrations. However, the detailed mechanics of catalysis is complex. Catalysts may affect the reaction environment favorably, or bind to the reagents to polarize bonds, e.g. acid catalysts for reactions of carbonyl compounds, or form specific intermediates that are not produced naturally, such as osmate esters in osmium tetroxide-catalyzed dihydroxylation of alkenes, or cause dissociation of reagents to reactive forms, such as chemisorbed hydrogen in catalytic hydrogenation.

Kinetically, catalytic reactions are typical chemical reactions; i.e. the reaction rate depends on the frequency of contact of the reactants in the rate-determining step. Usually, the catalyst participates in this slowest step, and rates are limited by amount of catalyst and its "activity". In heterogeneous catalysis, the diffusion of reagents to the surface and diffusion of products from the surface can be rate determining. A nanomaterial-based catalyst is an example of a heterogeneous catalyst. Analogous events associated with substrate binding and product dissociation apply to homogeneous catalysts.

Although catalysts are not consumed by the reaction itself, they may be inhibited, deactivated, or destroyed by secondary processes. In heterogeneous catalysis, typical secondary processes include coking where the catalyst becomes covered by polymeric side products. Additionally, heterogeneous catalysts can dissolve into the solution in a solid–liquid system or sublimate in a solid–gas system.

Background

The production of most industrially important chemicals involves catalysis. Similarly, most biochemically significant processes are catalysed. Research into catalysis is a major field in applied science and involves many areas of chemistry, notably organometallic chemistry and materials science. Catalysis is relevant to many aspects of environmental science, e.g. the catalytic converter in automobiles and the dynamics of the ozone hole. Catalytic reactions are preferred in environmentally friendly green chemistry due to the reduced amount of waste generated, as opposed to stoichiometric reactions in which all reactants are consumed and more side products are formed. Many transition metals and transition metal complexes are used in catalysis as well. Catalysts called enzymes are important in biology.

A catalyst works by providing an alternative reaction pathway to the reaction product. The rate of the reaction is increased as this alternative route has a lower activation energy than the reaction route not mediated by the catalyst. The disproportionation of hydrogen peroxide creates water and oxygen, as shown below.

$$2 H_2O_2 \rightarrow 2 H_2O + O_2$$

This reaction is preferable in the sense that the reaction products are more stable than the starting material, though the uncatalysed reaction is slow. In fact, the decomposition of hydrogen peroxide is so slow that hydrogen peroxide solutions are commercially available. This reaction is strongly affected by catalysts such as manganese dioxide, or the enzyme peroxidase in organisms. Upon the addition of a small amount of manganese dioxide, the hydrogen peroxide reacts rapidly. This effect is readily seen by the effervescence of oxygen. The manganese dioxide is not consumed in the reaction, and thus may be recovered unchanged, and re-used indefinitely. Accordingly, manganese dioxide *catalyses* this reaction.

General Principles

Units

Catalytic activity is usually denoted by the symbol z and measured in mol/s, a unit which was called katal and defined the SI unit for catalytic activity since 1999. Catalytic activity is not a kind of reaction rate, but a property of the catalyst under certain conditions, in relation to a specific chemical reaction. Catalytic activity of one katal (Symbol 1 kat = 1 mol/s) of a catalyst means an amount of that catalyst (substance, in Mol) that leads to a net reaction of one Mol per second of the reactants to the resulting reagents or other outcome which was intended for this chemical reaction. A catalyst may and usually will have different catalytic activity for distinct reactions.

Typical Mechanism

Catalysts generally react with one or more reactants to form intermediates that subsequently give the final reaction product, in the process regenerating the catalyst. The following is a typical reaction scheme, where C represents the catalyst, X and Y are reactants, and Z is the product of the reaction of X and Y:

$$X + C \rightarrow XC \, (1)$$

$$Y + XC \rightarrow XYC \, (2)$$

$$XYC \rightarrow CZ \, (3)$$

$$CZ \rightarrow C + Z \, (4)$$

Although the catalyst is consumed by reaction 1, it is subsequently produced by reaction 4, so it does not occur in the overall reaction equation:

$$X + Y \rightarrow Z$$

As a catalyst is regenerated in a reaction, often only small amounts are needed to increase the rate of the reaction. In practice, however, catalysts are sometimes consumed in secondary processes.

The catalyst does usually appear in the rate equation. For example, if the rate-determining step in the above reaction scheme is the first step X + C $\rightarrow$ XC, the catalyzed reaction will be second order with rate equation $v = k_{cat}[X][C]$, which is proportional to the catalyst concentration [C]. However [C] remains constant during the reaction so that the catalyzed reaction is pseudo-first order: $v = k_{obs}[X]$, where $k_{obs} = k_{cat}[C]$.

As an example of a detailed mechanism at the microscopic level, in 2008 Danish researchers first revealed the sequence of events when oxygen and hydrogen combine on the surface of titanium dioxide (TiO_2, or *titania*) to produce water. With a time-lapse series of scanning tunneling microscopy images, they determined the molecules undergo adsorption, dissociation and diffusion before reacting. The intermediate reaction states were: HO_2, H_2O_2, then H_3O_2 and the final reaction product (water molecule dimers), after which the water molecule desorbs from the catalyst surface.

Reaction Energetics

Generic potential energy diagram showing the effect of a catalyst in a hypothetical exothermic chemical reaction X + Y to give Z. The presence of the catalyst opens a different reaction pathway (shown in red) with a lower activation energy. The final result and the overall thermodynamics are the same.

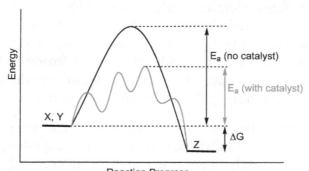

Catalysts work by providing an (alternative) mechanism involving a different transition state and lower activation energy. Consequently, more molecular collisions have the energy needed to reach the transition state. Hence, catalysts can enable reactions that would otherwise be blocked or slowed by a kinetic barrier. The catalyst may increase reaction rate or selectivity, or enable the reaction at lower temperatures. This effect can be illustrated with an energy profile diagram.

In the catalyzed elementary reaction, catalysts do not change the extent of a reaction: they have **no** effect on the chemical equilibrium of a reaction because the rate of both the forward and the reverse reaction are both affected. The second law of thermodynamics describes why a catalyst does not change the chemical equilibrium of a reaction. Suppose there was such a catalyst that shifted an equilibrium. Introducing the catalyst to the system would result in a reaction to move to the new equilibrium, producing energy. Production of energy is a necessary result since reactions are spontaneous only if Gibbs free energy is produced, and if there is no energy barrier, there is no need for a catalyst. Then, removing the catalyst would also result in reaction, producing energy; i.e. the addition and its reverse process, removal, would both produce energy. Thus, a catalyst that could change the equilibrium would be a perpetual motion machine, a contradiction to the laws of thermodynamics.

If a catalyst does change the equilibrium, then it must be consumed as the reaction proceeds, and thus it is also a reactant. Illustrative is the base-catalysed hydrolysis of esters, where the produced carboxylic acid immediately reacts with the base catalyst and thus the reaction equilibrium is shifted towards hydrolysis.

The SI derived unit for measuring the catalytic activity of a catalyst is the katal, which is moles per second. The productivity of a catalyst can be described by the turn over number (or TON) and the catalytic activity by the *turn over frequency* (TOF), which is the TON per time unit. The biochemical equivalent is the enzyme unit.

The catalyst stabilizes the transition state more than it stabilizes the starting material. It decreases the kinetic barrier by decreasing the *difference* in energy between starting material and transition state. It does not change the energy difference between starting materials and products (thermodynamic barrier), or the available energy (this is provided by the environment as heat or light).

Materials

The chemical nature of catalysts is as diverse as catalysis itself, although some generalizations can be made. Proton acids are probably the most widely used catalysts, especially for the many reactions involving water, including hydrolysis and its reverse. Multifunctional solids often are catalytically active, e.g. zeolites, alumina, higher-order oxides, graphitic carbon, nanoparticles, nanodots, and facets of bulk materials. Transition metals are often used to catalyze redox reactions (oxidation, hydrogenation). Examples are nickel, such as Raney nickel for hydrogenation, and vanadium(V) oxide for oxidation of sulfur dioxide into sulfur trioxide by the so-called contact process. Many catalytic processes, especially those used in organic synthesis, require "late transition metals", such as palladium, platinum, gold, ruthenium, rhodium, or iridium.

Some so-called catalysts are really precatalysts. Precatalysts convert to catalysts in the reaction. For example, Wilkinson's catalyst $RhCl(PPh_3)_3$ loses one triphenylphosphine ligand before entering the true catalytic cycle. Precatalysts are easier to store but are easily activated in situ. Because of this preactivation step, many catalytic reactions involve an induction period.

Chemical species that improve catalytic activity are called co-catalysts (cocatalysts) or promotors in cooperative catalysis.

Types

Catalysts can be heterogeneous or homogeneous, depending on whether a catalyst exists in the same phase as the substrate. Biocatalysts (enzymes) are often seen as a separate group.

Heterogeneous Catalysts

The microporous molecular structure of the zeolite ZSM-5 is exploited in catalysts used in refineries

Zeolites are extruded as pellets for easy handling in catalytic reactors.

Heterogeneous catalysts act in a different phase than the reactants. Most heterogeneous catalysts are solids that act on substrates in a liquid or gaseous reaction mixture. Diverse mechanisms for reactions on surfaces are known, depending on how the adsorption takes place (Langmuir-Hinshelwood, Eley-Rideal, and Mars-van Krevelen). The total surface area of solid has an important effect on the reaction rate. The smaller the catalyst particle size, the larger the surface area for a given mass of particles.

A heterogeneous catalyst has active sites, which are the atoms or crystal faces where the reaction actually occurs. Depending on the mechanism, the active site may be either a planar exposed metal surface, a crystal edge with imperfect metal valence or a complicated combination of the two. Thus, not only most of the volume, but also most of the surface of a heterogeneous catalyst may be catalytically inactive. Finding out the nature of the active site requires technically challenging research. Thus, empirical research for finding out new metal combinations for catalysis continues.

For example, in the Haber process, finely divided iron serves as a catalyst for the synthesis of ammonia from nitrogen and hydrogen. The reacting gases adsorb onto active sites on the iron particles. Once physically adsorbed, the reagents undergo chemisorption that results in dissociation into adsorbed atomic species, and new bonds between the resulting fragments form in part due to their close proximity. In this way the particularly strong triple bond in nitrogen is broken, which would be extremely uncommon in the gas phase due to its high activation energy. Thus, the activation energy of the overall reaction is lowered, and the rate of reaction increases. Another place where a heterogeneous catalyst is applied is in the oxidation of sulfur dioxide on vanadium(V) oxide for the production of sulfuric acid.

Heterogeneous catalysts are typically "supported," which means that the catalyst is dispersed on a second material that enhances the effectiveness or minimizes their cost. Supports prevent or reduce agglomeration and sintering of the small catalyst particles, exposing more surface area, thus catalysts have a higher specific activity (per gram) on a support. Sometimes the support is merely a surface on which the catalyst is spread to increase the surface area. More often, the support and the catalyst interact, affecting the catalytic reaction. Supports are porous materials with a high surface area, most commonly alumina, zeolites or various kinds of activated carbon. Specialized supports include silicon dioxide, titanium dioxide, calcium carbonate, and barium sulfate.

Electrocatalysts

In the context of electrochemistry, specifically in fuel cell engineering, various metal-containing catalysts are used to enhance the rates of the half reactions that comprise the fuel cell. One common type of fuel cell electrocatalyst is based upon nanoparticles of platinum that are supported on slightly larger carbon particles. When in contact with one of the electrodes in a fuel cell, this platinum increases the rate of oxygen reduction either to water, or to hydroxide or hydrogen peroxide.

Homogeneous Catalysts

Homogeneous catalysts function in the same phase as the reactants, but the mechanistic principles invoked in heterogeneous catalysis are generally applicable. Typically homogeneous catalysts are dissolved in a solvent with the substrates. One example of homogeneous catalysis involves the influence of H^+ on the esterification of carboxylic acids, such as the formation of methyl acetate from acetic acid and methanol. For inorganic chemists, homogeneous catalysis is often synonymous with organometallic catalysts.

Organocatalysis

Whereas transition metals sometimes attract most of the attention in the study of catalysis, small organic molecules without metals can also exhibit catalytic properties, as is apparent from the fact that many enzymes lack transition metals. Typically, organic catalysts require a higher loading (amount of catalyst per unit amount of reactant, expressed in mol% amount of substance) than transition metal(-ion)-based catalysts, but these catalysts are usually commercially available in bulk, helping to reduce costs. In the early 2000s, these organocatalysts were considered "new generation" and are competitive to traditional metal(-ion)-containing catalysts. Organocatalysts are supposed to operate akin to metal-free enzymes utilizing, e.g., non-covalent interactions such as hydrogen bonding. The discipline organocatalysis is divided in the application of covalent (e.g., proline, DMAP) and non-covalent (e.g., thiourea organocatalysis) organocatalysts referring to the preferred catalyst-substrate binding and interaction, respectively.

Enzymes and Biocatalysts

In biology, enzymes are protein-based catalysts in metabolism and catabolism. Most biocatalysts are enzymes, but other non-protein-based classes of biomolecules also exhibit catalytic properties including ribozymes, and synthetic deoxyribozymes.

Biocatalysts can be thought of as intermediate between homogeneous and heterogeneous catalysts, although strictly speaking soluble enzymes are homogeneous catalysts and membrane-bound enzymes are heterogeneous. Several factors affect the activity of enzymes (and other catalysts) including temperature, pH, concentration of enzyme, substrate, and products. A particularly important reagent in enzymatic reactions is water, which is the product of many bond-forming reactions and a reactant in many bond-breaking processes.

In biocatalysis, enzymes are employed to prepare many commodity chemicals including high-fructose corn syrup and acrylamide.

Some monoclonal antibodies whose binding target is a stable molecule which resembles the transition state of a chemical reaction can function as weak catalysts for that chemical reaction by lowering its activation energy. Such catalytic antibodies are sometimes called "abzymes".

Nanocatalysts

Nanocatalysts are nanomaterials with catalytic activities. They have been extensively explored for wide range of applications. Among them, the nanocatalysts with enzyme mimicking activities are collectively called as nanozymes.

Tandem Catalysis

In tandem catalysis two or more different catalysts are coupled in a one-pot reaction.

Autocatalysis

In autocatalysis the catalyst *is* a product of the overall reaction, in contrast to all other types of catalysis considered in this article. The simplest example of autocatalysis is a reaction of type A + B → 2 B, in one or in several steps. The overall reaction is just A → B, so that B is a product. But since B is also a reactant, it may be present in the rate equation and affect the reaction rate. As the reaction proceeds, the concentration of B increases and can accelerate the reaction as a catalyst. In effect, the reaction accelerates itself or is autocatalyzed.

A real example is the hydrolysis of an ester such as aspirin to a carboxylic acid and an alcohol. In the absence of added acid catalysts, the carboxylic acid product catalyzes the hydrolysis.

Significance

Estimates are that 90% of all commercially produced chemical products involve catalysts at some stage in the process of their manufacture. In 2005, catalytic processes generated about $900 billion in products worldwide. Catalysis is so pervasive that subareas are not readily classified. Some areas of particular concentration are surveyed below.

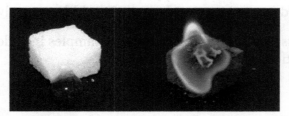

Left: Partially caramelised cube sugar, Right: burning cube sugar with ash as catalyst

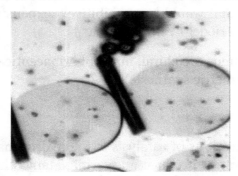

A Ti-Cr-Pt tube (~40 μm long) releases oxygen bubbles when immersed in hydrogen peroxide (via catalytic decomposition), forming a micropump.

Energy Processing

Petroleum refining makes intensive use of catalysis for alkylation, catalytic cracking (breaking long-chain hydrocarbons into smaller pieces), naphtha reforming and steam reforming (conversion of hydrocarbons into synthesis gas). Even the exhaust from the burning of fossil fuels is treated via catalysis: Catalytic converters, typically composed of platinum and rhodium, break down some of the more harmful byproducts of automobile exhaust.

$$2\ CO + 2\ NO \rightarrow 2\ CO_2 + N_2$$

With regard to synthetic fuels, an old but still important process is the Fischer-Tropsch synthesis of hydrocarbons from synthesis gas, which itself is processed via water-gas shift reactions, catalysed by iron. Biodiesel and related biofuels require processing via both inorganic and biocatalysts.

Fuel cells rely on catalysts for both the anodic and cathodic reactions.

Catalytic heaters generate flameless heat from a supply of combustible fuel.

Bulk Chemicals

Some of the largest-scale chemicals are produced via catalytic oxidation, often using oxygen. Examples include nitric acid (from ammonia), sulfuric acid (from sulfur dioxide to sulfur trioxide by the chamber process), terephthalic acid from p-xylene, and acrylonitrile from propane and ammonia.

Many other chemical products are generated by large-scale reduction, often via hydrogenation. The largest-scale example is ammonia, which is prepared via the Haber process from nitrogen. Methanol is prepared from carbon monoxide.

Bulk polymers derived from ethylene and propylene are often prepared via Ziegler-Natta catalysis. Polyesters, polyamides, and isocyanates are derived via acid-base catalysis.

Most carbonylation processes require metal catalysts, examples include the Monsanto acetic acid process and hydroformylation.

Fine Chemicals

Many fine chemicals are prepared via catalysis; methods include those of heavy industry as well as more specialized processes that would be prohibitively expensive on a large scale. Examples include the Heck reaction, and Friedel-Crafts reactions.

Because most bioactive compounds are chiral, many pharmaceuticals are produced by enantioselective catalysis (catalytic asymmetric synthesis).

Food Processing

One of the most obvious applications of catalysis is the hydrogenation (reaction with hydrogen gas) of fats using nickel catalyst to produce margarine. Many other foodstuffs are prepared via biocatalysis.

Environment

Catalysis impacts the environment by increasing the efficiency of industrial processes, but catalysis also plays a direct role in the environment. A notable example is the catalytic role of chlorine free radicals in the breakdown of ozone. These radicals are formed by the action of ultraviolet radiation on chlorofluorocarbons (CFCs).

$$Cl^{\cdot} + O_3 \rightarrow ClO^{\cdot} + O_2$$

$$ClO^{\cdot} + O^{\cdot} \rightarrow Cl^{\cdot} + O_2$$

History

In a general sense, anything that increases the rate of a process is a "catalyst". The concept of catalysis was invented by chemist Elizabeth Fulhame and described in a 1794 book, based on her novel work in oxidation-reduction experiments. The term *catalysis* was later used by Jöns Jakob Berzelius in 1835 to describe reactions that are accelerated by substances that remain unchanged after the reaction. Fulhame, who predated Berzelius, did work with water as opposed to metals in her reduction experiments. Other 18th century chemists who worked in catalysis were Eilhard Mitscherlich who referred to it as *contact* processes, and Johann Wolfgang Döbereiner who spoke of *contact action*. He developed Döbereiner's lamp, a lighter based on hydrogen and a platinum sponge, which became a commercial success in the 1820s that lives on today. Humphry Davy discovered the use of platinum in catalysis. In the 1880s, Wilhelm Ostwald at Leipzig University started a systematic investigation into reactions that were catalyzed by the presence of acids and bases, and found that chemical reactions occur at finite rates and that these rates can be used to determine the strengths of acids and bases. For this work, Ostwald was awarded the 1909 Nobel Prize in Chemistry.

Inhibitors, Poisons, and Promoters

Substances that reduce the action of catalysts are called catalyst inhibitors if reversible, and catalyst poisons if irreversible. Promoters are substances that increase the catalytic activity, even though they are not catalysts by themselves.

Inhibitors are sometimes referred to as "negative catalysts" since they decrease the reaction rate. However the term inhibitor is preferred since they do not work by introducing a reaction path with higher activation energy; this would not reduce the rate since the reaction would continue to occur by the non-catalyzed path. Instead they act either by deactivating catalysts, or by removing reaction intermediates such as free radicals.

The inhibitor may modify selectivity in addition to rate. For instance, in the reduction of alkynes to alkenes, a palladium (Pd) catalyst partly "poisoned" with lead(II) acetate ($Pb(CH_3CO_2)_2$) can be used. Without the deactivation of the catalyst, the alkene produced would be further reduced to alkane.

The inhibitor can produce this effect by, e.g., selectively poisoning only certain types of active sites. Another mechanism is the modification of surface geometry. For instance, in hydrogenation operations, large planes of metal surface function as sites of hydrogenolysis catalysis while sites catalyzing hydrogenation of unsaturates are smaller. Thus, a poison that covers surface randomly will tend to reduce the number of uncontaminated large planes but leave proportionally more smaller sites free, thus changing the hydrogenation vs. hydrogenolysis selectivity. Many other mechanisms are also possible.

Promoters can cover up surface to prevent production of a mat of coke, or even actively remove such material (e.g., rhenium on platinum in platforming). They can aid the dispersion of the catalytic material or bind to reagents.

Current Market

The global demand on catalysts in 2010 was estimated at approximately 29.5 billion USD. With the rapid recovery in automotive and chemical industry overall, the global catalyst market is expected to experience fast growth in the next years.

Materials Science

The interdisciplinarity field of materials science, also commonly termed materials science and engineering, involves the discovery and design of new materials, with an emphasis on solids. The intellectual origins of materials science stem from the Enlightenment, when researchers began to use analytical thinking from chemistry, physics, and engineering to understand ancient, phenomenological observations in metallurgy and mineralogy. Materials science still incorporates elements of physics, chemistry, and engineering. As such, the field was long considered by academic institutions as a sub-field of these related fields. Beginning in the 1940s, materials science began to be more widely recognized as a specific and distinct field of science and engineering, and major technical universities around the world created dedicated schools of the study.

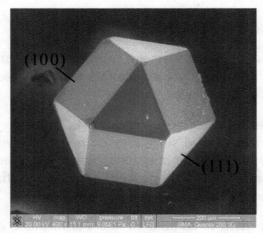

A diamond cuboctahedron showing seven crystallographic planes, imaged with scanning electron microscopy.

Materials science is a syncretic discipline hybridizing metallurgy, ceramics, solid-state physics, and chemistry. It is the first example of a new academic discipline emerging by fusion rather than fission.

Many of the most pressing scientific problems humans currently face are due to the limits of the materials that are available. Thus, breakthroughs in materials science are likely to affect the future of technology significantly.

Materials scientists emphasize understanding how the history of a material (its *processing*) influences its structure, and thus the material's properties and performance. The understanding of processing-structure-properties relationships is called the § materials paradigm. This paradigm is used to advance understanding in a variety of research areas, including nanotechnology, biomaterials, and metallurgy. Materials science is also an important part of forensic engineering and failure analysis - investigating materials, products, structures or components which fail or which do not operate or function as intended, causing personal injury or damage to property. Such investigations are key to understanding, for example, the causes of various aviation accidents and incidents.

History

The material of choice of a given era is often a defining point. Phrases such as Stone Age, Bronze Age, Iron Age, and Steel Age are great examples. Originally deriving from the manufacture of ceramics and its putative derivative metallurgy, materials science is one of the oldest forms of engineering and applied science. Modern materials science evolved directly from metallurgy, which itself evolved from mining and (likely) ceramics and the use of fire. A major breakthrough in the understanding of materials occurred in the late 19th century, when the American scientist Josiah Willard Gibbs demonstrated that the thermodynamic properties related to atomic structure in various phases are related to the physical properties of a material. Important elements of modern materials science are a product of the space race: the understanding and engineering of the metallic alloys, and silica and carbon materials, used in building space vehicles enabling the exploration of space. Materials science has driven, and been driven by, the development of revolutionary technologies such as rubbers, plastics, semiconductors, and biomaterials.

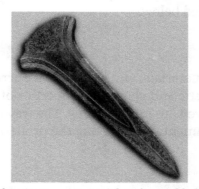

A late Bronze Age sword or dagger blade.

Before the 1960s (and in some cases decades after), many *materials science* departments were named *metallurgy* departments, reflecting the 19th and early 20th century emphasis on metals. The growth of materials science in the United States was catalyzed in part by the Advanced Research Projects Agency, which funded a series of university-hosted laboratories in the early 1960s "to expand the national program of basic research and training in the materials sciences." The field has since broadened to include every class of materials, including ceramics, polymers, semiconductors, magnetic materials, medical implant materials, biological materials, and nanomaterials.

Fundamentals

The materials paradigm represented in the form of a tetrahedron.

A material is defined as a substance (most often a solid, but other condensed phases can be included) that is intended to be used for certain applications. There are a myriad of materials around us—they can be found in anything from buildings to spacecraft. Materials can generally be divided into two classes: crystalline and non-crystalline. The traditional examples of materials are metals, semiconductors, ceramics and polymers. New and advanced materials that are being developed include nanomaterials and biomaterials, etc.

The basis of materials science involves studying the structure of materials, and relating them to their properties. Once a materials scientist knows about this structure-property correlation, they can then go on to study the relative performance of a material in a given application. The major determinants of the structure of a material and thus of its properties are its constituent chemical elements and the way in which it has been processed into its final form. These characteristics, taken together and related through the laws of thermodynamics and kinetics, govern a material's microstructure, and thus its properties.

Structure

As mentioned above, structure is one of the most important components of the field of materials science. Materials science examines the structure of materials from the atomic scale, all the way up to the macro scale. Characterization is the way materials scientists examine the structure of a material. This involves methods such as diffraction with X-rays, electrons, or neutrons, and various forms of spectroscopy and chemical analysis such as Raman spectroscopy, energy-dispersive spectroscopy (EDS), chromatography, thermal analysis, electron microscope analysis, etc. Structure is studied at various levels, as detailed below.

Atomic Structure

This deals with the atoms of the materials, and how they are arranged to give molecules, crystals, etc. Much of the electrical, magnetic and chemical properties of materials arise from this level of structure. The length scales involved are in angstroms. The way in which the atoms and molecules are bonded and arranged is fundamental to studying the properties and behavior of any material.

Nanostructure

Nanostructure deals with objects and structures that are in the 1—100 nm range. In many materials, atoms or molecules agglomerate together to form objects at the nanoscale. This causes many interesting electrical, magnetic, optical, and mechanical properties.

Buckminsterfullerene nanostructure.

In describing nanostructures it is necessary to differentiate between the number of dimensions on the nanoscale. Nanotextured surfaces have *one dimension* on the nanoscale, i.e., only the thickness of the surface of an object is between 0.1 and 100 nm. Nanotubes have *two dimensions* on the nanoscale, i.e., the diameter of the tube is between 0.1 and 100 nm; its length could be much greater. Finally, spherical nanoparticles have *three dimensions* on the nanoscale, i.e., the particle is between 0.1 and 100 nm in each spatial dimension. The terms nanoparticles and ultrafine particles (UFP) often are used synonymously although UFP can reach into the micrometre range. The term 'nanostructure' is often used when referring to magnetic technology. Nanoscale structure in biology is often called ultrastructure.

Materials which atoms and molecules form constituents in the nanoscale (i.e., they form nano-structure) are called nanomaterials. Nanomaterials are subject of intense research in the materials science community due to the unique properties that they exhibit.

Microstructure

Microstructure is defined as the structure of a prepared surface or thin foil of material as revealed by a microscope above 25× magnification. It deals with objects from 100 nm to a few cm. The microstructure of a material (which can be broadly classified into metallic, polymeric, ceramic and composite) can strongly influence physical properties such as strength, toughness, ductility, hardness, corrosion resistance, high/low temperature behavior, wear resistance, and so on. Most of the traditional materials (such as metals and ceramics) are microstructured.

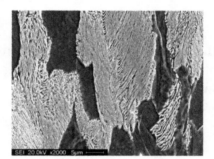

Microstructure of pearlite.

The manufacture of a perfect crystal of a material is physically impossible. For example, a crystalline material will contain defects such as precipitates, grain boundaries (Hall–Petch relationship), interstitial atoms, vacancies or substitutional atoms. The microstructure of materials reveals these defects, so that they can be studied.

Macro Structure

Macro structure is the appearance of a material in the scale millimeters to meters—it is the structure of the material as seen with the naked eye.

Crystallography

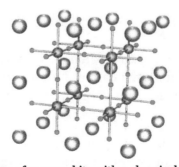

Crystal structure of a perovskite with a chemical formula ABX_3.

Crystallography is the science that examines the arrangement of atoms in crystalline solids. Crystallography is a useful tool for materials scientists. In single crystals, the effects of the crystalline arrangement of atoms is often easy to see macroscopically, because the natural shapes of crystals reflect the atomic structure. Further, physical properties are often controlled by crystalline defects. The understanding of crystal structures is an important prerequisite for understanding crystallographic defects. Mostly, materials do not occur as a single crystal, but in polycrystalline form, i.e., as an aggregate of small crystals with different orientations. Because of this, the powder diffraction method, which uses diffraction patterns of polycrystalline samples with a large number of crystals, plays an important role in structural determination. Most materials have a crystalline structure, but some important materials do not exhibit regular crystal structure. Polymers display varying degrees of crystallinity, and many are completely noncrystalline. Glass, some ceramics, and many natural materials are amorphous, not possessing any long-range order in their atomic arrangements. The study of polymers combines elements of chemical and statistical thermodynamics to give thermodynamic and mechanical, descriptions of physical properties.

Bonding

To obtain a full understanding of the material structure and how it relates to its properties, the materials scientist must study how the different atoms, ions and molecules are arranged and bonded to each other. This involves the study and use of quantum chemistry or quantum physics. Solid-state physics, solid-state chemistry and physical chemistry are also involved in the study of bonding and structure.

Properties

Materials exhibit myriad properties, including the following.

- Mechanical properties
- Chemical properties
- Electrical properties
- Thermal properties
- Optical properties
- Magnetic properties

The properties of a material determine its usability and hence its engineering application.

Synthesis and Processing

Synthesis and processing involves the creation of a material with the desired micro-nanostructure. From an engineering standpoint, a material cannot be used in industry if no economical production method for it has been developed. Thus, the processing of materials is vital to the field of materials science.

Different materials require different processing or synthesis methods. For example, the processing of metals has historically been very important and is studied under the branch of materials science named *physical metallurgy*. Also, chemical and physical methods are also used to synthesize other materials such as polymers, ceramics, thin films, etc. As of the early 21st century, new methods are being developed to synthesize nanomaterials such as graphene.

Thermodynamics

Thermodynamics is concerned with heat and temperature and their relation to energy and work. It defines macroscopic variables, such as internal energy, entropy, and pressure, that partly describe a body of matter or radiation. It states that the behavior of those variables is subject to general constraints, that are common to all materials, not the peculiar properties of particular materials. These general constraints are expressed in the four laws of thermodynamics. Thermodynamics describes the bulk behavior of the body, not the microscopic behaviors of the very large numbers of its microscopic constituents, such as molecules. The behavior of these microscopic particles is described by, and the laws of thermodynamics are derived from, statistical mechanics.

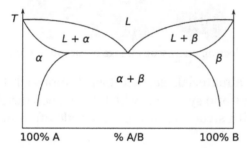

A phase diagram for a binary system displaying a eutectic point.

The study of thermodynamics is fundamental to materials science. It forms the foundation to treat general phenomena in materials science and engineering, including chemical reactions, magnetism, polarizability, and elasticity. It also helps in the understanding of phase diagrams and phase equilibrium.

Kinetics

Chemical kinetics is the study of the rates at which systems that are out of equilibrium change under the influence of various forces. When applied to materials science, it deals with how a material changes with time (moves from non-equilibrium to equilibrium state) due to application of a certain field. It details the rate of various processes evolving in materials including shape, size, composition and structure. Diffusion is important in the study of kinetics as this is the most common mechanism by which materials undergo change.

Kinetics is essential in processing of materials because, among other things, it details how the microstructure changes with application of heat.

In research

Materials science has received much attention from researchers. In most universities, many departments ranging from physics to chemistry to chemical engineering, along with materials science departments, are involved in materials research. Research in materials science is vibrant and consists of many avenues. The following list is in no way exhaustive. It serves only to highlight certain important research areas.

Nanomaterials

Nanomaterials describe, in principle, materials of which a single unit is sized (in at least one dimension) between 1 and 1000 nanometers (10^{-9} meter) but is usually 1—100 nm.

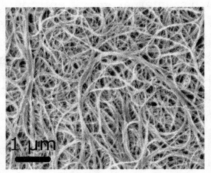

A scanning electron microscopy image of carbon nanotubes bundles

Nanomaterials research takes a materials science-based approach to nanotechnology, leveraging advances in materials metrology and synthesis which have been developed in support of microfabrication research. Materials with structure at the nanoscale often have unique optical, electronic, or mechanical properties.

The field of nanomaterials is loosely organized, like the traditional field of chemistry, into organic (carbon-based) nanomaterials such as fullerenes, and inorganic nanomaterials based on other elements, such as silicon. Examples of nanomaterials include fullerenes, carbon nanotubes, nanocrystals, etc.

Biomaterials

A biomaterial is any matter, surface, or construct that interacts with biological systems. As a science, *bio materials* is about fifty years old. The study of biomaterials is called *bio materials science*. It has experienced steady and strong growth over its history, with many companies investing large amounts of money into developing new products. Biomaterials science encompasses elements of medicine, biology, chemistry, tissue engineering, and materials science.

The iridescent nacre inside a nautilus shell.

Biomaterials can be derived either from nature or synthesized in a laboratory using a variety of chemical approaches using metallic components, polymers, ceramics, or composite materials.

They are often used and/or adapted for a medical application, and thus comprises whole or part of a living structure or biomedical device which performs, augments, or replaces a natural function. Such functions may be benign, like being used for a heart valve, or may be bioactive with a more interactive functionality such as hydroxylapatite coated hip implants. Biomaterials are also used everyday in dental applications, surgery, and drug delivery. For example, a construct with impregnated pharmaceutical products can be placed into the body, which permits the prolonged release of a drug over an extended period of time. A biomaterial may also be an autograft, allograft or xenograft used as an organ transplant material.

Electronic, Optical, and Magnetic

Semiconductors, metals, and ceramics are used today to form highly complex systems, such as integrated electronic circuits, optoelectronic devices, and magnetic and optical mass storage media. These materials form the basis of our modern computing world, and hence research into these materials is of vital importance.

Negative index metamaterial.

Semiconductors are a traditional example of these types of materials. They are materials that have properties that are intermediate between conductors and insulators. Their electrical conductivities are very sensitive to impurity concentrations, and this allows for the use of doping to achieve desirable electronic properties. Hence, semiconductors form the basis of the traditional computer.

This field also includes new areas of research such as superconducting materials, spintronics, metamaterials, etc. The study of these materials involves knowledge of materials science and solid-state physics or condensed matter physics.

Computational Science and Theory

With the increase in computing power, simulating the behavior of materials has become possible. This enables materials scientists to discover properties of materials formerly unknown, as well as to design new materials. Up until now, new materials were found by time consuming trial and error processes. But, now it is hoped that computational methods could drastically reduce that time, and allow tailoring materials properties. This involves simulating materials at all length scales, using methods such as density functional theory, molecular dynamics, etc.

In industry

Radical materials advances can drive the creation of new products or even new industries, but stable industries also employ materials scientists to make incremental improvements and troubleshoot issues with currently used materials. Industrial applications of materials science include materials design, cost-benefit tradeoffs in industrial production of materials, processing methods (casting, rolling, welding, ion implantation, crystal growth, thin-film deposition, sintering, glass-blowing, etc.), and analytic methods (characterization methods such as electron microscopy, X-ray diffraction, calorimetry, nuclear microscopy (HEFIB), Rutherford backscattering, neutron diffraction, small-angle X-ray scattering (SAXS), etc.).

Besides material characterization, the material scientist or engineer also deals with extracting materials and convering them into useful forms. Thus ingot casting, foundry methods, blast furnace extraction, and electrolytic extraction are all part of the required knowledge of a materials engineer. Often the presence, absence, or variation of minute quantities of secondary elements and compounds in a bulk material will greatly affect the final properties of the materials produced. For example, steels are classified based on 1/10 and 1/100 weight percentages of the carbon and other alloying elements they contain. Thus, the extracting and purifying methods used to extract iron in a blast furnace can affect the quality of steel that is produced.

Ceramics and Glasses

Another application of material science is the structures of ceramics and glass, typically associated with the most brittle materials. Bonding in ceramics and glasses uses covalent and ionic-covalent types with SiO_2 (silica or sand) as a fundamental building block. Ceramics are as soft as clay or as hard as stone and concrete. Usually, they are crystalline in form. Most glasses contain a metal oxide fused with silica. At high temperatures used to prepare glass, the material is a viscous liquid. The structure of glass forms into an amorphous state upon cooling. Windowpanes and eyeglasses are important examples. Fibers of glass are also available. Scratch resistant Corning Gorilla Glass is a well-known example of the application of materials science to drastically improve the properties of common components. Diamond and carbon in its graphite form are considered to be ceramics.

Si_3N_4 ceramic bearing parts

Engineering ceramics are known for their stiffness and stability under high temperatures, compression and electrical stress. Alumina, silicon carbide, and tungsten carbide are made from a fine powder of their constituents in a process of sintering with a binder. Hot pressing provides higher density material. Chemical vapor deposition can place a film of a ceramic on another material. Cermets are ceramic particles containing some metals. The wear resistance of tools is derived from cemented carbides with the metal phase of cobalt and nickel typically added to modify properties.

Composites

Filaments are commonly used for reinforcement in composite materials.

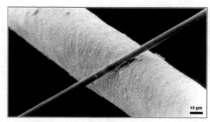

A 6 μm diameter carbon filament (running from bottom left to top right) siting atop the much larger human hair.

Another application of materials science in industry is making composite materials. These are structured materials composed of two or more macroscopic phases. Applications range from structural elements such as steel-reinforced concrete, to the thermal insulating tiles which play a key and integral role in NASA's Space Shuttle thermal protection system which is used to protect the surface of the shuttle from the heat of re-entry into the Earth's atmosphere. One example is reinforced Carbon-Carbon (RCC), the light gray material which withstands re-entry temperatures up to 1,510 °C (2,750 °F) and protects the Space Shuttle's wing leading edges and nose cap. RCC is a laminated composite material made from graphite rayon cloth and impregnated with a phenolic resin. After curing at high temperature in an autoclave, the laminate is pyrolized to convert the resin to carbon, impregnated with furfural alcohol in a vacuum chamber, and cured-pyrolized to convert the furfural alcohol to carbon. To provide oxidation resistance for reuse ability, the outer layers of the RCC are converted to silicon carbide.

Other examples can be seen in the "plastic" casings of television sets, cell-phones and so on. These plastic casings are usually a composite material made up of a thermoplastic matrix such as acrylonitrile butadiene styrene (ABS) in which calcium carbonate chalk, talc, glass fibers or carbon fibers have been added for added strength, bulk, or electrostatic dispersion. These additions may be termed reinforcing fibers, or dispersants, depending on their purpose.

Polymers

$$\left[\begin{array}{c} CH_3 \\ | \\ CH-CH_2 \end{array} \right]_n$$

The repeating unit of the polymer polypropylene

Polymers are chemical compounds made up of a large number of identical components linked together like chains. They are an important part of materials science. Polymers are the raw materials (the resins) used to make what are commonly called plastics and rubber. Plastics and rubber are really the final product, created after one or more polymers or additives have been added to a resin during processing, which is then shaped into a final form. Plastics which have been around, and which are in current widespread use, include polyethylene, polypropylene, polyvinyl chloride (PVC), polystyrene, nylons, polyesters, acrylics, polyurethanes, and polycarbonates and also rubbers which have been around are natural rubber, styrene butadiene rubber, chloroprene, and butadiene rubber. Plastics are generally classified as *commodity*, *specialty* and *engineering* plastics.

Polyvinyl chloride (PVC) is widely used, inexpensive, and annual production quantities are large. It lends itself to an vast array of applications, from artificial leather to electrical insulation and cabling, packaging, and containers. Its fabrication and processing are simple and well-established. The versatility of PVC is due to the wide range of plasticisers and other additives that it accepts. The term "additives" in polymer science refers to the chemicals and compounds added to the polymer base to modify its material properties.

Expanded polystyrene polymer packaging.

Polycarbonate would be normally considered an engineering plastic (other examples include PEEK, ABS). Such plastics are valued for their superior strengths and other special material properties. They are usually not used for disposable applications, unlike commodity plastics.

Specialty plastics are materials with unique characteristics, such as ultra-high strength, electrical conductivity, electro-fluorescence, high thermal stability, etc.

The dividing lines between the various types of plastics is not based on material but rather on their properties and applications. For example, polyethylene (PE) is a cheap, low friction polymer commonly used to make disposable bags for shopping and trash, and is considered a commodity plastic, whereas medium-density polyethylene (MDPE) is used for underground gas and water pipes, and another variety called ultra-high-molecular-weight polyethylene (UHMWPE) is an engineering plastic which is used extensively as the glide rails for industrial equipment and the low-friction socket in implanted hip joints.

Metal Alloys

The study of metal alloys is a significant part of materials science. Of all the metallic alloys in use today, the alloys of iron (steel, stainless steel, cast iron, tool steel, alloy steels) make up the largest proportion both by quantity and commercial value. Iron alloyed with various proportions of carbon gives low, mid and high carbon steels. An iron carbon alloy is only considered steel if the carbon level is between 0.01% and 2.00%. For the steels, the hardness and tensile strength of the steel is related to the amount of carbon present, with increasing carbon levels also leading to lower ductility and toughness. Heat treatment processes such as quenching and tempering can significantly change these properties however. Cast Iron is defined as an iron–carbon alloy with more than 2.00% but less than 6.67% carbon. Stainless steel is defined as a regular steel alloy with greater than 10% by weight alloying content of Chromium. Nickel and Molybdenum are typically also found in stainless steels.

Other significant metallic alloys are those of aluminium, titanium, copper and magnesium. Copper alloys have been known for a long time (since the Bronze Age), while the alloys of the other three metals have been relatively recently developed. Due to the chemical reactivity of these metals, the electrolytic extraction processes required were only developed relatively recently. The alloys of aluminium, titanium and magnesium are also known and valued for their high strength-to-weight ratios and, in the case of magnesium, their ability to provide electromagnetic shielding. These materials are ideal for situations where high strength-to-weight ratios are more important than bulk cost, such as in the aerospace industry and certain automotive engineering applications.

Wire rope made from steel alloy.

Semiconductors

The study of semiconductors is a significant part of materials science. A semiconductor is a material that has a resistivity between a metal and insulator. It's electronic properties can be greatly altered through intentionally introducing impurities, or doping. From these semiconductor materials, things such as diodes, transistors, light-emitting diodes (LEDs), and analog and digital electric circuits can be built, making them materials of interest in industry. Semiconductor devices have replaced thermionic devices (vacuum tubes) in most applications. Semiconductor devices are manufactured both as single discrete devices and as integrated circuits (ICs), which consist of a number—from a few to millions—of devices manufactured and interconnected on a single semiconductor substrate.

Of all the semiconductors in use today, silicon makes up the largest portion both by quantity and commercial value. Monocrystalline silicon is used to produce wafers used in the semiconductor and electronics industry. Second to silicon, gallium arsenide (GaAs) is the second most popular semiconductor used. Due to its higher electron mobility and saturation velocity compared to silicon, its a material of choice for high speed electronics applications. These superior properties are compelling reasons to use GaAs circuitry in mobile phones, satellite communications, microwave point-to-point links and higher frequency radar systems. Other semiconductor materials include germanium, silicon carbide, and gallium nitride and have various applications.

Relation to Other Fields

Materials science evolved—starting from the 1960s—because it was recognized that to create, discover and design new materials, one had to approach it in a unified manner. Thus, materials science and engineering emerged at the intersection of various fields such as metallurgy, solid state physics, chemistry, chemical engineering, mechanical engineering and electrical engineering.

The field is inherently interdisciplinary, and the materials scientists/engineers must be aware and make use of the methods of the physicist, chemist and engineer. The field thus, maintains close relationships with these fields. Also, many physicists, chemists and engineers also find themselves working in materials science.

The overlap between physics and materials science has led to the offshoot field of *materials physics*, which is concerned with the physical properties of materials. The approach is generally more macroscopic and applied than in condensed matter physics.

The field of materials science and engineering is important both from a scientific perspective, as well as from an engineering one. When discovering new materials, one encounters new phenomena that may not have been observed before. Hence, there is lot of science to be discovered when working with materials. Materials science also provides test for theories in condensed matter physics.

Materials are of the utmost importance for engineers, as the usage of the appropriate materials is crucial when designing systems. As a result, materials science is an increasingly important part of an engineer's education.

Emerging Technologies in Materials Science

Emerging technology	Status	Potentially marginalized technologies	Potential applications
Aerogel	Hypothetical, experiments, diffusion, early uses	Traditional insulation, glass	Improved insulation, insulative glass if it can be made clear, sleeves for oil pipelines, aerospace, high-heat & extreme cold applications
Amorphous metal	Experiments	Kevlar	Armor
Conductive polymers	Research, experiments, prototypes	Conductors	Lighter and cheaper wires, antistatic materials, organic solar cells
Femtotechnology, picotechnology	Hypothetical	Present nuclear	New materials; nuclear weapons, power
Fullerene	Experiments, diffusion	Synthetic diamond and carbon nanotubes (e.g., Buckypaper)	Programmable matter

Graphene	Hypothetical, experiments, diffusion, early uses	Silicon-based integrated circuit	Components with higher strength to weight ratios, transistors that operate at higher frequency, lower cost of display screens in mobile devices, storing hydrogen for fuel cell powered cars, filtration systems, longer-lasting and faster-charging batteries, sensors to diagnose diseases
High-temperature superconductivity	Cryogenic receiver front-end (CRFE) RF and microwave filter systems for mobile phone base stations; prototypes in dry ice; Hypothetical and experiments for higher temperatures	Copper wire, semiconductor integral circuits	No loss conductors, frictionless bearings, magnetic levitation, lossless high-capacity accumulators, electric cars, heat-free integral circuits and processors
LiTraCon	Experiments, already used to make Europe Gate	Glass	Building skyscrapers, towers, and sculptures like Europe Gate
Metamaterials	Hypothetical, experiments, diffusion	Classical optics	Microscopes, cameras, metamaterial cloaking, cloaking devices
Metal foam	Research, commercialization	Hulls	Space colonies, floating cities
Multi-function structures	Hypothetical, experiments, some prototypes, few commercial	Composite materials mostly	Wide range, e.g., self health monitoring, self healing material, morphing, ...
Nanomaterials: carbon nanotubes	Hypothetical, experiments, diffusion, early uses	Structural steel and aluminium	Stronger, lighter materials, space elevator
Programmable matter	Hypothetical, experiments	Coatings, catalysts	Wide range, e.g., claytronics, synthetic biology
Quantum dots	Research, experiments, prototypes	LCD, LED	Quantum dot laser, future use as programmable matter in display technologies (TV, projection), optical data communications (high-speed data transmission), medicine (laser scalpel)
Silicene	Hypothetical, research	Field-effect transistors	

Superalloy	Research, diffusion	Aluminum, titanium, composite materials	Aircraft jet engines
Synthetic diamond	early uses (drill bits, jewelry)	Silicon transistors	Electronics

Pigment

A pigment is a material that changes the color of reflected or transmitted light as the result of wavelength-selective absorption. This physical process differs from fluorescence, phosphorescence, and other forms of luminescence, in which a material emits light.

Natural ultramarine pigment in powdered form

Many materials selectively absorb certain wavelengths of light. Materials that humans have chosen and developed for use as pigments usually have special properties that make them ideal for coloring other materials. A pigment must have a high tinting strength relative to the materials it colors. It must be stable in solid form at ambient temperatures.

Synthetic ultramarine pigment is chemically identical to natural ultramarine

For industrial applications, as well as in the arts, permanence and stability are desirable properties. Pigments that are not permanent are called fugitive. Fugitive pigments fade over time, or with exposure to light, while some eventually blacken.

Pigments are used for coloring paint, ink, plastic, fabric, cosmetics, food, and other materials. Most pigments used in manufacturing and the visual arts are dry colorants, usually ground into a fine powder. This powder is added to a binder (or vehicle), a relatively neutral or colorless material that suspends the pigment and gives the paint its adhesion.

A distinction is usually made between a pigment, which is insoluble in its vehicle (resulting in a suspension), and a dye, which either is itself a liquid or is soluble in its vehicle (resulting in a solution). A colorant can act as either a pigment or a dye depending on the vehicle involved. In some cases, a pigment can be manufactured from a dye by precipitating a soluble dye with a metallic salt. The resulting pigment is called a lake pigment. The term biological pigment is used for all colored substances independent of their solubility.

In 2006, around 7.4 million tons of inorganic, organic and special pigments were marketed worldwide. Asia has the highest rate on a quantity basis followed by Europe and North America. By 2020, revenues will have risen to approx. US$34.2 billion. The global demand on pigments was roughly US$20.5 billion in 2009, around 1.5-2% up from the previous year. It is predicted to increase in a stable growth rate in the coming years. The worldwide sales are said to increase up to US$24.5 billion in 2015, and reach US$27.5 billion in 2018.

Physical Basis

Pigments appear the colors they are because they selectively reflect and absorb certain wavelengths of visible light. White light is a roughly equal mixture of the entire spectrum of visible light with a wavelength in a range from about 375 or 400 nanometers to about 760 or 780 nm. When this light encounters a pigment, parts of the spectrum are absorbed by the molecules or ions of the pigment. In organic pigments such as diazo or phthalocyanine compounds the light is absorbed by the conjugated systems of double bonds in the molecule. Some of the inorganic pigments such as vermilion (mercury sulfide) or cadmium yellow (cadmium sulfide) absorb light by transferring an electron from the negative ion (S^{2-}) to the positive ion (Hg^{2+} or Cd^{2+}). Such compounds are designated as charge-transfer complexes, with broad absorption bands that subtract most of the colors of the incident white light. The other wavelengths or parts of the spectrum are reflected or scattered. The new reflected light spectrum creates the appearance of a color. Pigments, unlike fluorescent or phosphorescent substances, can only subtract wavelengths from the source light, never add new ones.

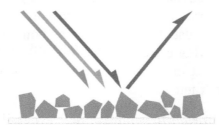

A wide variety of wavelengths (colors) encounter a pigment. This pigment absorbs red and green light, but reflects blue, creating the color blue.

The appearance of pigments is intimately connected to the color of the source light. Sunlight has a high color temperature, and a fairly uniform spectrum, and is considered a standard for white light. Artificial light sources tend to have great peaks in some parts of their spectrum, and deep valleys in others. Viewed under these conditions, pigments will appear different colors.

Color spaces used to represent colors numerically must specify their light source. Lab color measurements, unless otherwise noted, assume that the measurement was taken under a D65 light source, or "Daylight 6500 K", which is roughly the color temperature of sunlight.

Other properties of a color, such as its saturation or lightness, may be determined by the other substances that accompany pigments. Binders and fillers added to pure pigment chemicals also have their own reflection and absorption patterns, which can affect the final spectrum. Likewise, in pigment/binder mixtures, individual rays of light may not encounter pigment molecules, and may be reflected as is. These stray rays of source light contribute to a slightly less saturated color. Pure pigment allows very little white light to escape, producing a highly saturated color. A small quantity of pigment mixed with a lot of white binder, however, will appear desaturated and pale, due to the high quantity of escaping white light.

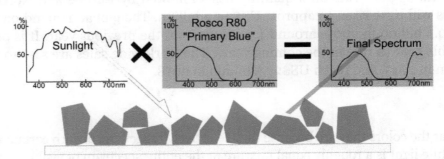

Sunlight encounters Rosco R80 "Primary Blue" pigment. The product of the source spectrum and the reflectance spectrum of the pigment results in the final spectrum, and the appearance of blue.

History

Naturally occurring pigments such as ochres and iron oxides have been used as colorants since prehistoric times. Archaeologists have uncovered evidence that early humans used paint for aesthetic purposes such as body decoration. Pigments and paint grinding equipment believed to be between 350,000 and 400,000 years old have been reported in a cave at Twin Rivers, near Lusaka, Zambia.

Before the Industrial Revolution, the range of color available for art and decorative uses was technically limited. Most of the pigments in use were earth and mineral pigments, or pigments of biological origin. Pigments from unusual sources such as botanical materials, animal waste, insects, and mollusks were harvested and traded over long distances. Some colors were costly or impossible to obtain, given the range of pigments that were available. Blue and purple came to be associated with royalty because of their rarity.

Biological pigments were often difficult to acquire, and the details of their production were kept secret by the manufacturers. Tyrian Purple is a pigment made from the mucus of one of several species of Murex snail. Production of Tyrian Purple for use as a fabric dye began as early as 1200 BCE by the Phoenicians, and was continued by the Greeks and Romans until 1453 CE, with the fall of Constantinople. The pigment was expensive and complex to produce, and items colored with it became associated with power and wealth. Greek historian Theopompus, writing in the 4th century BCE, reported that "purple for dyes fetched its weight in silver at Colophon [in Asia Minor]."

Mineral pigments were also traded over long distances. The only way to achieve a deep rich blue was by using a semi-precious stone, lapis lazuli, to produce a pigment known as ultramarine, and the best sources of lapis were remote. Flemish painter Jan van Eyck, working in the 15th century, did not ordinarily include blue in his paintings. To have one's portrait commissioned and painted with ultramarine blue was considered a great luxury. If a patron wanted blue, they were obliged to pay extra. When Van Eyck used lapis, he never blended it with other colors. Instead he applied it in pure form, almost as a decorative glaze. The prohibitive price of lapis lazuli forced artists to seek less expensive replacement pigments, both mineral (azurite, smalt) and biological (indigo).

Miracle of the Slave by Tintoretto (c. 1548). The son of a master dyer, Tintoretto used Carmine Red Lake pigment, derived from the cochineal insect, to achieve dramatic color effects.

Spain's conquest of a New World empire in the 16th century introduced new pigments and colors to peoples on both sides of the Atlantic. Carmine, a dye and pigment derived from a parasitic insect found in Central and South America, attained great status and value in Europe. Produced from harvested, dried, and crushed cochineal insects, carmine could be, and still is, used in fabric dye, food dye, body paint, or in its solid lake form, almost any kind of paint or cosmetic.

Natives of Peru had been producing cochineal dyes for textiles since at least 700 CE, but Europeans had never seen the color before. When the Spanish invaded the Aztec empire in what is now Mexico, they were quick to exploit the color for new trade opportunities. Carmine became the region's second most valuable export next to silver. Pigments produced from the cochineal insect gave the Catholic cardinals their vibrant robes and the English "Redcoats" their distinctive uniforms. The true source of the pigment, an insect, was kept secret until the 18th century, when biologists discovered the source.

Girl with a Pearl Earring by Johannes Vermeer (c. 1665).

While Carmine was popular in Europe, blue remained an exclusive color, associated with wealth and status. The 17th-century Dutch master Johannes Vermeer often made lavish use of lapis lazuli, along with Carmine and Indian yellow, in his vibrant paintings.

Development of Synthetic Pigments

The earliest known pigments were natural minerals. Natural iron oxides give a range of colors and are found in many Paleolithic and Neolithic cave paintings. Two examples include Red Ochre, anhydrous Fe_2O_3, and the hydrated Yellow Ochre ($Fe_2O_3.H_2O$). Charcoal, or carbon black, has also been used as a black pigment since prehistoric times.

Two of the first synthetic pigments were white lead (basic lead carbonate, $(PbCO_3)_2Pb(OH)_2$) and blue frit (Egyptian Blue). White lead is made by combining lead with vinegar (acetic acid, CH_3COOH) in the presence of CO_2. Blue frit is calcium copper silicate and was made from glass colored with a copper ore, such as malachite. These pigments were used as early as the second millennium BCE Later premodern additions to the range of synthetic pigments included vermillion, verdigris and lead-tin-yellow.

The Industrial and Scientific Revolutions brought a huge expansion in the range of synthetic pigments, pigments that are manufactured or refined from naturally occurring materials, available both for manufacturing and artistic expression. Because of the expense of Lapis Lazuli, much effort went into finding a less costly blue pigment.

Prussian Blue was the first modern synthetic pigment, discovered by accident in 1704. By the early 19th century, synthetic and metallic blue pigments had been added to the range of blues, including French ultramarine, a synthetic form of lapis lazuli, and the various forms of Cobalt and Cerulean Blue. In the early 20th century, organic chemistry added Phthalo Blue, a synthetic, organometallic pigment with overwhelming tinting power.

Self Portrait by Paul Cézanne. Working in the late 19th century, Cézanne had a palette of colors that earlier generations of artists could only have dreamed of.

Discoveries in color science created new industries and drove changes in fashion and taste. The discovery in 1856 of mauveine, the first aniline dye, was a forerunner for the development of hundreds of synthetic dyes and pigments like azo and diazo compounds which are the source of a wide spectrum of colors. Mauveine was discovered by an 18-year-old chemist named William Henry Perkin, who went on to exploit his discovery in industry and become wealthy. His success attracted a generation of followers, as young scientists went into organic chemistry to pursue riches. Within a few years, chemists had synthesized a substitute for madder in the production of Alizarin Crimson. By the closing decades of the 19th century, textiles, paints, and other commodities in colors such as red, crimson, blue, and purple had become affordable.

Development of chemical pigments and dyes helped bring new industrial prosperity to Germany and other countries in northern Europe, but it brought dissolution and decline elsewhere. In Spain's former New World empire, the production of cochineal colors employed thousands of low-paid workers. The Spanish monopoly on cochineal production had been worth a fortune until the early 19th century, when the Mexican War of Independence and other market changes disrupted production. Organic chemistry delivered the final blow for the cochineal color industry. When chemists created inexpensive substitutes for carmine, an industry and a way of life went into steep decline.

New Sources for Historic Pigments

Before the Industrial Revolution, many pigments were known by the location where they were produced. Pigments based on minerals and clays often bore the name of the city or region where they were mined. Raw Sienna and Burnt Sienna came from Siena, Italy, while Raw Umber and Burnt Umber came from Umbria. These pigments were among the easiest to synthesize, and chemists created modern colors based on the originals that were more consistent than colors mined from the original ore bodies. But the place names remained.

The Milkmaid by Johannes Vermeer (c. 1658). Vermeer was lavish in his choice of expensive pigments, including lead-tin-yellow, natural ultramarine and madder lake, as shown in this vibrant painting.

Historically and culturally, many famous natural pigments have been replaced with synthetic pigments, while retaining historic names. In some cases, the original color name has shifted in meaning, as a historic name has been applied to a popular modern color. By convention, a contemporary mixture of pigments that replaces a historical pigment is indicated by calling the resulting color a hue, but manufacturers are not always careful in maintaining this distinction. The following examples illustrate the shifting nature of historic pigment names:

- Indian Yellow was once produced by collecting the urine of cattle that had been fed only mango leaves. Dutch and Flemish painters of the 17th and 18th centuries favored it for its luminescent qualities, and often used it to represent sunlight. In the novel *Girl with a Pearl Earring*, Vermeer's patron remarks that Vermeer used "cow piss" to paint his wife. Since mango leaves are nutritionally inadequate for cattle, the practice of harvesting Indian Yellow was eventually declared to be inhumane. Modern hues of Indian Yellow are made from synthetic pigments.

- Ultramarine, originally the semi-precious stone lapis lazuli, has been replaced by an inexpensive modern synthetic pigment, French Ultramarine, manufactured from aluminium silicate with sulfur impurities. At the same time, Royal Blue, another name once given to tints produced from lapis lazuli, has evolved to signify a much lighter and brighter color, and is usually mixed from Phthalo Blue and titanium dioxide, or from inexpensive synthetic blue dyes. Since synthetic ultramarine is chemically identical with lapis lazuli, the "hue" designation is not used. French Blue, yet another historic name for ultramarine, was adopted by the textile and apparel industry as a color name in the 1990s, and was applied to a shade of blue that has nothing in common with the historic pigment ultramarine.

- Vermilion, a toxic mercury compound favored for its deep red-orange color by old master painters such as Titian, has been replaced in painters' palettes by various modern pigments, including cadmium reds. Although genuine Vermilion paint can still be purchased for fine arts and art conservation applications, few manufacturers make it, because of legal liability issues. Few artists buy it, because it has been superseded by modern pigments that are both less expensive and less toxic, as well as less reactive with other pigments. As a result, genuine Vermilion is almost unavailable. Modern vermilion colors are properly designated as Vermilion Hue to distinguish them from genuine Vermilion.

Titian used the historic pigment Vermilion to create the reds in the great fresco of Assunta, completed c. 1518.

Manufacturing and Industrial Standards

Pigments for sale at a market stall in Goa, India.

Before the development of synthetic pigments, and the refinement of techniques for extracting mineral pigments, batches of color were often inconsistent. With the development of a modern color industry, manufacturers and professionals have cooperated to create international standards for identifying, producing, measuring, and testing colors.

First published in 1905, the Munsell color system became the foundation for a series of color models, providing objective methods for the measurement of color. The Munsell system describes a color in three dimensions, hue, value (lightness), and chroma (color purity), where chroma is the difference from gray at a given hue and value.

By the middle years of the 20th century, standardized methods for pigment chemistry were available, part of an international movement to create such standards in industry. The International Organization for Standardization (ISO) develops technical standards for the manufacture of pigments and dyes. ISO standards define various industrial and chemical properties, and how to test for them. The principal ISO standards that relate to all pigments are as follows:

- ISO-787 General methods of test for pigments and extenders.

- ISO-8780 Methods of dispersion for assessment of dispersion characteristics.

Other ISO standards pertain to particular classes or categories of pigments, based on their chemical composition, such as ultramarine pigments, titanium dioxide, iron oxide pigments, and so forth.

Many manufacturers of paints, inks, textiles, plastics, and colors have voluntarily adopted the Colour Index International (CII) as a standard for identifying the pigments that they use in manufacturing particular colors. First published in 1925, and now published jointly on the web by the Society of Dyers and Colourists (United Kingdom) and the American Association of Textile Chemists and Colorists (USA), this index is recognized internationally as the authoritative reference on colorants. It encompasses more than 27,000 products under more than 13,000 generic color index names.

In the CII schema, each pigment has a generic index number that identifies it chemically, regardless of proprietary and historic names. For example, Phthalocyanine Blue BN has been known by a variety of generic and proprietary names since its discovery in the 1930s. In much of Europe, phthalocyanine blue is better known as Helio Blue, or by a proprietary name such as Winsor Blue. An American paint manufacturer, Grumbacher, registered an alternate spelling (Thanos Blue) as a trademark. Colour Index International resolves all these conflicting historic, generic, and proprietary names so that manufacturers and consumers can identify the pigment (or dye) used in a particular color product. In the CII, all phthalocyanine blue pigments are designated by a generic color index number as either PB15 or PB16, short for pigment blue 15 and pigment blue 16; these two numbers reflect slight variations in molecular structure that produce a slightly more greenish or reddish blue.

Scientific and Technical Issues

Selection of a pigment for a particular application is determined by cost, and by the physical properties and attributes of the pigment itself. For example, a pigment that is used to color glass must have very high heat stability in order to survive the manufacturing process; but, suspended in the glass vehicle, its resistance to alkali or acidic materials is not an issue. In artistic paint, heat stability is less important, while lightfastness and toxicity are greater concerns.

The following are some of the attributes of pigments that determine their suitability for particular manufacturing processes and applications:

- Lightfastness and sensitivity for damage from ultra violet light

- Heat stability

- Toxicity

- Tinting strength

- Staining

- Dispersion

- Opacity or transparency

- Resistance to alkalis and acids

- Reactions and interactions between pigments

Swatches

Swatches are used to communicate colors accurately. For different media like printing, computers, plastics, and textiles, different type of swatches are used. Generally, the medium which offers the broadest gamut of color shades is widely used across different media.

Printed Swatches

There are many reference standards providing printed swatches of color shades. PANTONE, RAL, Munsell etc. are widely used standards of color communication across different media like printing, plastics, and textiles.

Plastic Swatches

Companies manufacturing color masterbatches and pigments for plastics offer plastic swatches in injection molded color chips. These color chips are supplied to the designer or customer to choose and select the color for their specific plastic products.

Plastic swatches are available in various special effects like pearl, metallic, fluorescent, sparkle, mosaic etc. However, these effects are difficult to replicate on other media like print and computer display. wherein they have created plastic swatches on website by 3D modelling to including various special effects.

Computer Swatches

Pure pigments reflect light in a very specific way that cannot be precisely duplicated by the discrete light emitters in a computer display. However, by making careful measurements of pigments, close approximations can be made. The Munsell Color System provides a good conceptual explanation of what is missing. Munsell devised a system that provides an objective measure of color in three dimensions: hue, value (or lightness), and chroma. Computer displays in general are unable to show the true chroma of many pigments, but the hue and lightness can be reproduced with relative accuracy. However, when the gamma of a computer display deviates from the reference value, the hue is also systematically biased.

The following approximations assume a display device at gamma 2.2, using the sRGB color space. The further a display device deviates from these standards, the less accurate these swatches will be. Swatches are based on the average measurements of several lots of single-pigment watercolor paints, converted from Lab color space to sRGB color space for viewing on a computer display. Different brands and lots of the same pigment may vary in color. Furthermore, pigments have inherently complex reflectance spectra that will render their color appearance greatly different depending on the spectrum of the source illumination; a property called metamerism. Averaged measurements of pigment samples will only yield approximations of their true appearance under a specific source of illumination. Computer display systems use a technique called chromatic adaptation transforms to emulate the correlated color temperature of illumination sources, and cannot perfectly reproduce the intricate spectral combinations originally seen. In many cases, the perceived color of a pigment falls outside of the gamut of computer displays and a method called gamut mapping is used to approximate the true appearance. Gamut mapping trades off any one of lightness, hue, or saturation accuracy to render the color on screen, depending on the priority chosen in the conversion's ICC rendering intent.

#990024	PR106 - #E34234	#FFB02E
Tyrian red	Vermilion (genuine)	Indian yellow

PB29 - #003BAF	PB27 - #0B3E66
Ultramarine blue	Prussian blue

Biological Pigments

In biology, a pigment is any colored material of plant or animal cells. Many biological structures, such as skin, eyes, fur, and hair contain pigments (such as melanin). Animal skin coloration often comes about through specialized cells called chromatophores, which animals such as the octopus and chameleon can control to vary the animal's color. Many conditions affect the levels or nature of pigments in plant, animal, some protista, or fungus cells. For instance, the disorder called albinism affects the level of melanin production in animals.

Pigmentation in organisms serves many biological purposes, including camouflage, mimicry, aposematism (warning), sexual selection and other forms of signalling, photosynthesis (in plants), as well as basic physical purposes such as protection from sunburn.

Pigment color differs from structural color in that pigment color is the same for all viewing angles, whereas structural color is the result of selective reflection or iridescence, usually because of multilayer structures. For example, butterfly wings typically contain structural color, although many butterflies have cells that contain pigment as well.

Pigments by Elemental Composition

Transition metal compounds. From left to right, aqueous solutions of: Co(NO$_3$)$_2$ (red); K$_2$Cr$_2$O$_7$ (orange); K$_2$CrO$_4$ (yellow); NiCl$_2$ (turquoise); CuSO$_4$ (blue); KMnO$_4$ (purple).

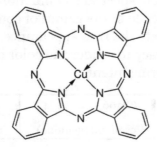

Phthalo Blue

Metal-based Pigments

- Cadmium pigments: cadmium yellow, cadmium red, cadmium green, cadmium orange, cadmium sulfoselenide

- Chromium pigments: chrome yellow and chrome green

- Cobalt pigments: cobalt violet, cobalt blue, cerulean blue, aureolin (cobalt yellow)

- Copper pigments: Azurite, Han purple, Han blue, Egyptian blue, Malachite, Paris green, Phthalocyanine Blue BN, Phthalocyanine Green G, verdigris, viridian

- Iron oxide pigments: sanguine, caput mortuum, oxide red, red ochre, Venetian red, Prussian blue

- Lead pigments: lead white, cremnitz white, Naples yellow, red lead, lead-tin-yellow

- Manganese pigments: manganese violet

- Mercury pigments: vermilion

- Titanium pigments: titanium yellow, titanium beige, titanium white, titanium black

- Zinc pigments: zinc white, zinc ferrite

Other Inorganic Pigments

- Carbon pigments: carbon black (including vine blac, lamp black), ivory black (bone char)

- Clay earth pigments (iron oxides): yellow ochre, raw sienna, burnt sienna, raw umber, burnt umber.

- Ultramarine pigments: ultramarine, ultramarine green shade

Biological and Organic

- Biological origins: alizarin (synthesized), alizarin crimson (synthesized), gamboge, cochineal red, rose madder, indigo, Indian yellow, Tyrian purple

- Non biological organic: quinacridone, magenta, phthalo green, phthalo blue, pigment red 170, diarylide yellow

References

- Nelson, D. L. and Cox, M. M. (2000) Lehninger, Principles of Biochemistry 3rd Ed. Worth Publishing: New York. ISBN 1-57259-153-6.

- Rayner-Canham, Marelene; Rayner-Canham, Geoffrey William (January 2001). Women in Chemistry: Their Changing Roles from Alchemical Times to the Mid-Twentieth Century. American Chemical Society. ISBN 9780841235229.

- Laidler, K.J. (1978) Physical Chemistry with Biological Applications, Benjamin/Cummings. pp. 415–417. ISBN 0805356800.

- Bender, Myron L; Komiyama, Makoto and Bergeron, Raymond J (1984) The Bioorganic Chemistry of Enzymatic Catalysis Wiley-Interscience, Hoboken, U.S. ISBN 0-471-05991-9

- Gage, John (1999). Color and Culture: Practice and Meaning from Antiquity to Abstraction. University of California Press. ISBN 0-520-22225-3.

- Meyer, Ralph (1991). The Artist's Handbook of Materials and Techniques, Fifth Edition. Viking. ISBN 0-670-83701-6.

- Lindlar, H. and Dubuis, R. (2016). "Palladium Catalyst for Partial Reduction of Acetylenes". Org. Synth. doi:10.15227/orgsyn.046.0089. ; Coll. Vol., 5, p. 880

- Wei, Hui; Wang, Erkang (2013-06-21). "Nanomaterials with enzyme-like characteristics (nanozymes): next-generation artificial enzymes". Chemical Society Reviews. 42 (14). doi:10.1039/C3CS35486E. ISSN 1460-4744.

- "Sto AG, Cabot Create Aerogel Insulation". Construction Digital. 15 November 2011. Retrieved 18 November 2011.

Carbon: A Comprehensive Study

One of the chemical elements is carbon. It is denoted by the symbol C and atomic number 6. Some of the allotropes of carbon are diamond and graphite. The other major components discussed are isotopes of carbon, the carbon cycle and the alpha and beta carbon. This chapter will not only provide an overview, it will also delve deep into the topics related to it.

Carbon

Carbon is a chemical element with symbol C and atomic number 6. On the periodic table, it is the first (row 2) of six elements in column (group 14), which have in common the composition of their outer electron shell. It is nonmetallic and tetravalent—making four electrons available to form covalent chemical bonds. Three isotopes occur naturally, ^{12}C and ^{13}C being stable while ^{14}C is radioactive, decaying with a half-life of about 5,730 years. Carbon is one of the few elements known since antiquity.

Carbon is the 15th most abundant element in the Earth's crust, and the fourth most abundant element in the universe by mass after hydrogen, helium, and oxygen. Carbon's abundance, its unique diversity of organic compounds, and its unusual ability to form polymers at the temperatures commonly encountered on Earth enables this element to serve as a common element of all known life. It is the second most abundant element in the human body by mass (about 18.5%) after oxygen.

The atoms of carbon can be bonded together in different ways, termed allotropes of carbon. The best known are graphite, diamond, and amorphous carbon. The physical properties of carbon vary widely with the allotropic form. For example, graphite is opaque and black while diamond is highly transparent. Graphite is soft enough to form a streak on paper, while diamond is the hardest naturally-occurring material known. Graphite is a good electrical conductor while diamond has a low electrical conductivity. Under normal conditions, diamond, carbon nanotubes, and graphene have the highest thermal conductivities of all known materials. All carbon allotropes are solids under normal conditions, with graphite being the most thermodynamically stable form. They are chemically resistant and require high temperature to react even with oxygen.

The most common oxidation state of carbon in inorganic compounds is +4, while +2 is found in carbon monoxide and transition metal carbonyl complexes. The largest sources of inorganic carbon are limestones, dolomites and carbon dioxide, but significant quantities occur in organic deposits of coal, peat, oil, and methane clathrates. Carbon forms a vast number of compounds, more than any other element, with almost ten million compounds described to date, and yet that number is but a fraction of the number of theoretically possible compounds under standard conditions.

Characteristics

The allotropes of carbon include graphite, one of the softest known substances, and diamond, the hardest naturally occurring substance. It bonds readily with other small atoms including other carbon atoms, and is capable of forming multiple stable covalent bonds with such atoms. Carbon is known to form almost ten million different compounds, a large majority of all chemical compounds. Carbon also has the highest sublimation point of all elements. At atmospheric pressure it has no melting point as its triple point is at 10.8 ± 0.2 MPa and 4,600 ± 300 K (~4,330 °C or 7,820 °F), so it sublimes at about 3,900 K.

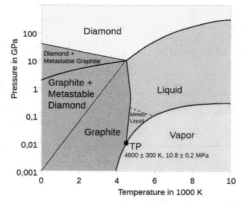

Theoretically predicted phase diagram of carbon

Carbon sublimes in a carbon arc which has a temperature of about 5,800 K (5,530 °C; 9,980 °F). Thus, irrespective of its allotropic form, carbon remains solid at higher temperatures than the highest melting point metals such as tungsten or rhenium. Although thermodynamically prone to oxidation, carbon resists oxidation more effectively than elements such as iron and copper that are weaker reducing agents at room temperature.

Carbon compounds form the basis of all known life on Earth, and the carbon-nitrogen cycle provides some of the energy produced by the Sun and other stars. Although it forms an extraordinary variety of compounds, most forms of carbon are comparatively unreactive under normal conditions. At standard temperature and pressure, it resists all but the strongest oxidizers. It does not react with sulfuric acid, hydrochloric acid, chlorine or any alkalis. At elevated temperatures, carbon reacts with oxygen to form carbon oxides, and will rob oxygen from metal oxides to leave the elemental metal. This exothermic reaction is used in the iron and steel industry to smelt iron and to control the carbon content of steel:

$$Fe_3O_4 + 4\ C_{(s)} \rightarrow 3\ Fe_{(s)} + 4\ CO_{(g)}$$

with sulfur to form carbon disulfide and with steam in the coal-gas reaction:

$$C_{(s)} + H_2O_{(g)} \rightarrow CO_{(g)} + H_{2(g)}.$$

Carbon combines with some metals at high temperatures to form metallic carbides, such as the iron carbide cementite in steel, and tungsten carbide, widely used as an abrasive and for making hard tips for cutting tools.

As of 2009, graphene appears to be the strongest material ever tested. The process of separating it from graphite will require some further technological development before it is economical for industrial processes.

The system of carbon allotropes spans a range of extremes:

Graphite is one of the softest materials known.	Synthetic nanocrystalline diamond is the hardest material known.
Graphite is a very good lubricant, displaying superlubricity.	Diamond is the ultimate abrasive.
Graphite is a conductor of electricity.	Diamond is an excellent electrical insulator, and has the highest breakdown electric field of any known material.
Some forms of graphite are used for thermal insulation (i.e. firebreaks and heat shields), but some other forms are good thermal conductors.	Diamond is the best known naturally occurring thermal conductor
Graphite is opaque.	Diamond is highly transparent.
Graphite crystallizes in the hexagonal system.	Diamond crystallizes in the cubic system.
Amorphous carbon is completely isotropic.	Carbon nanotubes are among the most anisotropic materials known.

Allotropes

Atomic carbon is a very short-lived species and, therefore, carbon is stabilized in various multi-atomic structures with different molecular configurations called allotropes. The three relatively well-known allotropes of carbon are amorphous carbon, graphite, and diamond. Once considered exotic, fullerenes are nowadays commonly synthesized and used in research; they include buckyballs, carbon nanotubes, carbon nanobuds and nanofibers. Several other exotic allotropes have also been discovered, such as lonsdaleite (questionable), glassy carbon, carbon nanofoam and linear acetylenic carbon (carbyne).

A large sample of glassy carbon.

The amorphous form is an assortment of carbon atoms in a non-crystalline, irregular, glassy state, which is essentially graphite but not held in a crystalline macrostructure. It is present as a powder, and is the main constituent of substances such as charcoal, lampblack (soot) and activated carbon. At normal pressures, carbon takes the form of graphite, in which each atom is bonded trigonally to three others in a plane composed of fused hexagonal rings, just like those in aromatic hydrocarbons. The resulting network is 2-dimensional, and the resulting flat sheets are stacked and loosely bonded through weak van der Waals forces. This gives graphite its softness and its cleaving properties (the sheets slip easily past one another). Because of the delocalization of one of the outer electrons of each atom to form a π-cloud, graphite conducts electricity, but only in the plane of each covalently bonded sheet. This results in a lower bulk electrical conductivity for carbon than for most metals. The delocalization also accounts for the energetic stability of graphite over diamond at room temperature.

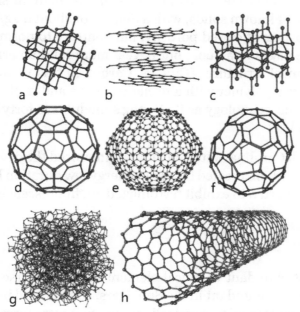

Some allotropes of carbon: a) diamond; b) graphite; c) lonsdaleite; d–f) fullerenes (C_{60}, C_{540}, C_{70}); g) amorphous carbon; h) carbon nanotube.

At very high pressures, carbon forms the more compact allotrope, diamond, having nearly twice the density of graphite. Here, each atom is bonded tetrahedrally to four others, forming a 3-dimensional network of puckered six-membered rings of atoms. Diamond has the same cubic structure as silicon and germanium, and because of the strength of the carbon-carbon bonds, it is the hardest naturally occurring substance measured by resistance to scratching. Contrary to the popular belief that *"diamonds are forever"*, they are thermodynamically unstable under normal conditions and transform into graphite. Due to a high activation energy barrier, the transition into graphite is so slow at normal temperature that it is unnoticeable. Under some conditions, carbon crystallizes as lonsdaleite, a hexagonal crystal lattice with all atoms covalently bonded and properties similar to those of diamond.

Fullerenes are a synthetic crystalline formation with a graphite-like structure, but in place of hexagons, fullerenes are formed of pentagons (or even heptagons) of carbon atoms. The missing (or additional) atoms warp the sheets into spheres, ellipses, or cylinders. The properties of fullerenes (split into buckyballs, buckytubes, and nanobuds) have not yet been fully analyzed and represent an intense area of research in nanomaterials. The names *"fullerene"* and *"buckyball"* are given after Richard Buckminster Fuller, popularizer of geodesic domes, which resemble the structure of fullerenes. The buckyballs are fairly large molecules formed completely of carbon bonded trigonally, forming spheroids (the best-known and simplest is the soccerball-shaped C_{60} buckminsterfullerene). Carbon nanotubes are structurally similar to buckyballs, except that each atom is bonded trigonally in a curved sheet that forms a hollow cylinder. Nanobuds were first reported in 2007 and are hybrid bucky tube/buckyball materials (buckyballs are covalently bonded to the outer wall of a nanotube) that combine the properties of both in a single structure.

Of the other discovered allotropes, carbon nanofoam is a ferromagnetic allotrope discovered in 1997. It consists of a low-density cluster-assembly of carbon atoms strung together in a loose three-dimensional web, in which the atoms are bonded trigonally in six- and seven-membered rings. It is among the lightest known solids, with a density of about 2 kg/m^3. Similarly, glassy carbon contains a high proportion of closed porosity, but contrary to normal graphite, the graphitic layers are not stacked like pages in a book, but have a more random arrangement. Linear acetylenic carbon has the chemical structure $-(C{:::}C)_n-$. Carbon in this modification is linear with sp orbital hybridization, and is a polymer with alternating single and triple bonds. This carbyne is of considerable interest to nanotechnology as its Young's modulus is forty times that of the hardest known material – diamond.

In 2015, a team at the North Carolina State University announced the development of another allotrope they have dubbed Q-carbon, created by a high energy low duration laser pulse on amorphous carbon dust. Q-carbon is reported to exhibit ferromagetism, fluorescence, and a hardness superior to diamonds.

Occurrence

Carbon is the fourth most abundant chemical element in the universe by mass after hydrogen, helium, and oxygen. Carbon is abundant in the Sun, stars, comets, and in the atmospheres of most planets. Some meteorites contain microscopic diamonds that were formed when the solar system was still a protoplanetary disk. Microscopic diamonds may also be formed by the intense pressure and high temperature at the sites of meteorite impacts.

Graphite ore. Penny is included for scale.

In 2014 NASA announced a greatly upgraded database for tracking polycyclic aromatic hydrocarbons (PAHs) in the universe. More than 20% of the carbon in the universe may be associated with PAHs, complex compounds of carbon and hydrogen without oxygen. These compounds figure in the PAH world hypothesis where they are hypothesized to have a role in abiogenesis and formation of life. PAHs seem to have been formed "a couple of billion years" after the Big Bang, are widespread throughout the universe, and are associated with new stars and exoplanets.

Raw diamond crystal.

It has been estimated that the solid earth as a whole contains 730 ppm of carbon, with 2000 ppm in the core and 120 ppm in the combined mantle and crust. Since the mass of the earth is 5.972×10^{24} kg, this would imply 4360 million gigatonnes of carbon. This is much more than the amount of carbon in the oceans or atmosphere (below).

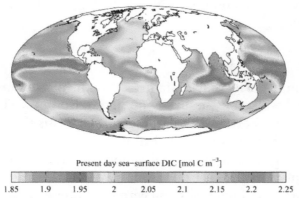

"Present day" (1990s) sea surface dissolved inorganic carbon concentration (from the GLODAP climatology)

In combination with oxygen in carbon dioxide, carbon is found in the Earth's atmosphere (approximately 810 gigatonnes of carbon) and dissolved in all water bodies (approximately 36,000 gigatonnes of carbon). Around 1,900 gigatonnes of carbon are present in the biosphere. Hydrocarbons (such as coal, petroleum, and natural gas) contain carbon as well. Coal "reserves" (not "resources") amount to around 900 gigatonnes with perhaps 18 000 Gt of resources. Oil reserves are around 150 gigatonnes. Proven sources of natural gas are about $175 \, 10^{12}$ cubic metres (containing about 105 gigatonnes of carbon), but studies estimate another $900 \, 10^{12}$ cubic metres of "unconventional" deposits such as shale gas, representing about 540 gigatonnes of carbon.

Carbon is also found in methane hydrates in polar regions and under the seas. Various estimates put this carbon between 500, 2500 Gt, or 3000 Gt.

In the past, quantities of hydrocarbons were greater. According to one source, in the period from 1751 to 2008 about 347 gigatonnes of carbon were released as carbon dioxide to the atmosphere from burning of fossil fuels. Another source puts the amount added to the atmosphere for the period since 1750 at 879 Gt, and the total going to the atmosphere, sea, and land (such as peat bogs) at almost 2000 Gt.

Carbon is a constituent (about 12% by mass) of the very large masses of carbonate rock (limestone, dolomite, marble and so on). Coal is very rich in carbon (anthracite contains 92–98%) and is the largest commercial source of mineral carbon, accounting for 4,000 gigatonnes or 80% of fossil fuel.

As for individual carbon allotropes, graphite is found in large quantities in the United States (mostly in New York and Texas), Russia, Mexico, Greenland, and India. Natural diamonds occur in the rock kimberlite, found in ancient volcanic "necks", or "pipes". Most diamond deposits are in Africa, notably in South Africa, Namibia, Botswana, the Republic of the Congo, and Sierra Leone. Diamond deposits have also been found in Arkansas, Canada, the Russian Arctic, Brazil, and in Northern and Western Australia. Diamonds are now also being recovered from the ocean floor off the Cape of Good Hope. Diamonds are found naturally, but about 30% of all industrial diamonds used in the U.S. are now manufactured.

Carbon-14 is formed in upper layers of the troposphere and the stratosphere at altitudes of 9–15 km by a reaction that is precipitated by cosmic rays. Thermal neutrons are produced that collide with the nuclei of nitrogen-14, forming carbon-14 and a proton.

Carbon-rich asteroids are relatively preponderant in the outer parts of the asteroid belt in our solar system. These asteroids have not yet been directly sampled by scientists. The asteroids can be used in hypothetical space-based carbon mining, which may be possible in the future, but is currently technologically impossible.

Isotopes

Isotopes of carbon are atomic nuclei that contain six protons plus a number of neutrons (varying from 2 to 16). Carbon has two stable, naturally occurring isotopes. The isotope carbon-12 (^{12}C) forms 98.93% of the carbon on Earth, while carbon-13 (^{13}C) forms the remaining 1.07%. The concentration of ^{12}C is further increased in biological materials because biochemical reactions discriminate against ^{13}C. In 1961, the International Union of Pure and Applied Chemistry (IUPAC) adopted the isotope carbon-12 as the basis for atomic weights. Identification of carbon in nuclear magnetic resonance (NMR) experiments is done with the isotope ^{13}C.

Carbon-14 (^{14}C) is a naturally occurring radioisotope, created in the upper atmosphere (lower stratosphere and upper troposphere) by interaction of nitrogen with cosmic rays. It is found in trace amounts on Earth of up to 1 part per trillion (0.0000000001%), mostly confined to the atmosphere and superficial deposits, particularly of peat and other organic materials. This isotope decays by 0.158 MeV β^- emission. Because of its relatively short half-life of 5730 years, ^{14}C is virtually absent in ancient rocks. The amount of ^{14}C in the atmosphere and in living organisms is almost constant, but decreases predictably in their bodies after death. This principle is used in radiocarbon dating, invented in 1949, which has been used extensively to determine the age of carbonaceous materials with ages up to about 40,000 years.

There are 15 known isotopes of carbon and the shortest-lived of these is ^{8}C which decays through proton emission and alpha decay and has a half-life of 1.98739×10^{-21} s. The exotic ^{19}C exhibits a nuclear halo, which means its radius is appreciably larger than would be expected if the nucleus were a sphere of constant density.

Formation in Stars

Formation of the carbon atomic nucleus requires a nearly simultaneous triple collision of alpha particles (helium nuclei) within the core of a giant or supergiant star which is known as the triple-alpha process, as the products of further nuclear fusion reactions of helium with hydrogen or another helium nucleus produce lithium-5 and beryllium-8 respectively, both of which are highly unstable and decay almost instantly back into smaller nuclei. This happens in conditions of temperatures over 100 megakelvin and helium concentration that the rapid expansion and cooling of the early universe prohibited, and therefore no significant carbon was created during the Big Bang.

According to current physical cosmology theory, carbon is formed in the interiors of stars in the horizontal branch by the collision and transformation of three helium nuclei. When those stars die as supernova, the carbon is scattered into space as dust. This dust becomes component material for the formation of second or third-generation star systems with accreted planets. The Solar System is one such star system with an abundance of carbon, enabling the existence of life as we know it.

The CNO cycle is an additional fusion mechanisms that powers stars, wherein carbon operates as a catalyst.

Rotational transitions of various isotopic forms of carbon monoxide (for example, ^{12}CO, ^{13}CO, and ^{18}CO) are detectable in the submillimeter wavelength range, and are used in the study of newly forming stars in molecular clouds.

Carbon Cycle

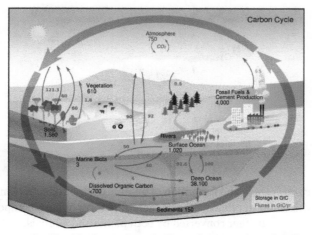

Diagram of the carbon cycle. The black numbers indicate how much carbon is stored in various reservoirs, in billions tonnes ("GtC" stands for gigatonnes of carbon; figures are circa 2004). The purple numbers indicate how much carbon moves between reservoirs each year. The sediments, as defined in this diagram, do not include the ~70 million GtC of carbonate rock and kerogen.

Under terrestrial conditions, conversion of one element to another is very rare. Therefore, the amount of carbon on Earth is effectively constant. Thus, processes that use carbon must obtain it from somewhere and dispose of it somewhere else. The paths of carbon in the environment form the carbon cycle. For example, photosynthetic plants draw carbon dioxide from the atmosphere (or seawater) and build it into biomass, as in the Calvin cycle, a process of carbon fixation. Some of this biomass is eaten by animals, while some carbon is exhaled by animals as carbon dioxide. The carbon cycle is considerably more complicated than this short loop; for example, some carbon dioxide is dissolved in the oceans; if bacteria do not consume it, dead plant or animal matter may become petroleum or coal, which releases carbon when burned.

Compounds

Organic Compounds

Carbon can form very long chains of interconnecting C-C bonds, a property that is called catenation. Carbon-carbon bonds are strong and stable. Through catenation, carbon forms a countless number of compounds. A tally of unique compounds shows that more contain carbon that those that do not. A similar claim can be made for hydrogen because most organic compounds also contain hydrogen.

Structural formula of methane, the simplest possible organic compound.

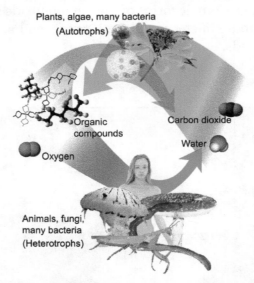

Correlation between the *carbon cycle* and formation of organic compounds. In plants, carbon dioxide formed by carbon fixation can join with water in photosynthesis (green) to form organic compounds, which can be used and further converted by both plants and animals.

The simplest form of an organic molecule is the hydrocarbon—a large family of organic molecules that are composed of hydrogen atoms bonded to a chain of carbon atoms. Chain length, side chains and functional groups all affect the properties of organic molecules.

Carbon occurs in all known organic life and is the basis of organic chemistry. When united with hydrogen, it forms various hydrocarbons that are important to industry as refrigerants, lubricants, solvents, as chemical feedstock for the manufacture of plastics and petrochemicals, and as fossil fuels.

When combined with oxygen and hydrogen, carbon can form many groups of important biological compounds including sugars, lignans, chitins, alcohols, fats, and aromatic esters, carotenoids and terpenes. With nitrogen it forms alkaloids, and with the addition of sulfur also it forms antibiotics, amino acids, and rubber products. With the addition of phosphorus to these other elements, it forms DNA and RNA, the chemical-code carriers of life, and adenosine triphosphate (ATP), the most important energy-transfer molecule in all living cells.

Inorganic Compounds

Commonly carbon-containing compounds which are associated with minerals or which do not contain hydrogen or fluorine, are treated separately from classical organic compounds; the definition is not rigid. Among these are the simple oxides of carbon. The most prominent oxide is carbon dioxide (CO_2). This was once the principal constituent of the paleoatmosphere, but is a minor component of the Earth's atmosphere today. Dissolved in water, it forms carbonic acid ($H2CO3$), but as most compounds with multiple single-bonded oxygens on a single carbon it is unstable. Through this intermediate, though, resonance-stabilized carbonate ions are produced. Some important minerals are carbonates, notably calcite. Carbon disulfide ($CS2$) is similar.

The other common oxide is carbon monoxide (CO). It is formed by incomplete combustion, and is a colorless, odorless gas. The molecules each contain a triple bond and are fairly polar, resulting in a tendency to bind permanently to hemoglobin molecules, displacing oxygen, which has a lower binding affinity. Cyanide (CN^-), has a similar structure, but behaves much like a halide ion (pseudohalogen). For example, it can form the nitride cyanogen molecule ($(CN)_2$), similar to diatomic halides. Other uncommon oxides are carbon suboxide ($C3O2$), the unstable dicarbon monoxide (C_2O), carbon trioxide (CO_3), cyclopentanepentone (C_5O_5), cyclohexanehexone (C_6O_6), and mellitic anhydride ($C_{12}O_9$).

With reactive metals, such as tungsten, carbon forms either carbides (C^{4-}), or acetylides ($C2-2$) to form alloys with high melting points. These anions are also associated with methane and acetylene, both very weak acids. With an electronegativity of 2.5, carbon prefers to form covalent bonds. A few carbides are covalent lattices, like carborundum (SiC), which resembles diamond.

Organometallic Compounds

Organometallic compounds by definition contain at least one carbon-metal bond. A wide range of such compounds exist; major classes include simple alkyl-metal compounds (for example, tetraethyllead), η^2-alkene compounds (for example, Zeise's salt), and η^3-allyl compounds (for example, allylpalladium chloride dimer); metallocenes containing cyclopentadienyl ligands (for example, ferrocene); and transition metal carbene complexes. Many metal carbonyls exist (for example, tetracarbonylnickel); some workers consider the carbon monoxide ligand to be purely inorganic, and not organometallic.

While carbon is understood to exclusively form four bonds, an interesting compound containing an octahedral hexacoordinated carbon atom has been reported. The cation of the compound is $[(Ph_3PAu)_6C]^{2+}$. This phenomenon has been attributed to the aurophilicity of the gold ligands.

History and Etymology

The English name *carbon* comes from the Latin *carbo* for coal and charcoal, whence also comes the French *charbon*, meaning charcoal. In German, Dutch and Danish, the names for carbon are *Kohlenstoff*, *koolstof* and *kulstof* respectively, all literally meaning coal-substance.

Antoine Lavoisier in his youth

Carbon was discovered in prehistory and was known in the forms of soot and charcoal to the earliest human civilizations. Diamonds were known probably as early as 2500 BCE in China, while carbon in the form of charcoal was made around Roman times by the same chemistry as it is today, by heating wood in a pyramid covered with clay to exclude air.

Carl Wilhelm Scheele

In 1722, René Antoine Ferchault de Réaumur demonstrated that iron was transformed into steel through the absorption of some substance, now known to be carbon. In 1772, Antoine Lavoisier showed that diamonds are a form of carbon; when he burned samples of charcoal and diamond and found that neither produced any water and that both released the same amount of carbon dioxide per gram. In 1779, Carl Wilhelm Scheele showed that graphite, which had been thought of as a form of lead, was instead identical with charcoal but with a small admixture of iron, and that it gave "aerial acid" (his name for carbon dioxide) when oxidized with nitric acid. In 1786, the French scientists Claude Louis Berthollet, Gaspard Monge and C. A. Vandermonde confirmed that graphite was mostly carbon by oxidizing it in oxygen in much the same way Lavoisier had done with diamond. Some iron again was left, which the French scientists thought was necessary to the graphite structure. In their publication they proposed the name *carbone* for the element in graphite which was given off as a gas upon burning graphite. Antoine Lavoisier then listed carbon as an element in his 1789 textbook.

A new allotrope of carbon, fullerene, that was discovered in 1985 includes nanostructured forms such as buckyballs and nanotubes. Their discoverers – Robert Curl, Harold Kroto and Richard Smalley – received the Nobel Prize in Chemistry in 1996. The resulting renewed interest in new forms lead to the discovery of further exotic allotropes, including glassy carbon, and the realization that "amorphous carbon" is not strictly amorphous.

Production

Graphite

Commercially viable natural deposits of graphite occur in many parts of the world, but the most important sources economically are in China, India, Brazil and North Korea. Graphite deposits are of metamorphic origin, found in association with quartz, mica and feldspars in schists, gneisses and metamorphosed sandstones and limestone as lenses or veins, sometimes of a metre or more in thickness. Deposits of graphite in Borrowdale, Cumberland, England were at first of sufficient size and purity that, until the 19th century, pencils were made simply by sawing blocks of natural graphite into strips before encasing the strips in wood. Today, smaller deposits of graphite are obtained by crushing the parent rock and floating the lighter graphite out on water.

There are three types of natural graphite—amorphous, flake or crystalline flake, and vein or lump. Amorphous graphite is the lowest quality and most abundant. Contrary to science, in industry "amorphous" refers to very small crystal size rather than complete lack of crystal structure. Amorphous is used for lower value graphite products and is the lowest priced graphite. Large amorphous graphite deposits are found in China, Europe, Mexico and the United States. Flake graphite is less common and of higher quality than amorphous; it occurs as separate plates that crystallized in metamorphic rock. Flake graphite can be four times the price of amorphous. Good quality flakes can be processed into expandable graphite for many uses, such as flame retardants. The foremost deposits are found in Austria, Brazil, Canada, China, Germany and Madagascar. Vein or lump graphite is the rarest, most valuable, and highest quality type of natural graphite. It occurs in veins along intrusive contacts in solid lumps, and it is only commercially mined in Sri Lanka.

According to the USGS, world production of natural graphite was 1.1 million tonnes in 2010, to which China contributed 800,000 t, India 130,000 t, Brazil 76,000 t, North Korea 30,000 t and Canada 25,000 t. No natural graphite was reported mined in the United States, but 118,000 t of synthetic graphite with an estimated value of $998 million was produced in 2009.

Diamond

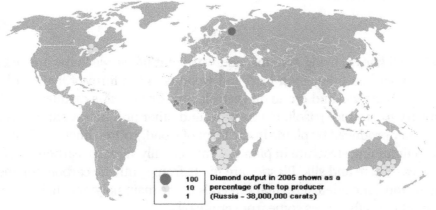

Diamond output in 2005

The diamond supply chain is controlled by a limited number of powerful businesses, and is also highly concentrated in a small number of locations around the world.

Only a very small fraction of the diamond ore consists of actual diamonds. The ore is crushed, during which care has to be taken in order to prevent larger diamonds from being destroyed in this process and subsequently the particles are sorted by density. Today, diamonds are located in the diamond-rich density fraction with the help of X-ray fluorescence, after which the final sorting steps are done by hand. Before the use of X-rays became commonplace, the separation was done with grease belts; diamonds have a stronger tendency to stick to grease than the other minerals in the ore.

Historically diamonds were known to be found only in alluvial deposits in southern India. India led the world in diamond production from the time of their discovery in approximately the 9th century BCE to the mid-18th century AD, but the commercial potential of these sources had been exhausted by the late 18th century and at that time India was eclipsed by Brazil where the first non-Indian diamonds were found in 1725.

Diamond production of primary deposits (kimberlites and lamproites) only started in the 1870s after the discovery of the Diamond fields in South Africa. Production has increased over time and now an accumulated total of 4.5 billion carats have been mined since that date. About 20% of that amount has been mined in the last 5 years alone, and during the last ten years 9 new mines have started production while 4 more are waiting to be opened soon. Most of these mines are located in Canada, Zimbabwe, Angola, and one in Russia.

In the United States, diamonds have been found in Arkansas, Colorado and Montana. In 2004, a startling discovery of a microscopic diamond in the United States led to the January 2008 bulk-sampling of kimberlite pipes in a remote part of Montana.

Today, most commercially viable diamond deposits are in Russia, Botswana, Australia and the Democratic Republic of Congo. In 2005, Russia produced almost one-fifth of the global diamond output, reports the British Geological Survey. Australia has the richest diamantiferous pipe with production reaching peak levels of 42 metric tons (41 long tons; 46 short tons) per year in the 1990s. There are also commercial deposits being actively mined in the Northwest Territories of Canada, Siberia (mostly in Yakutia territory; for example, Mir pipe and Udachnaya pipe), Brazil, and in Northern and Western Australia.

Applications

Carbon is essential to all known living systems, and without it life as we know it could not exist. The major economic use of carbon other than food and wood is in the form of hydrocarbons, most notably the fossil fuel methane gas and crude oil (petroleum). Crude oil is distilled in refineries by the petrochemical industry to produce gasoline, kerosene, and other products. Cellulose is a natural, carbon-containing polymer produced by plants in the form of wood, cotton, linen, and hemp. Cellulose is used primarily for maintaining structure in plants. Commercially valuable carbon polymers of animal origin include wool, cashmere and silk. Plastics are made from synthetic carbon polymers, often with oxygen and nitrogen atoms included at regular intervals in the main polymer chain. The raw materials for many of these synthetic substances come from crude oil.

The uses of carbon and its compounds are extremely varied. It can form alloys with iron, of which the most common is carbon steel. Graphite is combined with clays to form the 'lead' used in pencils used for writing and drawing. It is also used as a lubricant and a pigment, as a molding material in glass manufacture, in electrodes for dry batteries and in electroplating and electroforming, in brushes for electric motors and as a neutron moderator in nuclear reactors.

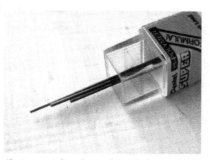

Pencil leads for mechanical pencils are made of graphite (often mixed with a clay or synthetic binder).

Charcoal is used as a drawing material in artwork, barbecue grilling, iron smelting, and in many other applications. Wood, coal and oil are used as fuel for production of energy and heating. Gem quality diamond is used in jewelry, and industrial diamonds are used in drilling, cutting and polishing tools for machining metals and stone. Plastics are made from fossil hydrocarbons, and carbon fiber, made by pyrolysis of synthetic polyester fibers is used to reinforce plastics to form advanced, lightweight composite materials.

Sticks of vine and compressed charcoal.

Carbon fiber is made by pyrolysis of extruded and stretched filaments of polyacrylonitrile (PAN) and other organic substances. The crystallographic structure and mechanical properties of the fiber depend on the type of starting material, and on the subsequent processing. Carbon fibers made from PAN have structure resembling narrow filaments of graphite, but thermal processing may re-order the structure into a continuous rolled sheet. The result is fibers with higher specific tensile strength than steel.

A cloth of woven carbon fibres

Carbon black is used as the black pigment in printing ink, artist's oil paint and water colours, carbon paper, automotive finishes, India ink and laser printer toner. Carbon black is also used as a filler in rubber products such as tyres and in plastic compounds. Activated charcoal is used as an absorbent and adsorbent in filter material in applications as diverse as gas masks, water purification, and kitchen extractor hoods, and in medicine to absorb toxins, poisons, or gases from the digestive system. Carbon is used in chemical reduction at high temperatures. Coke is used to reduce iron ore into iron (smelting). Case hardening of steel is achieved by heating finished steel components in carbon powder. Carbides of silicon, tungsten, boron and titanium, are among the hardest known materials, and are used as abrasives in cutting and grinding tools. Carbon compounds make up most of the materials used in clothing, such as natural and synthetic textiles and leather, and almost all of the interior surfaces in the built environment other than glass, stone and metal.

Diamonds

The diamond industry falls into two categories: one dealing with gem-grade diamonds and the other, with industrial-grade diamonds. While a large trade in both types of diamonds exists, the two markets act in dramatically different ways.

Unlike precious metals such as gold or platinum, gem diamonds do not trade as a commodity: there is a substantial mark-up in the sale of diamonds, and there is not a very active market for resale of diamonds.

Industrial diamonds are valued mostly for their hardness and heat conductivity, with the gemological qualities of clarity and color being mostly irrelevant. About 80% of mined diamonds (equal to about 100 million carats or 20 tonnes annually) are unsuitable for use as gemstones are relegated for industrial use (known as *bort*). synthetic diamonds, invented in the 1950s, found almost immediate industrial applications; 3 billion carats (600 tonnes) of synthetic diamond is produced annually.

The dominant industrial use of diamond is in cutting, drilling, grinding, and polishing. Most of these applications do not require large diamonds; in fact, most diamonds of gem-quality except for their small size can be used industrially. Diamonds are embedded in drill tips or saw blades, or ground into a powder for use in grinding and polishing applications. Specialized applications include use in laboratories as containment for high pressure experiments, high-performance bearings, and limited use in specialized windows. With the continuing advances in the production of synthetic diamonds, new applications are becoming feasible. Garnering much excitement is the possible use of diamond as a semiconductor suitable for microchips, and because of its exceptional heat conductance property, as a heat sink in electronics.

Precautions

Pure carbon has extremely low toxicity to humans and can be handled and even ingested safely in the form of graphite or charcoal. It is resistant to dissolution or chemical attack, even in the acidic contents of the digestive tract. Consequently, once it enters into the body's tissues it is likely to remain there indefinitely. Carbon black was probably one of the first pigments to be used for tattooing, and Ötzi the Iceman was found to have carbon tattoos that survived during his life and for 5200 years after his death. Inhalation of coal dust or soot (carbon black) in large quantities can be dangerous, irritating lung tissues and causing the congestive lung disease, coalworker's pneumoconiosis. Diamond dust used as an abrasive can harmful if ingested or inhaled. Microparticles of carbon are produced in diesel engine exhaust fumes, and may accumulate in the lungs. In these examples, the harm may result from contaminants (e.g., organic chemicals, heavy metals) rather than from the carbon itself.

Worker at carbon black plant in Sunray, Texas (photo by John Vachon, 1942)

Carbon generally has low toxicity to life on Earth; but carbon nanoparticles are deadly to *Drosophila*.

Carbon may burn vigorously and brightly in the presence of air at high temperatures. Large accumulations of coal, which have remained inert for hundreds of millions of years in the absence of oxygen, may spontaneously combust when exposed to air in coal mine waste tips, ship cargo holds and coal bunkers, and storage dumps.

In nuclear applications where graphite is used as a neutron moderator, accumulation of Wigner energy followed by a sudden, spontaneous release may occur. Annealing to at least 250 °C can release the energy safely, although in the Windscale fire the procedure went wrong, causing other reactor materials to combust.

The great variety of carbon compounds include such lethal poisons as tetrodotoxin, the lectin ricin from seeds of the castor oil plant *Ricinus communis*, cyanide (CN^-), and carbon monoxide; and such essentials to life as glucose and protein.

Bonding to Carbon

Chemical bonds to carbon	
Core organic chemistry	Many uses in chemistry
Academic research, but no widespread use	Bond unknown

CH																	He
CLi	CBe											CB	CC	CN	CO	CF	Ne
CNa	CMg											CAl	CSi	CP	CS	CCl	CAr
CK	CCa	CSc	CTi	CV	CCr	CMn	CFe	CCo	CNi	CCu	CZn	CGa	CGe	CAs	CSe	CBr	CKr
CRb	CSr	CY	CZr	CNb	CMo	CTc	CRu	CRh	CPd	CAg	CCd	CIn	CSn	CSb	CTe	CI	CXe
CCs	CBa		CHf	CTa	CW	CRe	COs	CIr	CPt	CAu	CHg	CTl	CPb	CBi	CPo	CAt	Rn
Fr	CRa		Rf	Db	CSg	Bh	Hs	Mt	Ds	Rg	Cn	Nh	Fl	Mc	Lv	Ts	Og

↓

CLa	CCe	CPr	CNd	CPm	CSm	CEu	CGd	CTb	CDy	CHo	CEr	CTm	CYb	CLu
Ac	CTh	CPa	CU	CNp	CPu	CAm	CCm	CBk	CCf	CEs	Fm	Md	No	Lr

Allotropes of Carbon

Carbon is capable of forming many allotropes due to its valency. Well-known forms of carbon include diamond and graphite. In recent decades many more allotropes and forms of carbon have been discovered and researched including ball shapes such as buckminsterfullerene and sheets such as graphene. Larger scale structures of carbon include nanotubes, nanobuds and nanoribbons. Other unusual forms of carbon exist at very high temperature or extreme pressures.

Diamond

Diamond is a well known allotrope of carbon. The hardness and high dispersion of light of diamond make it useful for both industrial applications and jewelry. Diamond is the hardest known natural mineral. This makes it an excellent abrasive and makes it hold polish and luster extremely well. No known naturally occurring substance can cut (or even scratch) a diamond, except another diamond.

The market for industrial-grade diamonds operates much differently from its gem-grade counterpart. Industrial diamonds are valued mostly for their hardness and heat conductivity, making many of the gemological characteristics of diamond, including clarity and color, mostly irrelevant. This helps explain why 80% of mined diamonds (equal to about 100 million carats or 20 tonnes annually) are unsuitable for use as gemstones and known as *bort*, are destined for industrial use.

In addition to mined diamonds, synthetic diamonds found industrial applications almost immediately after their invention in the 1950s; another 400 million carats (80 tonnes) of synthetic diamonds are produced annually for industrial use which is nearly four times the mass of natural diamonds mined over the same period.

The dominant industrial use of diamond is in cutting, drilling (drill bits), grinding (diamond edged cutters), and polishing. Most uses of diamonds in these technologies do not require large diamonds; in fact, most diamonds that are gem-quality can find an industrial use. Diamonds are embedded in drill tips or saw blades, or ground into a powder for use in grinding and polishing applications (due to its extraordinary hardness). Specialized applications include use in laboratories as containment for high pressure experiments, high-performance bearings, and limited use in specialized windows.

With the continuing advances being made in the production of synthetic diamond, future applications are beginning to become feasible. Garnering much excitement is the possible use of diamond as a semiconductor suitable to build microchips from, or the use of diamond as a heat sink in electronics. Significant research efforts in Japan, Europe, and the United States are under way to capitalize on the potential offered by diamond's unique material properties, combined with increased quality and quantity of supply starting to become available from synthetic diamond manufacturers.

Each carbon atom in a diamond is covalently bonded to four other carbons in a tetrahedron. These tetrahedrons together form a 3-dimensional network of six-membered carbon rings (similar to cyclohexane), in the chair conformation, allowing for zero bond angle strain. This stable network of covalent bonds and hexagonal rings, is the reason that diamond is so strong. Although graphite is the most stable allotrope of carbon under standard laboratory conditions (273 or 298 K, 1 atm), a recent computational study indicated that under idealized conditions ($T = 0$, $p = 0$), diamond is the most stable allotrope by 1.1 kJ/mol compared to graphite.

Graphite

Graphite, named by Abraham Gottlob Werner in 1789, is one of the most common allotropes of carbon. Unlike diamond, graphite is an electrical conductor. Thus, it can be used in, for instance, electrical arc lamp electrodes. Likewise, under standard conditions, graphite is the most stable form of carbon. Therefore, it is used in thermochemistry as the standard state for defining the heat of formation of carbon compounds.

Graphite conducts electricity, due to delocalization of the pi bond electrons above and below the planes of the carbon atoms. These electrons are free to move, so are able to conduct electricity. However, the electricity is only conducted along the plane of the layers. In diamond, all four outer electrons of each carbon atom are 'localised' between the atoms in covalent bonding. The movement of electrons is restricted and diamond does not conduct an electric current. In graphite, each carbon atom uses only 3 of its 4 outer energy level electrons in covalently bonding to three other carbon atoms in a plane. Each carbon atom contributes one electron to a delocalised system of electrons that is also a part of the chemical bonding. The delocalised electrons are free to move throughout the plane. For this reason, graphite conducts electricity along the planes of carbon atoms, but does not conduct in a direction at right angles to the plane.

Graphite powder is used as a dry lubricant. Although it might be thought that this industrially important property is due entirely to the loose interlamellar coupling between sheets in the structure, in fact in a vacuum environment (such as in technologies for use in space), graphite was found to be a very poor lubricant. This fact led to the discovery that graphite's lubricity is due to adsorbed air and water between the layers, unlike other layered dry lubricants such as molybdenum disulfide. Recent studies suggest that an effect called superlubricity can also account for this effect.

When a large number of crystallographic defects bind these planes together, graphite loses its lubrication properties and becomes what is known as pyrolytic carbon, a useful material in blood-contacting implants such as prosthetic heart valves.

Graphite is the most stable allotrope of carbon. Contrary to popular belief, high-purity graphite does not readily burn, even at elevated temperatures. For this reason, it is used in nuclear reactors and for high-temperature crucibles for melting metals. At very high temperatures and pressures (roughly 2000 °C and 5 GPa), it can be transformed into diamond.

Natural and crystalline graphites are not often used in pure form as structural materials due to their shear-planes, brittleness and inconsistent mechanical properties.

In its pure glassy (isotropic) synthetic forms, pyrolytic graphite and carbon fiber graphite are extremely strong, heat-resistant (to 3000 °C) materials, used in reentry shields for missile nosecones, solid rocket engines, high temperature reactors, brake shoes and electric motor brushes.

Intumescent or expandable graphites are used in fire seals, fitted around the perimeter of a fire door. During a fire the graphite intumesces (expands and chars) to resist fire penetration and prevent the spread of fumes. A typical start expansion temperature (SET) is between 150 and 300 °C.

Density: graphite's specific gravity is 2.3, which makes it lighter than diamonds.

Chemical activity: it is slightly more reactive than diamond. This is because the reactants are able to penetrate between the hexagonal layers of carbon atoms in graphite. It is unaffected by ordinary solvents, dilute acids, or fused alkalis. However, chromic acid oxidises it to carbon dioxide.

Graphene

A single layer of graphite is called graphene and has extraordinary electrical, thermal, and physical properties. It can be produced by epitaxy on an insulating or conducting substrate or by mechanical exfoliation (repeated peeling) from graphite. Its applications may include replacing silicon in high-performance electronic devices.

Amorphous Carbon

Amorphous carbon is the name used for carbon that does not have any crystalline structure. As with all glassy materials, some short-range order can be observed, but there is no long-range pattern of atomic positions. While entirely amorphous carbon can be produced, most amorphous carbon actually contains microscopic crystals of graphite-like, or even diamond-like carbon.

Coal and soot or carbon black are informally called amorphous carbon. However, they are products of pyrolysis (the process of decomposing a substance by the action of heat), which does not produce true amorphous carbon under normal conditions.

Nanocarbons

Buckminsterfullerenes

The *buckminsterfullerenes*, or usually just *fullerenes* or *buckyballs* for short, were discovered in 1985 by a team of scientists from Rice University and the University of Sussex, three of whom were awarded the 1996 Nobel Prize in Chemistry. They are named for the resemblance of their allotropic structure to the geodesic structures devised by the scientist and architect Richard Buckminster "Bucky" Fuller. Fullerenes are molecules of varying sizes composed entirely of carbon, which take the form of a hollow sphere, ellipsoid, or tube.

As of the early twenty-first century, the chemical and physical properties of fullerenes are still under heavy study, in both pure and applied research labs. In April 2003, fullerenes were under study for potential medicinal use — binding specific antibiotics to the structure to target resistant bacteria and even target certain cancer cells such as melanoma.

Carbon Nanotubes

Carbon nanotubes, also called buckytubes, are cylindrical carbon molecules with novel properties that make them potentially useful in a wide variety of applications (e.g., nano-electronics, optics, materials applications, etc.). They exhibit extraordinary strength, unique electrical properties, and are efficient conductors of heat. Inorganic nanotubes have also been synthesized. A nanotube is a member of the fullerene structural family, which also includes buckyballs. Whereas buckyballs are spherical in shape, a nanotube is cylindrical, with at least one end typically capped with a hemisphere of the buckyball structure. Their name is derived from their size, since the diameter of a nanotube is on the order of a few nanometers (approximately 50,000 times smaller than the width of a human hair), while they can be up to several centimeters in length. There are two main types of nanotubes: single-walled nanotubes (SWNTs) and multi-walled nanotubes (MWNTs).

Carbon Nanobuds

Carbon nanobuds are a newly discovered allotrope of carbon in which fullerene like "buds" are covalently attached to the outer sidewalls of the carbon nanotubes. This hybrid material has useful properties of both fullerenes and carbon nanotubes. For instance, they have been found to be exceptionally good field emitters.

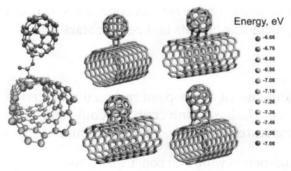

Computer models of stable nanobud structures

Glassy Carbon

Glassy carbon or vitreous carbon is a class of non-graphitizing carbon widely used as an electrode material in electrochemistry, as well as for high-temperature crucibles and as a component of some prosthetic devices.

A large sample of glassy carbon.

It was first produced by Bernard Redfern in the mid-1950s at the laboratories of The Carborundum Company, Manchester, UK. He had set out to develop a polymer matrix to mirror a diamond structure and discovered a resole (phenolic) resin that would, with special preparation, set without a catalyst. Using this resin the first glassy carbon was produced.

The preparation of glassy carbon involves subjecting the organic precursors to a series of heat treatments at temperatures up to 3000 °C. Unlike many non-graphitizing carbons, they are impermeable to gases and are chemically extremely inert, especially those prepared at very high temperatures. It has been demonstrated that the rates of oxidation of certain glassy carbons in oxygen, carbon dioxide or water vapour are lower than those of any other carbon. They are also highly resistant to attack by acids. Thus, while normal graphite is reduced to a powder by a mixture of concentrated sulfuric and nitric acids at room temperature, glassy carbon is unaffected by such treatment, even after several months.

Atomic and Diatomic Carbon

Under certain conditions, carbon can be found in its atomic form. It is formed by passing large electric currents through carbon under very low pressures. It is extremely unstable, but it is an intermittent product used in the creation of carbenes.

Diatomic carbon can also be found under certain conditions. It is often detected via spectroscopy in extraterrestrial bodies, including comets and certain stars.

Carbon Nanofoam

Carbon nanofoam is the fifth known allotrope of carbon discovered in 1997 by Andrei V. Rode and co-workers at the Australian National University in Canberra. It consists of a low-density cluster-assembly of carbon atoms strung together in a loose three-dimensional web.

Each cluster is about 6 nanometers wide and consists of about 4000 carbon atoms linked in graphite-like sheets that are given negative curvature by the inclusion of heptagons among the regular hexagonal pattern. This is the opposite of what happens in the case of buckminsterfullerenes, in which carbon sheets are given positive curvature by the inclusion of pentagons.

The large-scale structure of carbon nanofoam is similar to that of an aerogel, but with 1% of the density of previously produced carbon aerogels – only a few times the density of air at sea level. Unlike carbon aerogels, carbon nanofoam is a poor electrical conductor.

Carbide-derived Carbon

Carbide-derived carbon (CDC) is a family of carbon materials with different surface geometries and carbon ordering that are produced via selective removal of metals from metal carbide precursors, such as TiC, SiC, Ti_3AlC_2, Mo_2C, etc. This synthesis is accomplished using chlorine treatment, hydrothermal synthesis, or high-temperature selective metal desorption under vacuum. Depending on the synthesis method, carbide precursor, and reaction parameters, multiple carbon allotropes can be achieved, including endohedral particles composed of predominantly amorphous carbon, carbon nanotubes, epitaxial graphene, nanocrystalline diamond, onion-like carbon, and graphitic ribbons, barrels, and horns. These structures exhibit high porosity and specific surface areas, with highly tunable pore diameters, making them promising materials for supercapacitor-based energy storage, water filtration and capacitive desalinization, catalyst support, and cytokine removal.

Lonsdaleite (Hexagonal Diamond)

Lonsdaleite is a hexagonal allotrope of the carbon allotrope diamond, believed to form from graphite present in meteorites upon their impact to Earth. The great heat and stress of the impact transforms the graphite into diamond, but retains graphite's hexagonal crystal lattice. Hexagonal diamond has also been synthesized in the laboratory, by compressing and heating graphite either in a static press or using explosives. It can also be produced by the thermal decomposition of a polymer, poly(hydridocarbyne), at atmospheric pressure, under inert gas atmosphere (e.g. argon, nitrogen), starting at temperature 110 °C (230 °F).

Linear Acetylenic Carbon (LAC)

A one-dimensional carbon polymer with the structure $-(C:::C)_n-$.

Other Possible Forms

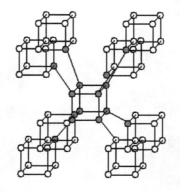

Crystal structure of C_8 cubic carbon

- Chaoite is a mineral believed to have been formed in meteorite impacts. It has been described as slightly harder than graphite with a reflection colour of grey to white. However, the existence of carbyne phases is disputed.

- Metallic carbon: Theoretical studies have shown that there are regions in the phase diagram, at extremely high pressures, where carbon has metallic character.

- bcc-carbon: At ultrahigh pressures of above 1000 GPa, diamond is predicted to transform into the so-called C_8 structure, a body-centered cubic structure with 8 atoms in the unit cell. This cubic carbon phase might have importance in astrophysics. Its structure is known in one of the metastable phases of silicon and is similar to cubane. Superdense and superhard material resembling this phase has been synthesized and published in 1979 and 2008. The structure of this phase was proposed in 2012 as carbon sodalite.

- bct-carbon: Body-centered tetragonal carbon proposed by theorists in 2010

- M-carbon: Monoclinic C-centered carbon was first thought to have been created in 1963 by compressing graphite at room temperature. Its structure was theorized in 2006, then in 2009 it was related to those experimental observations. Many structural candidates, including bct-carbon, were proposed to be equally compatible with experimental data available at the time, until in 2012 it was theoretically proven that this structure is kinetically likeliest to form from graphite. High-resolution data appeared shortly after, demonstrating that among all structure candidates only M-carbon is compatible with experiment.

- Q-carbon: Ferromagnetic carbon discovered in 2015.

- T-carbon: Every carbon atom in diamond is replaced with a carbon tetrahedron (hence 'T-carbon'). This was proposed by theorists in 2011.

- There is an evidence that white dwarf stars have a core of crystallized carbon and oxygen nuclei. The largest of these found in the universe so far, BPM 37093, is located 50 light-years (4.7×10^{14} km) away in the constellation Centaurus. A news release from the Harvard-Smithsonian Center for Astrophysics described the 2,500-mile (4,000 km)-wide stellar core as a *diamond*, and it was named as *Lucy*, after the Beatles' song "Lucy in the Sky With Diamonds"; however, it is more likely an exotic form of carbon.

- Prismane C_8 is a theoretically-predicted metastable carbon allotrope comprising an atomic cluster of eight carbon atoms, with the shape of an elongated triangular bipyramid—a six-atom triangular prism with two more atoms above and below its bases.

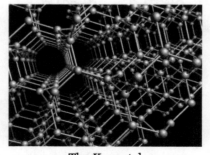

The K_4 crystal

- The Laves graph or K_4 crystal is a theoretically-predicted three-dimensional crystalline metastable carbon structure in which each carbon atom is bonded to three others, at 120° angles (like graphite), but where the bond planes of adjacent lie at an angle of 70.5°, rather than coinciding

- Penta-graphene

- Haeckelites Ordered arrangements of pentagons, hexagons, and heptagons which can either be flat or tubular.

- Phagraphene Graphene allotrope with distorted Dirac cones.

Variability of Carbon

The system of carbon allotropes spans an astounding range of extremes, considering that they are all merely structural formations of the same element.

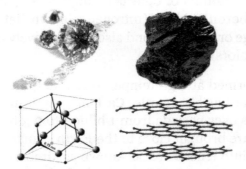

Diamond and graphite are two allotropes of carbon: pure forms of the same element that differ in structure.

Between diamond and graphite:

- Diamond crystallizes in the cubic system but graphite crystallizes in the hexagonal system.

- Diamond is clear and transparent, but graphite is black and opaque.

- Diamond is the hardest mineral known (10 on the Mohs scale), but graphite is one of the softest (1–2 on Mohs scale).

- Diamond is the ultimate abrasive, but graphite is soft and is a very good lubricant.

- Diamond is an excellent electrical insulator, but graphite is a conductor of electricity.

- Diamond is an excellent thermal conductor, but some forms of graphite are used for thermal insulation (for example heat shields and firebreaks).

- At standard temperature and pressure, graphite is the thermodynamically stable form. Thus diamonds do not exist forever. The conversion from diamond to graphite, however, has a very high activation energy and is therefore extremely slow.

Despite the hardness of diamonds, the chemical bonds that hold the carbon atoms in diamonds together are actually weaker than those that hold together graphite. The difference is that in diamond, the bonds form an inflexible three-dimensional lattice. In graphite, the atoms are tightly bonded into sheets, but the sheets can slide easily over each other, making graphite soft.

Diamond

Diamond is a metastable allotrope of carbon, where the carbon atoms are arranged in a variation of the face-centered cubic crystal structure called a diamond lattice. Diamond is less stable than graphite, but the conversion rate from diamond to graphite is negligible at standard conditions. Diamond is renowned as a material with superlative physical qualities, most of which originate from the strong covalent bonding between its atoms. In particular, diamond has the highest hardness and thermal conductivity of any bulk material. Those properties determine the major industrial application of diamond in cutting and polishing tools and the scientific applications in diamond knives and diamond anvil cells.

Because of its extremely rigid lattice, it can be contaminated by very few types of impurities, such as boron and nitrogen. Small amounts of defects or impurities (about one per million of lattice atoms) color diamond blue (boron), yellow (nitrogen), brown (lattice defects), green (radiation exposure), purple, pink, orange or red. Diamond also has relatively high optical dispersion (ability to disperse light of different colors).

Most natural diamonds are formed at high temperature and pressure at depths of 140 to 190 kilometers (87 to 118 mi) in the Earth's mantle. Carbon-containing minerals provide the carbon source, and the growth occurs over periods from 1 billion to 3.3 billion years (25% to 75% of the age of the Earth). Diamonds are brought close to the Earth's surface through deep volcanic eruptions by magma, which cools into igneous rocks known as kimberlites and lamproites. Diamonds can also be produced synthetically in a HPHT method which approximately simulates the conditions in the Earth's mantle. An alternative, and completely different growth technique is chemical vapor deposition (CVD). Several non-diamond materials, which include cubic zirconia and silicon carbide and are often called diamond simulants, resemble diamond in appearance and many properties. Special gemological techniques have been developed to distinguish natural diamonds, synthetic diamonds, and diamond simulants.

History

Diamonds have been treasured as gemstones since their use as religious icons in ancient India. Their usage in engraving tools also dates to early human history. The popularity of diamonds has risen since the 19th century because of increased supply, improved cutting and polishing techniques, growth in the world economy, and innovative and successful advertising campaigns.

In 1772, Antoine Lavoisier used a lens to concentrate the rays of the sun on a diamond in an atmosphere of oxygen, and showed that the only product of the combustion was carbon dioxide, proving that diamond is composed of carbon. Later in 1797, Smithson Tennant repeated and expanded that experiment. By demonstrating that burning diamond and graphite releases the same amount of gas, he established the chemical equivalence of these substances.

The most familiar uses of diamonds today are as gemstones used for adornment, a use which dates back into antiquity, and as industrial abrasives for cutting hard materials. The dispersion of white light into spectral colors is the primary gemological characteristic of gem diamonds. In the 20th century, experts in gemology developed methods of grading diamonds and other gemstones based on the characteristics most important to their value as a gem. Four characteristics, known informally as the *four Cs*, are now commonly used as the basic descriptors of diamonds: these are *carat* (its weight), *cut* (quality of the cut is graded according to proportions, symmetry and polish), *color* (how close to white or colorless; for fancy diamonds how intense is its hue), and *clarity* (how free is it from inclusions). A large, flawless diamond is known as a paragon.

Natural History

The formation of natural diamond requires very specific conditions—exposure of carbon-bearing materials to high pressure, ranging approximately between 45 and 60 kilobars (4.5 and 6 GPa), but at a comparatively low temperature range between approximately 900 and 1,300 °C (1,650 and 2,370 °F). These conditions are met in two places on Earth; in the lithospheric mantle below relatively stable continental plates, and at the site of a meteorite strike.

Formation in Cratons

The conditions for diamond formation to happen in the lithospheric mantle occur at considerable depth corresponding to the requirements of temperature and pressure. These depths are estimated between 140 and 190 kilometers (87 and 118 mi) though occasionally diamonds have crystallized at depths about 300 km (190 mi). The rate at which temperature changes with increasing depth into the Earth varies greatly in different parts of the Earth. In particular, under oceanic plates the temperature rises more quickly with depth, beyond the range required for diamond formation at the depth required. The correct combination of temperature and pressure is only found in the thick, ancient, and stable parts of continental plates where regions of lithosphere known as *cratons* exist. Long residence in the cratonic lithosphere allows diamond crystals to grow larger.

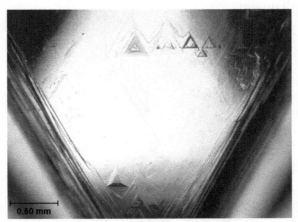

One face of an uncut octahedral diamond, showing trigons (of positive and negative relief) formed by natural chemical etching

Through studies of carbon isotope ratios (similar to the methodology used in carbon dating, except with the stable isotopes C-12 and C-13), it has been shown that the carbon found in diamonds comes from both inorganic and organic sources. Some diamonds, known as *harzburgitic*, are formed from inorganic carbon originally found deep in the Earth's mantle. In contrast, *eclogitic* diamonds contain organic carbon from organic detritus that has been pushed down from the surface of the Earth's crust through subduction before transforming into diamond. These two different source of carbon have measurably different ^{13}C:^{12}C ratios. Diamonds that have come to the Earth's surface are generally quite old, ranging from under 1 billion to 3.3 billion years old. This is 22% to 73% of the age of the Earth.

Diamonds occur most often as euhedral or rounded octahedra and twinned octahedra known as *macles*. As diamond's crystal structure has a cubic arrangement of the atoms, they have many facets that belong to a cube, octahedron, rhombicosidodecahedron, tetrakis hexahedron or disdyakis dodecahedron. The crystals can have rounded off and unexpressive edges and can be elongated. Sometimes they are found grown together or form double "twinned" crystals at the surfaces of the octahedron. These different shapes and habits of some diamonds result from differing external circumstances. Diamonds (especially those with rounded crystal faces) are commonly found coated in *nyf*, an opaque gum-like skin.

Transport from Mantle

Diamond-bearing rock is carried from the mantle to the Earth's surface by deep-origin volcanic eruptions. The magma for such a volcano must originate at a depth where diamonds can be formed—150 km (93 mi) or more (three times or more the depth of source magma for most volcanoes). This is a relatively rare occurrence. These typically small surface volcanic craters extend downward in formations known as volcanic pipes. The pipes contain material that was transported toward the surface by volcanic action, but was not ejected before the volcanic activity ceased. During eruption these pipes are open to the surface, resulting in open circulation; many xenoliths of surface rock and even wood and fossils are found in volcanic pipes. Diamond-bearing volcanic pipes are closely related to the oldest, coolest regions of continental crust (cratons). This is because cratons are very thick, and their lithospheric mantle extends to great enough depth that diamonds are stable. Not all pipes contain diamonds, and even fewer contain enough diamonds to make mining economically viable.

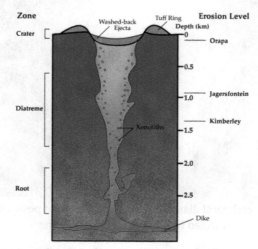

Schematic diagram of a volcanic pipe

The magma in volcanic pipes is usually one of two characteristic types, which cool into igneous rock known as either kimberlite or lamproite. The magma itself does not contain diamond; instead, it acts as an elevator that carries deep-formed rocks (xenoliths), minerals (xenocrysts), and fluids upward. These rocks are characteristically rich in magnesium-bearing olivine, pyroxene, and amphibole minerals which are often altered to serpentine by heat and fluids during and after eruption. Certain *indicator minerals* typically occur within diamantiferous kimberlites and are used as mineralogical tracers by prospectors, who follow the indicator trail back to the volcanic pipe which may contain diamonds. These minerals are rich in chromium (Cr) or titanium (Ti), elements which impart bright colors to the minerals. The most common indicator minerals are chromium garnets (usually bright red chromium-pyrope, and occasionally green ugrandite-series garnets), eclogitic garnets, orange titanium-pyrope, red high-chromium spinels, dark chromite, bright green chromium-diopside, glassy green olivine, black picroilmenite, and magnetite. Kimberlite deposits are known as *blue ground* for the deeper serpentinized part of the deposits, or as *yellow ground* for the near surface smectite clay and carbonate weathered and oxidized portion.

Once diamonds have been transported to the surface by magma in a volcanic pipe, they may erode out and be distributed over a large area. A volcanic pipe containing diamonds is known as a *primary source* of diamonds. *Secondary sources* of diamonds include all areas where a significant number of diamonds have been eroded out of their kimberlite or lamproite matrix, and accumulated because of water or wind action. These include alluvial deposits and deposits along existing and ancient shorelines, where loose diamonds tend to accumulate because of their size and density. Diamonds have also rarely been found in deposits left behind by glaciers (notably in Wisconsin and Indiana); in contrast to alluvial deposits, glacial deposits are minor and are therefore not viable commercial sources of diamond.

Space Diamonds

Not all diamonds found on Earth originated on Earth. Primitive interstellar meteorites were found to contain carbon possibly in the form of diamond. A type of diamond called carbonado that is found in South America and Africa may have been deposited there via an asteroid impact (not formed from the impact) about 3 billion years ago. These diamonds may have formed in the intrastellar environment, but as of 2008, there was no scientific consensus on how carbonado diamonds originated.

Diamonds can also form under other naturally occurring high-pressure conditions. Very small diamonds of micrometer and nanometer sizes, known as *microdiamonds* or *nanodiamonds* respectively, have been found in meteorite impact craters. Such impact events create shock zones of high pressure and temperature suitable for diamond formation. Impact-type microdiamonds can be used as an indicator of ancient impact craters. Popigai crater in Russia may have the world's largest diamond deposit, estimated at trillions of carats, and formed by an asteroid impact.

Scientific evidence indicates that white dwarf stars have a core of crystallized carbon and oxygen nuclei. The largest of these found in the universe so far, BPM 37093, is located 50 light-years (4.7×10^{14} km) away in the constellation Centaurus. A news release from the Harvard-Smithsonian Center for Astrophysics described the 2,500-mile (4,000 km)-wide stellar core as a *diamond.*

Material Properties

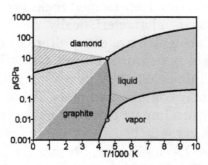

Theoretically predicted phase diagram of carbon

A diamond is a transparent crystal of tetrahedrally bonded carbon atoms in a covalent network lattice (sp^3) that crystallizes into the diamond lattice which is a variation of the face centered cubic structure. Diamonds have been adapted for many uses because of the material's exceptional physical characteristics. Most notable are its extreme hardness and thermal conductivity ($900-2320$ $W \cdot m^{-1} \cdot K^{-1}$), as well as wide bandgap and high optical dispersion. Above 1700 °C (1973 K/3583 °F) in vacuum or oxygen-free atmosphere, diamond converts to graphite; in air, transformation starts at ~700 °C. Diamond's ignition point is 720 – 800 °C in oxygen and 850 – 1000 °C in air. Naturally occurring diamonds have a density ranging from 3.15–3.53 g/cm^3, with pure diamond close to 3.52 g/cm^3. The chemical bonds that hold the carbon atoms in diamonds together are weaker than those in graphite. In diamonds, the bonds form an inflexible three-dimensional lattice, whereas in graphite, the atoms are tightly bonded into sheets, which can slide easily over one another, making the overall structure weaker. In a diamond, each carbon atom is surrounded by neighboring four carbon atoms forming a tetrahedral shaped unit.

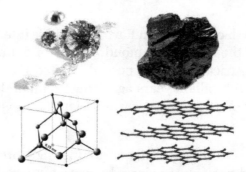

Diamond and graphite are two allotropes of carbon: pure forms of the same element that differ in structure.

Hardness

Diamond is the hardest known natural material on both the Vickers and the Mohs scale. Diamond's hardness has been known since antiquity, and is the source of its name.

Diamond hardness depends on its purity, crystalline perfection and orientation: hardness is higher for flawless, pure crystals oriented to the <111> direction (along the longest diagonal of the cubic diamond lattice). Therefore, whereas it might be possible to scratch some diamonds with other materials, such as boron nitride, the hardest diamonds can only be scratched by other diamonds and nanocrystalline diamond aggregates.

The hardness of diamond contributes to its suitability as a gemstone. Because it can only be scratched by other diamonds, it maintains its polish extremely well. Unlike many other gems, it is well-suited to daily wear because of its resistance to scratching—perhaps contributing to its popularity as the preferred gem in engagement or wedding rings, which are often worn every day.

The extreme hardness of diamond in certain orientations makes it useful in materials science, as in this pyramidal diamond embedded in the working surface of a Vickers hardness tester.

The hardest natural diamonds mostly originate from the Copeton and Bingara fields located in the New England area in New South Wales, Australia. These diamonds are generally small, perfect to semiperfect octahedra, and are used to polish other diamonds. Their hardness is associated with the crystal growth form, which is single-stage crystal growth. Most other diamonds show more evidence of multiple growth stages, which produce inclusions, flaws, and defect planes in the crystal lattice, all of which affect their hardness. It is possible to treat regular diamonds under a combination of high pressure and high temperature to produce diamonds that are harder than the diamonds used in hardness gauges.

Somewhat related to hardness is another mechanical property *toughness*, which is a material's ability to resist breakage from forceful impact. The toughness of natural diamond has been measured as 7.5–10 MPa·m$^{1/2}$. This value is good compared to other ceramic materials, but poor compared to most engineering materials such as engineering alloys, which typically exhibit toughnesses over 100 MPa·m$^{1/2}$. As with any material, the macroscopic geometry of a diamond contributes to its resistance to breakage. Diamond has a cleavage plane and is therefore more fragile in some orientations than others. Diamond cutters use this attribute to cleave some stones, prior to faceting. "Impact toughness" is one of the main indexes to measure the quality of synthetic industrial diamonds.

Pressure Resistance

Used in so-called diamond anvil experiments to create high-pressure environments, diamonds are able to withstand crushing pressures in excess of 600 gigapascals (6 million atmospheres).

Electrical Conductivity

Other specialized applications also exist or are being developed, including use as semiconductors: some blue diamonds are natural semiconductors, in contrast to most diamonds, which are excellent electrical insulators. The conductivity and blue color originate from boron impurity. Boron substitutes for carbon atoms in the diamond lattice, donating a hole into the valence band.

Substantial conductivity is commonly observed in nominally undoped diamond grown by chemical vapor deposition. This conductivity is associated with hydrogen-related species adsorbed at the surface, and it can be removed by annealing or other surface treatments.

Surface Property

Diamonds are naturally lipophilic and hydrophobic, which means the diamonds' surface cannot be wet by water but can be easily wet and stuck by oil. This property can be utilized to extract diamonds using oil when making synthetic diamonds. However, when diamond surfaces are chemically modified with certain ions, they are expected to become so hydrophilic that they can stabilize multiple layers of water ice at human body temperature.

The surface of diamonds is partially oxidized. The oxidized surface can be reduced by heat treatment under hydrogen flow. That is to say, this heat treatment partially removes oxygen-containing functional groups. But diamonds (sp^3C) are unstable against high temperature (above about 400 °C (752 °F)) under atmospheric pressure. The structure gradually changes into sp^2C above this temperature. Thus, diamonds should be reduced under this temperature.

Chemical Stability

Diamonds are not very reactive. Under room temperature diamonds do not react with any chemical reagents including strong acids and bases. A diamond's surface can only be oxidized at temperatures above about 850 °C (1,560 °F) in air. Diamond also reacts with fluorine gas above about 700 °C (1,292 °F).

Color

Diamond has a wide bandgap of 5.5 eV corresponding to the deep ultraviolet wavelength of 225 nanometers. This means pure diamond should transmit visible light and appear as a clear colorless crystal. Colors in diamond originate from lattice defects and impurities. The diamond crystal lattice is exceptionally strong and only atoms of nitrogen, boron and hydrogen can be introduced into diamond during the growth at significant concentrations (up to atomic percents). Transition metals nickel and cobalt, which are commonly used for growth of synthetic diamond by high-pressure high-temperature techniques, have been detected in diamond as individual atoms; the maximum concentration is 0.01% for nickel and even less for cobalt. Virtually any element can be introduced to diamond by ion implantation.

Brown diamonds at the National Museum of Natural History in Washington, D.C.

Nitrogen is by far the most common impurity found in gem diamonds and is responsible for the yellow and brown color in diamonds. Boron is responsible for the blue color. Color in diamond has two additional sources: irradiation (usually by alpha particles), that causes the color in green diamonds; and plastic deformation of the diamond crystal lattice. Plastic deformation is the cause of color in some brown and perhaps pink and red diamonds. In order of rarity, yellow diamond is followed by brown, colorless, then by blue, green, black, pink, orange, purple, and red. "Black", or Carbonado, diamonds are not truly black, but rather contain numerous dark inclusions that give the gems their dark appearance. Colored diamonds contain impurities or structural defects that cause the coloration, while pure or nearly pure diamonds are transparent and colorless. Most diamond impurities replace a carbon atom in the crystal lattice, known as a carbon flaw. The most common impurity, nitrogen, causes a slight to intense yellow coloration depending upon the type and concentration of nitrogen present. The Gemological Institute of America (GIA) classifies low saturation yellow and brown diamonds as diamonds in the *normal color range,* and applies a grading scale from "D" (colorless) to "Z" (light yellow). Diamonds of a different color, such as blue, are called *fancy colored* diamonds, and fall under a different grading scale.

The most famous colored diamond, Hope Diamond in 1974.

In 2008, the Wittelsbach Diamond, a 35.56-carat (7.112 g) blue diamond once belonging to the King of Spain, fetched over US$24 million at a Christie's auction. In May 2009, a 7.03-carat (1.406 g) blue diamond fetched the highest price per carat ever paid for a diamond when it was sold at auction for 10.5 million Swiss francs (6.97 million euro or US$9.5 million at the time). That record was however beaten the same year: a 5-carat (1.0 g) vivid pink diamond was sold for $10.8 million in Hong Kong on December 1, 2009.

Identification

Diamonds can be identified by their high thermal conductivity. Their high refractive index is also indicative, but other materials have similar refractivity. Diamonds cut glass, but this does not positively identify a diamond because other materials, such as quartz, also lie above glass on the Mohs scale and can also cut it. Diamonds can scratch other diamonds, but this can result in damage to one or both stones. Hardness tests are infrequently used in practical gemology because of their potentially destructive nature. The extreme hardness and high value of diamond means that gems are typically polished slowly using painstaking traditional techniques and greater attention to detail than is the case with most other gemstones; these tend to result in extremely flat, highly polished facets with exceptionally sharp facet edges. Diamonds also possess an extremely high refractive index and fairly high dispersion. Taken together, these factors affect the overall appearance of a polished diamond and most diamantaires still rely upon skilled use of a loupe (magnifying glass) to identify diamonds 'by eye'.

Industry

The diamond industry can be separated into two distinct categories: one dealing with gem-grade diamonds and another for industrial-grade diamonds. Both markets value diamonds differently.

A round brilliant cut diamond set in a ring

Gem-grade Diamonds

A large trade in gem-grade diamonds exists. Although most gem-grade diamonds are sold newly polished, there is a well-established market for resale of polished diamonds (e.g. pawnbroking, auctions, second-hand jewelry stores, diamantaires, bourses, etc.). One hallmark of the trade in gem-quality diamonds is its remarkable concentration: wholesale trade and diamond cutting is limited to just a few locations; in 2003, 92% of the world's diamonds were cut and polished in Surat, India. Other important centers of diamond cutting and trading are the Antwerp diamond district in Belgium, where the International Gemological Institute is based, London, the Diamond District in New York City, the Diamond Exchange District in Tel Aviv, and Amsterdam. One contributory factor is the geological nature of diamond deposits: several large primary kimberlite-pipe mines each account for significant portions of market share (such as the Jwaneng mine in Botswana, which is a single large-pit mine that can produce between 12,500,000 carats (2,500 kg) to 15,000,000 carats (3,000 kg) of diamonds per year). Secondary alluvial diamond deposits, on the other hand, tend to be fragmented amongst many different operators because they can be dispersed over many hundreds of square kilometers (e.g., alluvial deposits in Brazil).

The production and distribution of diamonds is largely consolidated in the hands of a few key players, and concentrated in traditional diamond trading centers, the most important being Antwerp, where 80% of all rough diamonds, 50% of all cut diamonds and more than 50% of all rough, cut and industrial diamonds combined are handled. This makes Antwerp a de facto "world diamond capital". The city of Antwerp also hosts the Antwerpsche Diamantkring, created in 1929 to become the first and biggest diamond bourse dedicated to rough diamonds. Another important diamond center is New York City, where almost 80% of the world's diamonds are sold, including auction sales.

The De Beers company, as the world's largest diamond mining company, holds a dominant position in the industry, and has done so since soon after its founding in 1888 by the British imperialist Cecil Rhodes. De Beers is currently the world's largest operator of diamond production facilities (mines) and distribution channels for gem-quality diamonds. The Diamond Trading Company (DTC) is a subsidiary of De Beers and markets rough diamonds from De Beers-operated mines. De Beers and its subsidiaries own mines that produce some 40% of annual world diamond production. For most of the 20th century over 80% of the world's rough diamonds passed through De Beers, but by 2001–2009 the figure had decreased to around 45%, and by 2013 the company's market share had further decreased to around 38% in value terms and even less by volume. De Beers sold off the vast majority of its diamond stockpile in the late 1990s – early 2000s and the remainder largely represents working stock (diamonds that are being sorted before sale). This was well documented in the press but remains little known to the general public.

As a part of reducing its influence, De Beers withdrew from purchasing diamonds on the open market in 1999 and ceased, at the end of 2008, purchasing Russian diamonds mined by the largest Russian diamond company Alrosa. As of January 2011, De Beers states that it only sells diamonds from the following four countries: Botswana, Namibia, South Africa and Canada. Alrosa had to suspend their sales in October 2008 due to the global energy crisis, but the company reported that it had resumed selling rough diamonds on the open market by October 2009. Apart from Alrosa, other important diamond mining companies include BHP Billiton, which is the world's largest mining company; Rio Tinto Group, the owner of Argyle (100%), Diavik (60%), and Murowa (78%) diamond mines; and Petra Diamonds, the owner of several major diamond mines in Africa.

Diamond polisher in Amsterdam

Further down the supply chain, members of The World Federation of Diamond Bourses (WFDB) act as a medium for wholesale diamond exchange, trading both polished and rough diamonds. The WFDB consists of independent diamond bourses in major cutting centers such as Tel Aviv, Antwerp, Johannesburg and other cities across the USA, Europe and Asia. In 2000, the WFDB and The International Diamond Manufacturers Association established the World Diamond Council to prevent the trading of diamonds used to fund war and inhumane acts. WFDB's additional activities include sponsoring the World Diamond Congress every two years, as well as the establishment of the *International Diamond Council* (IDC) to oversee diamond grading.

Once purchased by Sightholders (which is a trademark term referring to the companies that have a three-year supply contract with DTC), diamonds are cut and polished in preparation for sale as gemstones ('industrial' stones are regarded as a by-product of the gemstone market; they are used for abrasives). The cutting and polishing of rough diamonds is a specialized skill that is concentrated in a limited number of locations worldwide. Traditional diamond cutting centers are Antwerp, Amsterdam, Johannesburg, New York City, and Tel Aviv. Recently, diamond cutting centers have been established in China, India, Thailand, Namibia and Botswana. Cutting centers with lower cost of labor, notably Surat in Gujarat, India, handle a larger number of smaller carat diamonds, while smaller quantities of larger or more valuable diamonds are more likely to be handled in Europe or North America. The recent expansion of this industry in India, employing low cost labor, has allowed smaller diamonds to be prepared as gems in greater quantities than was previously economically feasible.

Diamonds which have been prepared as gemstones are sold on diamond exchanges called *bourses*. There are 28 registered diamond bourses in the world. Bourses are the final tightly controlled step in the diamond supply chain; wholesalers and even retailers are able to buy relatively small lots of diamonds at the bourses, after which they are prepared for final sale to the consumer. Diamonds can be sold already set in jewelry, or sold unset ("loose"). According to the Rio Tinto Group, in 2002 the diamonds produced and released to the market were valued at US\$9 billion as rough diamonds, US\$14 billion after being cut and polished, US\$28 billion in wholesale diamond jewelry, and US\$57 billion in retail sales.

Cutting

The Darya-I-Nur Diamond—an example of unusual diamond cut and jewelry arrangement

Mined rough diamonds are converted into gems through a multi-step process called "cutting". Diamonds are extremely hard, but also brittle and can be split up by a single blow. Therefore, diamond cutting is traditionally considered as a delicate procedure requiring skills, scientific knowledge, tools and experience. Its final goal is to produce a faceted jewel where the specific angles between the facets would optimize the diamond luster, that is dispersion of white light, whereas the number and area of facets would determine the weight of the final product. The weight reduction upon cutting is significant and can be of the order of 50%. Several possible shapes are considered, but the final decision is often determined not only by scientific, but also practical considerations. For example, the diamond might be intended for display or for wear, in a ring or a necklace, singled or surrounded by other gems of certain color and shape. Some of them may be considered as classical, such as round, pear, marquise, oval, hearts and arrows diamonds, etc. Some of them are special, produced by certain companies, for example, Phoenix, Cushion, Sole Mio diamonds, etc.

The most time-consuming part of the cutting is the preliminary analysis of the rough stone. It needs to address a large number of issues, bears much responsibility, and therefore can last years in case of unique diamonds. The following issues are considered:

- The hardness of diamond and its ability to cleave strongly depend on the crystal orientation. Therefore, the crystallographic structure of the diamond to be cut is analyzed using X-ray diffraction to choose the optimal cutting directions.

- Most diamonds contain visible non-diamond inclusions and crystal flaws. The cutter has to decide which flaws are to be removed by the cutting and which could be kept.

- The diamond can be split by a single, well calculated blow of a hammer to a pointed tool, which is quick, but risky. Alternatively, it can be cut with a diamond saw, which is a more reliable but tedious procedure.

After initial cutting, the diamond is shaped in numerous stages of polishing. Unlike cutting, which is a responsible but quick operation, polishing removes material by gradual erosion and is extremely time consuming. The associated technique is well developed; it is considered as a routine and can be performed by technicians. After polishing, the diamond is reexamined for possible flaws, either remaining or induced by the process. Those flaws are concealed through various diamond enhancement techniques, such as repolishing, crack filling, or clever arrangement of the stone in the jewelry. Remaining non-diamond inclusions are removed through laser drilling and filling of the voids produced.

Marketing

Marketing has significantly affected the image of diamond as a valuable commodity.

N. W. Ayer & Son, the advertising firm retained by De Beers in the mid-20th century, succeeded in reviving the American diamond market. And the firm created new markets in countries where no diamond tradition had existed before. N. W. Ayer's marketing included product placement, advertising focused on the diamond product itself rather than the De Beers brand, and associations with celebrities and royalty. Without advertising the De Beers brand, De Beers was advertising its competitors' diamond products as well, but this was not a concern as De Beers dominated the diamond market throughout the 20th century. De Beers' market share dipped temporarily to 2nd place in the global market below Alrosa in the aftermath of the global economic crisis of 2008, down to less than 29% in terms of carats mined, rather than sold. The campaign lasted for decades but was effectively discontinued by early 2011. De Beers still advertises diamonds, but the advertising now mostly promotes its own brands, or licensed product lines, rather than completely "generic" diamond products. The campaign was perhaps best captured by the slogan "a diamond is forever". This slogan is now being used by De Beers Diamond Jewelers, a jewelry firm which is a 50%/50% joint venture between the De Beers mining company and LVMH, the luxury goods conglomerate.

Brown-colored diamonds constituted a significant part of the diamond production, and were predominantly used for industrial purposes. They were seen as worthless for jewelry (not even being assessed on the diamond color scale). After the development of Argyle diamond mine in Australia in 1986, and marketing, brown diamonds have become acceptable gems. The change was mostly due to the numbers: the Argyle mine, with its 35,000,000 carats (7,000 kg) of diamonds per year, makes about one-third of global production of natural diamonds; 80% of Argyle diamonds are brown.

Industrial-grade Diamonds

Industrial diamonds are valued mostly for their hardness and thermal conductivity, making many of the gemological characteristics of diamonds, such as the 4 Cs, irrelevant for most applications. 80% of mined diamonds (equal to about 135,000,000 carats (27,000 kg) annually), are unsuitable for use as gemstones, and used industrially. In addition to mined diamonds, synthetic diamonds found industrial applications almost immediately after their invention in the 1950s; another 570,000,000 carats (114,000 kg) of synthetic diamond is produced annually for industrial use (in 2004; in 2014 it's 4,500,000,000 carats (900,000 kg), 90% of which is produced in China). Approximately 90% of diamond grinding grit is currently of synthetic origin.

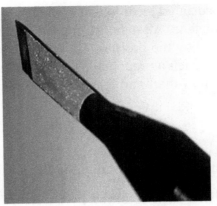

A scalpel with synthetic diamond blade

The boundary between gem-quality diamonds and industrial diamonds is poorly defined and partly depends on market conditions (for example, if demand for polished diamonds is high, some lower-grade stones will be polished into low-quality or small gemstones rather than being sold for industrial use). Within the category of industrial diamonds, there is a sub-category comprising the lowest-quality, mostly opaque stones, which are known as bort.

Close-up photograph of an angle grinder blade with tiny diamonds shown embedded in the metal

Industrial use of diamonds has historically been associated with their hardness, which makes diamond the ideal material for cutting and grinding tools. As the hardest known naturally occurring material, diamond can be used to polish, cut, or wear away any material, including other diamonds. Common industrial applications of this property include diamond-tipped drill bits and saws, and the use of diamond powder as an abrasive. Less expensive industrial-grade diamonds, known as bort, with more flaws and poorer color than gems, are used for such purposes. Diamond is not suitable for machining ferrous alloys at high speeds, as carbon is soluble in iron at the high temperatures created by high-speed machining, leading to greatly increased wear on diamond tools compared to alternatives.

Specialized applications include use in laboratories as containment for high pressure experiments, high-performance bearings, and limited use in specialized windows. With the continuing advances being made in the production of synthetic diamonds, future applications are becoming feasible. The high thermal conductivity of diamond makes it suitable as a heat sink for integrated circuits in electronics.

Mining

Approximately 130,000,000 carats (26,000 kg) of diamonds are mined annually, with a total value of nearly US$9 billion, and about 100,000 kg (220,000 lb) are synthesized annually.

Roughly 49% of diamonds originate from Central and Southern Africa, although significant sources of the mineral have been discovered in Canada, India, Russia, Brazil, and Australia. They are mined from kimberlite and lamproite volcanic pipes, which can bring diamond crystals, originating from deep within the Earth where high pressures and temperatures enable them to form, to the surface. The mining and distribution of natural diamonds are subjects of frequent controversy such as concerns over the sale of *blood diamonds* or *conflict diamonds* by African paramilitary groups. The diamond supply chain is controlled by a limited number of powerful businesses, and is also highly concentrated in a small number of locations around the world.

Only a very small fraction of the diamond ore consists of actual diamonds. The ore is crushed, during which care is required not to destroy larger diamonds, and then sorted by density. Today, diamonds are located in the diamond-rich density fraction with the help of X-ray fluorescence, after which the final sorting steps are done by hand. Before the use of X-rays became commonplace, the separation was done with grease belts; diamonds have a stronger tendency to stick to grease than the other minerals in the ore.

Siberia's Udachnaya diamond mine

Historically, diamonds were found only in alluvial deposits in Guntur and Krishna district of the Krishna River delta in Southern India. India led the world in diamond production from the time of their discovery in approximately the 9th century BC to the mid-18th century AD, but the commercial potential of these sources had been exhausted by the late 18th century and at that time India was eclipsed by Brazil where the first non-Indian diamonds were found in 1725. Currently, one of the most prominent Indian mines is located at Panna.

Diamond extraction from primary deposits (kimberlites and lamproites) started in the 1870s after the discovery of the Diamond Fields in South Africa. Production has increased over time and now an accumulated total of 4,500,000,000 carats (900,000 kg) have been mined since that date. Twenty percent of that amount has been mined in the last five years, and during the last 10 years, nine new mines have started production; four more are waiting to be opened soon. Most of these mines are located in Canada, Zimbabwe, Angola, and one in Russia.

In the U.S., diamonds have been found in Arkansas, Colorado, Wyoming, and Montana. In 2004, the discovery of a microscopic diamond in the U.S. led to the January 2008 bulk-sampling of kimberlite pipes in a remote part of Montana. The Crater of Diamonds State Park in Arkansas is open to the public, and is the only mine in the world where members of the public can dig for diamonds.

Today, most commercially viable diamond deposits are in Russia (mostly in Sakha Republic, for example Mir pipe and Udachnaya pipe), Botswana, Australia (Northern and Western Australia) and the Democratic Republic of the Congo. In 2005, Russia produced almost one-fifth of the global diamond output, according to the British Geological Survey. Australia boasts the richest diamantiferous pipe, with production from the Argyle diamond mine reaching peak levels of 42 metric tons per year in the 1990s. There are also commercial deposits being actively mined in the Northwest Territories of Canada and Brazil. Diamond prospectors continue to search the globe for diamond-bearing kimberlite and lamproite pipes.

Political Issues

In some of the more politically unstable central African and west African countries, revolutionary groups have taken control of diamond mines, using proceeds from diamond sales to finance their operations. Diamonds sold through this process are known as *conflict diamonds* or *blood diamonds*. Major diamond trading corporations continue to fund and fuel these conflicts by doing business with armed groups.

Unsustainable diamond mining in Sierra Leone

In response to public concerns that their diamond purchases were contributing to war and human rights abuses in central and western Africa, the United Nations, the diamond industry and diamond-trading nations introduced the Kimberley Process in 2002. The Kimberley Process aims to ensure that conflict diamonds do not become intermixed with the diamonds not controlled by such rebel groups. This is done by requiring diamond-producing countries to provide proof that the money they make from selling the diamonds is not used to fund criminal or revolutionary activities. Although the Kimberley Process has been moderately successful in limiting the number of conflict diamonds entering the market, some still find their way in. According to the International Diamond Manufacturers Association, conflict diamonds constitute 2–3% of all diamonds traded. Two major flaws still hinder the effectiveness of the Kimberley Process: (1) the relative ease of smuggling diamonds across African borders, and (2) the violent nature of diamond mining in nations that are not in a technical state of war and whose diamonds are therefore considered "clean".

The Canadian Government has set up a body known as the Canadian Diamond Code of Conduct to help authenticate Canadian diamonds. This is a stringent tracking system of diamonds and helps protect the "conflict free" label of Canadian diamonds.

Synthetics, Simulants, and Enhancements

Synthetics

Synthetic diamonds are diamonds manufactured in a laboratory, as opposed to diamonds mined from the Earth. The gemological and industrial uses of diamond have created a large demand for rough stones. This demand has been satisfied in large part by synthetic diamonds, which have been manufactured by various processes for more than half a century. However, in recent years it has become possible to produce gem-quality synthetic diamonds of significant size. It is possible to make colorless synthetic gemstones that, on a molecular level, are identical to natural stones and so visually similar that only a gemologist with special equipment can tell the difference.

Synthetic diamonds of various colors grown by the high-pressure high-temperature technique

The majority of commercially available synthetic diamonds are yellow and are produced by so-called *high-pressure high-temperature* (HPHT) processes. The yellow color is caused by nitrogen impurities. Other colors may also be reproduced such as blue, green or pink, which are a result of the addition of boron or from irradiation after synthesis.

Colorless gem cut from diamond grown by chemical vapor deposition

Another popular method of growing synthetic diamond is chemical vapor deposition (CVD). The growth occurs under low pressure (below atmospheric pressure). It involves feeding a mixture of gases (typically 1 to 99 methane to hydrogen) into a chamber and splitting them to chemically active radicals in a plasma ignited by microwaves, hot filament, arc discharge, welding torch or laser. This method is mostly used for coatings, but can also produce single crystals several millimeters in size.

As of 2010, nearly all 5,000 million carats (1,000 tonnes) of synthetic diamonds produced per year are for industrial use. Around 50% of the 133 million carats of natural diamonds mined per year end up in industrial use. Mining companies' expenses average $40 to $60 per carat for natural colorless diamonds, while synthetic manufacturers' expenses average $2,500 per carat for synthetic, gem-quality colorless diamonds. However, a purchaser is more likely to encounter a synthetic when looking for a fancy-colored diamond because nearly all synthetic diamonds are fancy-colored, while only 0.01% of natural diamonds are.

Simulants

A diamond simulant is a non-diamond material that is used to simulate the appearance of a diamond, and may be referred to as diamante. Cubic zirconia is the most common. The gemstone moissanite (silicon carbide) can be treated as a diamond simulant, though more costly to produce than cubic zirconia. Both are produced synthetically.

Gem-cut synthetic silicon carbide set in a ring

Enhancements

Diamond enhancements are specific treatments performed on natural or synthetic diamonds (usually those already cut and polished into a gem), which are designed to better the gemological characteristics of the stone in one or more ways. These include laser drilling to remove inclusions, application of sealants to fill cracks, treatments to improve a white diamond's color grade, and treatments to give fancy color to a white diamond.

Coatings are increasingly used to give a diamond simulant such as cubic zirconia a more "diamond-like" appearance. One such substance is diamond-like carbon—an amorphous carbonaceous material that has some physical properties similar to those of the diamond. Advertising suggests that such a coating would transfer some of these diamond-like properties to the coated stone, hence enhancing the diamond simulant. Techniques such as Raman spectroscopy should easily identify such a treatment.

Identification

Early diamond identification tests included a scratch test relying on the superior hardness of diamond. This test is destructive, as a diamond can scratch another diamond, and is rarely used nowadays. Instead, diamond identification relies on its superior thermal conductivity. Electronic thermal probes are widely used in the gemological centers to separate diamonds from their imitations. These probes consist of a pair of battery-powered thermistors mounted in a fine copper tip. One thermistor functions as a heating device while the other measures the temperature of the copper tip: if the stone being tested is a diamond, it will conduct the tip's thermal energy rapidly enough to produce a measurable temperature drop. This test takes about 2–3 seconds.

Whereas the thermal probe can separate diamonds from most of their simulants, distinguishing between various types of diamond, for example synthetic or natural, irradiated or non-irradiated, etc., requires more advanced, optical techniques. Those techniques are also used for some diamonds simulants, such as silicon carbide, which pass the thermal conductivity test. Optical techniques can distinguish between natural diamonds and synthetic diamonds. They can also identify the vast majority of treated natural diamonds. "Perfect" crystals (at the atomic lattice level) have never been found, so both natural and synthetic diamonds always possess characteristic imperfections, arising from the circumstances of their crystal growth, that allow them to be distinguished from each other.

Laboratories use techniques such as spectroscopy, microscopy and luminescence under shortwave ultraviolet light to determine a diamond's origin. They also use specially made instruments to aid them in the identification process. Two screening instruments are the *DiamondSure* and the *DiamondView*, both produced by the DTC and marketed by the GIA.

Several methods for identifying synthetic diamonds can be performed, depending on the method of production and the color of the diamond. CVD diamonds can usually be identified by an orange fluorescence. D-J colored diamonds can be screened through the Swiss Gemmological Institute's Diamond Spotter. Stones in the D-Z color range can be examined through the DiamondSure UV/visible spectrometer, a tool developed by De Beers. Similarly, natural diamonds usually have minor imperfections and flaws, such as inclusions of foreign material, that are not seen in synthetic diamonds.

Screening devices based on diamond type detection can be used to make a distinction between diamonds that are certainly natural and diamonds that are potentially synthetic. Those potentially synthetic diamonds require more investigation in a specialized lab. Examples of commercial screening devices are D-Screen (WTOCD/HRD Antwerp) and Alpha Diamond Analyzer (Bruker/HRD Antwerp).

Stolen Diamonds

Occasionally large thefts of diamonds take place. In February 2013 armed robbers carried out a raid at Brussels Airport and escaped with gems estimated to be worth $50m (£32m; 37m euros). The gang broke through a perimeter fence and raided the cargo hold of a Swiss-bound plane. The gang have since been arrested and large amounts of cash and diamonds recovered.

The identification of stolen diamonds presents a set of difficult problems. Rough diamonds will have a distinctive shape depending on whether their source is a mine or from an alluvial environment such as a beach or river - alluvial diamonds have smoother surfaces than those that have been mined. Determining the provenance of cut and polished stones is much more complex.

The Kimberley Process was developed to monitor the trade in rough diamonds and prevent their being used to fund violence. Before exporting, rough diamonds are certificated by the government of the country of origin. Some countries, such as Venezuela, are not party to the agreement. The Kimberley Process does not apply to local sales of rough diamonds within a country.

Diamonds may be etched by laser with marks invisible to the naked eye. Lazare Kaplan, a US-based company, developed this method. However, whatever is marked on a diamond can readily be removed.

Graphite

Graphite, archaically referred to as plumbago, is a crystalline form of carbon, a semimetal, a native element mineral, and one of the allotropes of carbon. Graphite is the most stable form of carbon under standard conditions. Therefore, it is used in thermochemistry as the standard state for defining the heat of formation of carbon compounds. Graphite may be considered the highest grade of coal, just above anthracite and alternatively called meta-anthracite, although it is not normally used as fuel because it is difficult to ignite.

Types and Varieties

There are three principal types of natural graphite, each occurring in different types of ore deposit:

- Crystalline flake graphite (or flake graphite for short) occurs as isolated, flat, plate-like particles with hexagonal edges if unbroken and when broken the edges can be irregular or angular;

- *Amorphous graphite*: very fine flake graphite is sometimes called amorphous in the trade;

- Lump graphite (also called vein graphite) occurs in fissure veins or fractures and appears as massive platy intergrowths of fibrous or acicular crystalline aggregates, and is probably hydrothermal in origin.

- Highly ordered pyrolytic graphite or more correctly highly *oriented* pyrolytic graphite (HOPG) refers to graphite with an angular spread between the graphite sheets of less than 1°.

- The name "graphite fiber" is also sometimes used to refer to carbon fiber or carbon fiber-reinforced polymer.

Occurrence

Graphite occurs in metamorphic rocks as a result of the reduction of sedimentary carbon compounds during metamorphism. It also occurs in igneous rocks and in meteorites. Minerals associated with graphite include quartz, calcite, micas and tourmaline. In meteorites it occurs with troilite and silicate minerals. Small graphitic crystals in meteoritic iron are called cliftonite.

Graphite output in 2005

According to the United States Geological Survey (USGS), world production of natural graphite in 2012 was 1,100,000 tonnes, of which the following major exporters are: China (750 kt), India (150 kt), Brazil (75 kt), North Korea (30 kt) and Canada (26 kt). Graphite is not mined in the United States, but U.S. production of synthetic graphite in 2010 was 134 kt valued at $1.07 billion.

Properties

Structure

Graphite has a layered, planar structure. The individual layers are called graphene. In each layer, the carbon atoms are arranged in a honeycomb lattice with separation of 0.142 nm, and the distance between planes is 0.335 nm. Atoms in the plane are bonded covalently, with only three of the four potential bonding sites satisfied. The fourth electron is free to migrate in the plane, making graphite electrically conductive. However, it does not conduct in a direction at right angles to the plane. Bonding between layers is via weak van der Waals bonds, which allows layers of graphite to be easily separated, or to slide past each other.

The two known forms of graphite, *alpha* (hexagonal) and *beta* (rhombohedral), have very similar physical properties, except the graphene layers stack slightly differently. The alpha graphite may be either flat or buckled. The alpha form can be converted to the beta form through mechanical treatment and the beta form reverts to the alpha form when it is heated above 1300 °C.

Graphite's unit cell

Scanning tunneling microscope image of graphite surface atoms

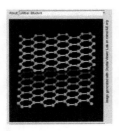

Ball-and-stick model of graphite (two graphene layers)

Animated view of the unit cell in three layers of graphene (note that this is a slightly different unit cell from the one to the left)

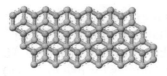

Plane view of layer stacking

Side view of layer stacking

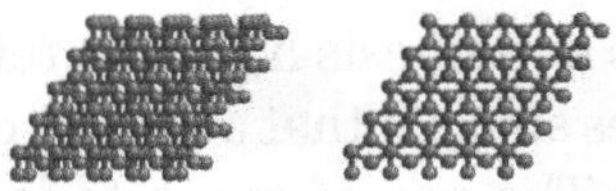

Rotating graphite stereogram

Other Properties

The acoustic and thermal properties of graphite are highly anisotropic, since phonons propagate quickly along the tightly-bound planes, but are slower to travel from one plane to another. Graphite's high thermal stability and electrical conductivity facilitate its widespread use as electrodes and refractories in high temperature material processing applications. However, in oxygen containing atmospheres graphite readily oxidizes to form CO_2 at temperatures of 700 °C and above.

Graphite plates and sheets, 10–15 cm high, Mineral specimen from Kimmirut, Baffin Island.

Graphite is an electric conductor, consequently, useful in such applications as arc lamp electrodes. It can conduct electricity due to the vast electron delocalization within the carbon layers (a phenomenon called aromaticity). These valence electrons are free to move, so are able to conduct electricity. However, the electricity is primarily conducted within the plane of the layers. The conductive properties of powdered graphite allows its use as pressure sensor in carbon microphones.

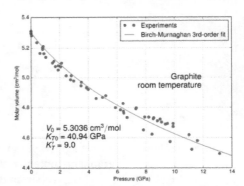

Molar volume vs. pressure at room temperature.

Graphite and graphite powder are valued in industrial applications for their self-lubricating and dry lubricating properties. There is a common belief that graphite's lubricating properties are solely due to the loose interlamellar coupling between sheets in the structure. However, it has been shown that in a vacuum environment (such as in technologies for use in space), graphite degrades as a lubricant, due to the hypoxic conditions This observation led to the hypothesis that the lubrication is due to the presence of fluids between the layers, such as air and water, which are naturally adsorbed from the environment. This hypothesis has been refuted by studies showing that air and water are not absorbed. Recent studies suggest that an effect called superlubricity can also account for graphite's lubricating properties. The use of graphite is limited by its tendency to facilitate pitting corrosion in some stainless steel, and to promote galvanic corrosion between dissimilar metals (due to its electrical conductivity). It is also corrosive to aluminium in the presence of moisture. For this reason, the US Air Force banned its use as a lubricant in aluminium aircraft, and discouraged its use in aluminium-containing automatic weapons. Even graphite pencil marks on aluminium parts may facilitate corrosion. Another high-temperature lubricant, hexagonal boron nitride, has the same molecular structure as graphite. It is sometimes called *white graphite*, due to its similar properties.

When a large number of crystallographic defects bind these planes together, graphite loses its lubrication properties and becomes what is known as pyrolytic graphite. It is also highly anisotropic, and diamagnetic, thus it will float in mid-air above a strong magnet. If it is made in a fluidized bed at 1000–1300 °C then it is isotropic turbostratic, and is used in blood contacting devices like mechanical heart valves and is called pyrolytic carbon, and is not diamagnetic. Pyrolytic graphite, and pyrolytic carbon are often confused but are very different materials.

Natural and crystalline graphites are not often used in pure form as structural materials, due to their shear-planes, brittleness and inconsistent mechanical properties.

History of Natural Graphite Use

In the 4th millennium B.C., during the Neolithic Age in southeastern Europe, the Marița culture used graphite in a ceramic paint for decorating pottery.

Some time before 1565 (some sources say as early as 1500), an enormous deposit of graphite was discovered on the approach to Grey Knotts from the hamlet of Seathwaite in Borrowdale parish, Cumbria, England, which the locals found very useful for marking sheep. During the reign of Elizabeth I (1533–1603), Borrowdale graphite was used as a refractory material to line moulds for cannonballs, resulting in rounder, smoother balls that could be fired farther, contributing to the strength of the English navy. This particular deposit of graphite was extremely pure and soft, and could easily be broken into sticks. Because of its military importance, this unique mine and its production were strictly controlled by the Crown.

Other Names

Historically, graphite was called black lead or plumbago. Plumbago was commonly used in its massive mineral form. Both of these names arise from confusion with the similar-appearing lead ores, particularly galena. The Latin word for lead, *plumbum*, gave its name to the English term for this grey metallic-sheened mineral and even to the leadworts or plumbagos, plants with flowers that resemble this colour.

The term *black lead* usually refers to a powdered or processed graphite, matte black in color.

Abraham Gottlob Werner coined the name *graphite* ("writing stone") in 1789. He attempted to clear up the confusion between molybdena, plumbago and blacklead after Carl Wilhelm Scheele in 1778 proved that there are at least three different minerals. Scheele's analysis showed that the chemical compounds molybdenum sulfide (molybdenite), lead(II) sulfide (galena) and graphite were three different soft black minerals.

Uses of Natural Graphite

Natural graphite is mostly consumed for refractories, batteries, steelmaking, expanded graphite, brake linings, foundry facings and lubricants. Graphene, which occurs naturally in graphite, has unique physical properties and is among the strongest substances known. However, the process of separating it from graphite will require more technological development.

Refractories

This end-use began before 1900 with the graphite crucible used to hold molten metal; this is now a minor part of refractories. In the mid-1980s, the carbon-magnesite brick became important, and a bit later the alumina-graphite shape. Currently the order of importance is alumina-graphite shapes, carbon-magnesite brick, monolithics (gunning and ramming mixes), and then crucibles.

Crucibles began using very large flake graphite, and carbon-magnesite brick requiring not quite so large flake graphite; for these and others there is now much more flexibility in size of flake required, and amorphous graphite is no longer restricted to low-end refractories. Alumina-graphite shapes are used as continuous casting ware, such as nozzles and troughs, to convey the molten steel from ladle to mold, and carbon magnesite bricks line steel converters and electric arc furnaces to withstand extreme temperatures. Graphite Blocks are also used in parts of blast furnace linings where the high thermal conductivity of the graphite is critical. High-purity monolithics are often used as a continuous furnace lining instead of the carbon-magnesite bricks.

The US and European refractories industry had a crisis in 2000–2003, with an indifferent market for steel and a declining refractory consumption per tonne of steel underlying firm buyouts and many plant closures. Many of the plant closures resulted from the acquisition of Harbison-Walker Refractories by RHI AG and some plants had their equipment auctioned off. Since much of the lost capacity was for carbon-magnesite brick, graphite consumption within refractories area moved towards alumina-graphite shapes and monolithics, and away from the brick. The major source of carbon-magnesite brick is now imports from China. Almost all of the above refractories are used to make steel and account for 75% of refractory consumption; the rest is used by a variety of industries, such as cement.

According to the USGS, US natural graphite consumption in refractories was 12,500 tonnes in 2010.

Batteries

The use of graphite in batteries has been increasing in the last 30 years. Natural and synthetic graphite are used to construct the anode of all major battery technologies. The lithium-ion battery utilizes roughly twice the amount of graphite than lithium carbonate.

The demand for batteries, primarily nickel-metal-hydride and lithium-ion batteries, has caused a growth in graphite demand in the late 1980s and early 1990s. This growth was driven by portable electronics, such as portable CD players and power tools. Laptops, mobile phones, tablet, and smartphone products have increased the demand for batteries. Electric vehicle batteries are anticipated to increase graphite demand. As an example, a lithium-ion battery in a fully electric Nissan Leaf contains nearly 40 kg of graphite.

Steelmaking

Natural graphite in this end use mostly goes into carbon raising in molten steel, although it can be used to lubricate the dies used to extrude hot steel. Supplying carbon raisers is very competitive, therefore subject to cut-throat pricing from alternatives such as synthetic graphite powder, petroleum coke, and other forms of carbon. A carbon raiser is added to increase the carbon content of the steel to the specified level. An estimate based on USGS US graphite consumption statistics indicates that 10,500 tonnes were used in this fashion in 2005.

Brake Linings

Natural amorphous and fine flake graphite are used in brake linings or brake shoes for heavier (nonautomotive) vehicles, and became important with the need to substitute for asbestos. This use has been important for quite some time, but nonasbestos organic (NAO) compositions are beginning to reduce graphite's market share. A brake-lining industry shake-out with some plant closures has not been beneficial, nor has an indifferent automotive market. According to the USGS, US natural graphite consumption in brake linings was 6,510 tonnes in 2005.

Foundry Facings and Lubricants

A foundry facing mold wash is a water-based paint of amorphous or fine flake graphite. Painting the inside of a mold with it and letting it dry leaves a fine graphite coat that will ease separation of the object cast after the hot metal has cooled. Graphite lubricants are specialty items for use at very high or very low temperatures, as forging die lubricant, an antiseize agent, a gear lubricant for mining machinery, and to lubricate locks. Having low-grit graphite, or even better no-grit graphite (ultra high purity), is highly desirable. It can be used as a dry powder, in water or oil, or as colloidal graphite (a permanent suspension in a liquid). An estimate based on USGS graphite consumption statistics indicates that 2,200 tonnes was used in this fashion in 2005.

Pencils

The ability to leave marks on paper and other objects gave graphite its name, given in 1789 by German mineralogist Abraham Gottlob Werner. It stems from *graphein*, meaning *to write/draw* in Ancient Greek.

Graphite pencils

From the 16th Century, pencils were made with leads of English natural graphite, but modern pencil lead is most commonly a mix of powdered graphite and clay; it was invented by Nicolas-Jacques

Conté in 1795. It is chemically unrelated to the metal lead, whose ores had a similar appearance, hence the continuation of the name. Plumbago is another older term for natural graphite used for drawing, typically as a lump of the mineral without a wood casing. The term plumbago drawing is normally restricted to 17th and 18th century works, mostly portraits.

Today, pencils are still a small but significant market for natural graphite. Around 7% of the 1.1 million tonnes produced in 2011 was used to make pencils. Low-quality amorphous graphite is used and sourced mainly from China.

Other Uses

Natural graphite has found uses in zinc-carbon batteries, in electric motor brushes, and various specialized applications. Graphite of various hardness or softness results in different qualities and tones when used as an artistic medium. Railroads would often mix powdered graphite with waste oil or linseed oil to create a heat resistant protective coating for the exposed portions of a steam locomotive's boiler, such as the smokebox or lower part of the firebox.

Expanded Graphite

Expanded graphite is made by immersing natural flake graphite in a bath of chromic acid, then concentrated sulfuric acid, which forces the crystal lattice planes apart, thus expanding the graphite. The expanded graphite can be used to make graphite foil or used directly as "hot top" compound to insulate molten metal in a ladle or red-hot steel ingots and decrease heat loss, or as firestops fitted around a fire door or in sheet metal collars surrounding plastic pipe (during a fire, the graphite expands and chars to resist fire penetration and spread), or to make high-performance gasket material for high-temperature use. After being made into graphite foil, the foil is machined and assembled into the bipolar plates in fuel cells. The foil is made into heat sinks for laptop computers which keeps them cool while saving weight, and is made into a foil laminate that can be used in valve packings or made into gaskets. Old-style packings are now a minor member of this grouping: fine flake graphite in oils or greases for uses requiring heat resistance. A GAN estimate of current US natural graphite consumption in this end use is 7,500 tonnes.

Intercalated Graphite

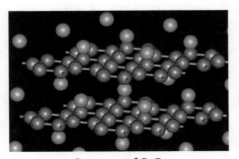

Structure of CaC_6

Graphite forms intercalation compounds with some metals and small molecules. In these compounds, the host molecule or atom gets "sandwiched" between the graphite layers, resulting in a type of compounds with variable stoichiometry. A prominent example of an intercalation compound is potassium graphite, denoted by the formula KC_8. Graphite intercalation compounds are superconductors. The highest transition temperature (by June 2009) $T_c = 11.5$ K is achieved in CaC_6, and it further increases under applied pressure (15.1 K at 8 GPa).

Uses of Synthetic Graphite

Invention of A Process to Produce Synthetic Graphite

A process to make synthetic graphite was invented accidentally by Edward Goodrich Acheson (1856–1931). In the mid-1890s, Acheson discovered that overheating carborundum, which he is also credited with discovering, produced almost pure graphite. While studying the effects of high temperature on carborundum, he had found that silicon vaporizes at about 4,150 °C (7,500 °F), leaving the carbon behind in graphitic carbon. This graphite was another major discovery for him, and it became extremely valuable and helpful as a lubricant.

In 1896 Acheson received a patent for his method of synthesizing graphite, and in 1897 started commercial production. The Acheson Graphite Co. was formed in 1899. In 1928 this company was merged with National Carbon Company (now GrafTech International). Acheson also developed a variety of colloidal graphite products including Oildag and Aquadag. These were later manufactured by the Acheson Colloids Co. (now Acheson Industries, a unit of Henkel AG).

Scientific Research

Highly oriented pyrolytic graphite (HOPG) is the highest-quality synthetic form of graphite. It is used in scientific research, in particular, as a length standard for scanner calibration of scanning probe microscope.

Electrodes

Graphite electrodes carry the electricity that melts scrap iron and steel, and sometimes direct-reduced iron (DRI), in electric arc furnaces, which are the vast majority of steel furnaces. They are made from petroleum coke after it is mixed with coal tar pitch. They are then extruded and shaped, baked to carbonize the binder (pitch), and finally graphitized by heating it to temperatures approaching 3000 °C, at which the carbon atoms arrange into graphite. They can vary in size up to 11 feet long and 30 inches in diameter. An increasing proportion of global steel is made using electric arc furnaces, and the electric arc furnace itself is getting more efficient, making more steel per tonne of electrode. An estimate based on USGS data indicates that graphite electrode consumption was 197,000 tonnes in 2005.

Electrolytic aluminium smelting also uses graphitic carbon electrodes. On a much smaller scale, synthetic graphite electrodes are used in electrical discharge machining (EDM), commonly to make injection molds for plastics.

Powder and Scrap

The powder is made by heating powdered petroleum coke above the temperature of graphitization, sometimes with minor modifications. The graphite scrap comes from pieces of unusable electrode material (in the manufacturing stage or after use) and lathe turnings, usually after crushing and sizing. Most synthetic graphite powder goes to carbon raising in steel (competing with natural graphite), with some used in batteries and brake linings. According to the USGS, US synthetic graphite powder and scrap production was 95,000 tonnes in 2001 (latest data).

Neutron Moderator

Special grades of synthetic graphite also find use as a matrix and neutron moderator within nuclear reactors. Its low neutron cross-section also recommends it for use in proposed fusion reactors. Care must be taken that reactor-grade graphite is free of neutron absorbing materials such as boron, widely used as the seed electrode in commercial graphite deposition systems—this caused the failure of the Germans' World War II graphite-based nuclear reactors. Since they could not isolate the difficulty they were forced to use far more expensive heavy water moderators. Graphite used for nuclear reactors is often referred to as nuclear graphite.

Other Uses

Graphite (carbon) fiber and carbon nanotubes are also used in carbon fiber reinforced plastics, and in heat-resistant composites such as reinforced carbon-carbon (RCC). Commercial structures made from carbon fiber graphite composites include fishing rods, golf club shafts, bicycle frames, sports car body panels, the fuselage of the Boeing 787 Dreamliner and pool cue sticks and have been successfully employed in reinforced concrete, The mechanical properties of carbon fiber graphite-reinforced plastic composites and grey cast iron are strongly influenced by the role of graphite in these materials. In this context, the term "(100%) graphite" is often loosely used to refer to a pure mixture of carbon reinforcement and resin, while the term "composite" is used for composite materials with additional ingredients.

Modern smokeless powder is coated in graphite to prevent the buildup of static charge.

Graphite has been used in at least three radar absorbent materials. It was mixed with rubber in Sumpf and Schornsteinfeger, which were used on U-boat snorkels to reduce their radar cross section. It was also used in tiles on early F-117 Nighthawk (1983)s.

Graphite Mining, Beneficiation, and Milling

Graphite is mined by both open pit and underground methods. Graphite usually needs beneficiation. This may be carried out by hand-picking the pieces of gangue (rock) and hand-screening the product or by crushing the rock and floating out the graphite. Beneficiation by flotation encounters the difficulty that graphite is very soft and "marks" (coats) the particles of gangue. This makes the "marked" gangue particles float off with the graphite, yielding impure concentrate. There are two ways of obtaining a commercial concentrate or product: repeated regrinding and floating (up to seven times) to purify the concentrate, or by acid leaching (dissolving) the gangue with hydrofluoric acid (for a silicate gangue) or hydrochloric acid (for a carbonate gangue).

In milling, the incoming graphite products and concentrates can be ground before being classified (sized or screened), with the coarser flake size fractions (below 8 mesh, 8–20 mesh, 20–50 mesh) carefully preserved, and then the carbon contents are determined. Some standard blends can be prepared from the different fractions, each with a certain flake size distribution and carbon content. Custom blends can also be made for individual customers who want a certain flake size distribution and carbon content. If flake size is unimportant, the concentrate can be ground more freely. Typical end products include a fine powder for use as a slurry in oil drilling and coatings for foundry molds, carbon raiser in the steel industry (Synthetic graphite powder and powdered petroleum coke can also be used as carbon raiser). Environmental impacts from graphite mills consist of air pollution including fine particulate exposure of workers and also soil contamination from powder spillages leading to heavy metal contamination of soil.

Occupational Safety

People can be exposed to graphite in the workplace by breathing it in, skin contact, and eye contact.

United States

The Occupational Safety and Health Administration (OSHA) has set the legal limit (permissible exposure limit) for graphite exposure in the workplace as a time weighted average (TWA) of 15 million particles per cubic foot (1.5 mg/m^3) over an 8-hour workday. The National Institute for Occupational Safety and Health (NIOSH) has set a recommended exposure limit (REL) of TWA 2.5 mg/m^3 respirable dust over an 8-hour workday. At levels of 1250 mg/m^3, graphite is immediately dangerous to life and health.

Graphite Recycling

The most common way of recycling graphite occurs when synthetic graphite electrodes are either manufactured and pieces are cut off or lathe turnings are discarded, or the electrode (or other) are used all the way down to the electrode holder. A new electrode replaces the old one, but a sizeable piece of the old electrode remains. This is crushed and sized, and the resulting graphite powder is mostly used to raise the carbon content of molten steel. Graphite-containing refractories are sometimes also recycled, but often not because of their graphite: the largest-volume items, such as carbon-magnesite bricks that contain only 15–25% graphite, usually contain too little graphite. However, some recycled carbon-magnesite brick is used as the basis for furnace-repair materials, and also crushed carbon-magnesite brick is used in slag conditioners. While crucibles have a high graphite content, the volume of crucibles used and then recycled is very small.

A high-quality flake graphite product that closely resembles natural flake graphite can be made from steelmaking kish. Kish is a large-volume near-molten waste skimmed from the molten iron feed to a basic oxygen furnace, and consists of a mix of graphite (precipitated out of the supersaturated iron), lime-rich slag, and some iron. The iron is recycled on site, leaving a mixture of graphite and slag. The best recovery process uses hydraulic classification (which utilizes a flow of water to separate minerals by specific gravity: graphite is light and settles nearly last) to get a 70% graphite rough concentrate. Leaching this concentrate with hydrochloric acid gives a 95% graphite product with a flake size ranging from 10 mesh down.

Isotopes of Carbon

Carbon (C) has 15 known isotopes, from ^{8}C to ^{22}C, of which ^{12}C and ^{13}C are stable. The longest-lived radioisotope is ^{14}C, with a half-life of 5,700 years. This is also the only carbon radioisotope found in nature—trace quantities are formed cosmogenically by the reaction $^{14}N + {}^1n \rightarrow {}^{14}C + {}^1H$. The most stable artificial radioisotope is ^{11}C, which has a half-life of 20.334 minutes. All other radioisotopes have half-lives under 20 seconds, most less than 200 milliseconds. The least stable isotope is ^8C, with a half-life of 2.0×10^{-21} s. Averaging over natural abundances, the relative atomic mass for carbon is 12.0107(8).

Carbon-11

Carbon-11 or ^{11}C is a radioactive isotope of carbon that decays to boron-11. This decay mainly occurs due to positron emission; however, around 0.19–0.23% of the time, it is a result of electron capture. It has a half-life of 20.334 minutes.

$$^{11}C \rightarrow {}^{11}B + e^+ + ve + 0.96 \text{ MeV}$$

$$^{11}C + e^- \rightarrow {}^{11}B + ve + 1.98 \text{ MeV It}$$

is produced from nitrogen in a cyclotron by the reaction

$$^{14}N + p \rightarrow {}^{11}C + {}^4He$$

Carbon-11 is commonly used as a radioisotope for the radioactive labeling of molecules in positron emission tomography. Among the many molecules used in this context is the radioligand [11C] DASB.

Natural Isotopes

There are three naturally occurring isotopes of carbon: 12, 13, and 14. ^{12}C and ^{13}C are stable, occurring in a natural proportion of approximately 99:1. ^{14}C is produced by thermal neutrons from cosmic radiation in the upper atmosphere, and is transported down to earth to be absorbed by living biological material. Isotopically, ^{14}C constitutes a negligible part; but, since it is radioactive with a half-life of 5,700 years, it is radiometrically detectable. Since dead tissue doesn't absorb ^{14}C, the amount of ^{14}C is one of the methods used within the field of archeology for radiometric dating of biological material.

Paleoclimate

^{12}C and ^{13}C are measured as the isotope ratio $\delta^{13}C$ in benthic foraminifera and used as a proxy for nutrient cycling and the temperature dependent air-sea exchange of CO_2 (ventilation) (Lynch-Stieglitz et al., 1995). Plants find it easier to use the lighter isotopes (^{12}C) when they convert sunlight and carbon dioxide into food. So, for example, large blooms of plankton (free-floating organisms) absorb large amounts of ^{12}C from the oceans. Originally, the ^{12}C was mostly incorporated into the seawater from the atmosphere. If the oceans that the plankton live in are stratified (meaning that there are layers of warm water near the top, and colder water deeper down), then the surface water does not mix very much with the deeper waters, so that when the plankton dies, it sinks and takes away ^{12}C from the surface, leaving the surface layers relatively rich in ^{13}C. Where cold waters well up from the depths (such as in the North Atlantic), the water carries ^{12}C back up with it. So, when the ocean was less stratified than today, there was much more ^{12}C in the skeletons of surface-dwelling species. Other indicators of past climate include the presence of tropical species, coral growths rings, etc.

Tracing Food Sources and Diets

The quantities of the different isotopes can be measured by mass spectrometry and compared to a standard; the result (e.g. the delta of the ^{13}C = δ^{13}C) is expressed as parts per thousand (‰).

$$\delta^{13}c = \left(\frac{\left(\frac{^{13}c}{^{12}c} \right)_{sample}}{\left(\frac{^{13}c}{^{12}c} \right)_{standard}} - 1 \right) \times 1000^{o}/_{oo}$$

Stable carbon isotopes in carbon dioxide are utilized differentially by plants during photosynthesis. Grasses in temperate climates (barley, rice, wheat, rye and oats, plus sunflower, potato, tomatoes, peanuts, cotton, sugar beet, and most trees and their nuts/fruits, roses and Kentucky bluegrass) follow a C3 photosynthetic pathway that will yield δ^{13}C values averaging about −26.5‰. Grasses in hot arid climates (maize in particular, but also millet, sorghum, sugar cane and crabgrass) follow a C4 photosynthetic pathway that produces δ^{13}C values averaging about −12.5‰.

It follows that eating these different plants will affect the δ^{13}C values in the consumer's body tissues. If an animal (or human) eats only C3 plants, their δ^{13}C values will be from −18.5 to −22.0‰ in their bone collagen and −14.5‰ in the hydroxylapatite of their teeth and bones.

In contrast, C4 feeders will have bone collagen with a value of −7.5‰ and hydroxylapatite value of −0.5‰.

In actual case studies, millet and maize eaters can easily be distinguished from rice and wheat eaters. Studying how these dietary preferences are distributed geographically through time can illuminate migration paths of people and dispersal paths of different agricultural crops. However, human groups have often mixed C3 and C4 plants (northern Chinese historically subsisted on wheat and millet), or mixed plant and animal groups together (for example, southeastern Chinese subsisting on rice and fish).

Table

nuclide symbol	Z(p)	N(n)	isotopic mass (u)	half-life	decay mode (s)	daughter isotope (s)[n 1]	nuclear spin	representative isotopic composition (mole fraction)	range of natural variation (mole fraction)
^{8}C	6	2	8.037675(25)	2.0(4) × 10^{-21} s [230(50) keV]	2p	6 Be [n 2]	0+		
^{9}C	6	3	9.0310367(23)	126.5(9) ms	β^{+} (60%)	9 B [n 3]	(3/2−)		
					β^{+}, p (23%)	8 Be [n 4]			
					β^{+}, α (17%)	5 Li [n 5]			
^{10}C	6	4	10.0168532(4)	19.290(12) s	β^{+}	10 B	0+		

^{11}C[n 6]	6	5	11.0114336(10)	20.334(24) min	β+ (99.79%)	^{11}B	3/2−		
					EC (.21%)	^{11}B			
^{12}C	6	6	12 exactly[n 7]	**Stable**			0+	0.9893(8)	0.98853–0.99037
^{13}C[n 8]	6	7	13.0033548378(10)	**Stable**			1/2−	0.0107(8)	0.00963–0.01147
^{14}C[n 9]	6	8	14.003241989(4)	5,730 years	β−	^{14}N	0+	Trace[n 10]	$<10^{-12}$
^{15}C	6	9	15.0105993(9)	2.449(5) s	β−	^{15}N	1/2+		
^{16}C	6	10	16.014701(4)	0.747(8) s	β−, n (97.9%)	^{15}N	0+		
					β− (2.1%)	^{16}N			
^{17}C	6	11	17.022586(19)	193(5) ms	β− (71.59%)	^{17}N	(3/2+)		
					β−, n (28.41%)	^{16}N			
^{18}C	6	12	18.02676(3)	92(2) ms	β− (68.5%)	^{18}N	0+		
					β−, n (31.5%)	^{17}N			
^{19}C[n 11]	6	13	19.03481(11)	46.2(23) ms	β−, n (47.0%)	^{18}N	(1/2+)		
					β− (46.0%)	^{19}N			
					β−, 2n (7%)	^{17}N			
^{20}C	6	14	20.04032(26)	16(3) ms [14(+6-5) ms]	β−, n (72.0%)	^{19}N	0+		
					β− (28.0%)	^{20}N			
^{21}C	6	15	21.04934(54)#	<30 ns	n	^{20}C	(1/2+)#		
^{22}C[n 12]	6	16	22.05720(97)#	6.2(13) ms [6.1(+14-12) ms]	β−	^{22}N	0+		

1. Bold for stable isotopes

2. Subsequently decays by double proton emission to ^4He for a net reaction of $^8C \rightarrow {}^4He + 4^1H$

3. Immediately decays by proton emission to ^8Be, which immediately decays to two ^4He atoms for a net reaction of $^9C \rightarrow 2^4He + ^1H + e^+$

4. Immediately decays into two ^4He atoms for a net reaction of $^9C \rightarrow 2^4He + ^1H + e^+$

5. Immediately decays by proton emission to ^4He for a net reaction of $^9C \rightarrow 2^4He + ^1H + e^+$

6. Used for labeling molecules in PET scans

7. The unified atomic mass unit is defined as 1/12 the mass of an unbound atom of carbon-12 at ground state

8. Ratio of ^{12}C to ^{13}C used to measure biological productivity in ancient times and differing types of photosynthesis

9. Has an important use in radiodating

10. Primarily cosmogenic, produced by neutrons striking atoms of ^{14}N ($^{14}N + ^1n \rightarrow ^{14}C + ^1H$)

11. Has 1 halo neutron

12. Has 2 halo neutrons

Carbon Cycle

The carbon cycle is the biogeochemical cycle by which carbon is exchanged among the biosphere, pedosphere, geosphere, hydrosphere, and atmosphere of the Earth. Along with the nitrogen cycle and the water cycle, the carbon cycle comprises a sequence of events that are key to making the Earth capable of sustaining life; it describes the movement of carbon as it is recycled and reused throughout the biosphere, including carbon sinks.

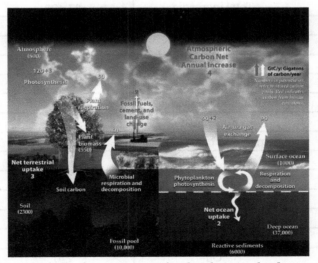

This diagram of the fast carbon cycle shows the movement of carbon between land, atmosphere, and oceans in billions of tons per year. Yellow numbers are natural fluxes, red are human contributions, white indicate stored carbon. Note this diagram does not account for volcanic and tectonic activity, which also sequesters and releases carbon.

The *global carbon budget* is the balance of the exchanges (incomes and losses) of carbon between the carbon reservoirs or between one specific loop (e.g., atmosphere <-> biosphere) of the carbon cycle. An examination of the carbon budget of a pool or reservoir can provide information about whether the pool or reservoir is functioning as a source or sink for carbon dioxide. The carbon cycle was initially discovered by Joseph Priestley and Antoine Lavoisier, and popularized by Humphry Davy.

Global Climate

Carbon-based molecules are crucial for life on Earth, because it is the main component of biological compounds. Carbon is also a major component of many minerals. Carbon also exists in various forms in the atmosphere. Carbon dioxide (CO_2) is partly responsible for the greenhouse effect and is the most important human-contributed greenhouse gas.

In the past two centuries, human activities have seriously altered the global carbon cycle, most significantly in the atmosphere. Although carbon dioxide levels have changed naturally over the past several thousand years, human emissions of carbon dioxide into the atmosphere exceed natural fluctuations. Changes in the amount of atmospheric CO_2 are considerably altering weather patterns and indirectly influencing oceanic chemistry. Current carbon dioxide levels in the atmosphere exceed measurements from the last 420,000 years and levels are rising faster than ever recorded, making it of critical importance to better understand how the carbon cycle works and what its effects are on the global climate.

Main Components

Carbon pools in the major reservoirs on earth.	
Pool	**Quantity (gigatons)**
Atmosphere	720
Oceans (total)	38,400
Total inorganic	37,400
Total organic	1,000
Surface layer	670
Deep layer	36,730
Lithosphere	
Sedimentary carbonates	> 60,000,000
Kerogens	15,000,000
Terrestrial biosphere (total)	2,000
Living biomass	600 - 1,000
Dead biomass	1,200
Aquatic biosphere	1 - 2
Fossil fuels (total)	4,130
Coal	3,510
Oil	230
Gas	140
Other (peat)	250

The global carbon cycle is now usually divided into the following major reservoirs of carbon interconnected by pathways of exchange:

- The atmosphere

- The terrestrial biosphere

- The oceans, including dissolved inorganic carbon and living and non-living marine biota

- The sediments, including fossil fuels, fresh water systems and non-living organic material.

- The Earth's interior, carbon from the Earth's mantle and crust. These carbon stores interact with the other components through geological processes

The carbon exchanges between reservoirs occur as the result of various chemical, physical, geological, and biological processes. The ocean contains the largest active pool of carbon near the surface of the Earth. The natural flows of carbon between the atmosphere, ocean, terrestrial ecosystems, and sediments is fairly balanced, so that carbon levels would be roughly stable without human influence.

Atmosphere

Carbon in the Earth's atmosphere exists in two main forms: carbon dioxide and methane. Both of these gases absorb and retain heat in the atmosphere and are partially responsible for the greenhouse effect. Methane produces a large greenhouse effect per volume as compared to carbon dioxide, but it exists in much lower concentrations and is more short-lived than carbon dioxide, making carbon dioxide the more important greenhouse gas of the two.

Epiphytes on electric wires. This kind of plant takes both CO_2 and water from the atmosphere for living and growing.

Carbon dioxide leaves the atmosphere through photosynthesis, thus entering the terrestrial and oceanic biospheres. Carbon dioxide also dissolves directly from the atmosphere into bodies of water (oceans, lakes, etc.), as well as dissolving in precipitation as raindrops fall through the atmosphere. When dissolved in water, carbon dioxide reacts with water molecules and forms carbonic acid, which contributes to ocean acidity. It can then be absorbed by rocks through weathering. It also can acidify other surfaces it touches or be washed into the ocean.

Human activities over the past two centuries have significantly increased the amount of carbon in the atmosphere, mainly in the form of carbon dioxide, both by modifying ecosystems' ability to extract carbon dioxide from the atmosphere and by emitting it directly, e.g., by burning fossil fuels and manufacturing concrete.

Terrestrial Biosphere

The terrestrial biosphere includes the organic carbon in all land-living organisms, both alive and dead, as well as carbon stored in soils. About 500 gigatons of carbon are stored above ground in plants and other living organisms, while soil holds approximately 1,500 gigatons of carbon. Most carbon in the terrestrial biosphere is organic carbon, while about a third of soil carbon is stored in inorganic forms, such as calcium carbonate. Organic carbon is a major component of all organisms living on earth. Autotrophs extract it from the air in the form of carbon dioxide, converting it into organic carbon, while heterotrophs receive carbon by consuming other organisms.

A portable soil respiration system measuring soil CO_2 flux

Because carbon uptake in the terrestrial biosphere is dependent on biotic factors, it follows a diurnal and seasonal cycle. In CO_2 measurements, this feature is apparent in the Keeling curve. It is strongest in the northern hemisphere, because this hemisphere has more land mass than the southern hemisphere and thus more room for ecosystems to absorb and emit carbon.

Carbon leaves the terrestrial biosphere in several ways and on different time scales. The combustion or respiration of organic carbon releases it rapidly into the atmosphere. It can also be exported into the oceans through rivers or remain sequestered in soils in the form of inert carbon. Carbon stored in soil can remain there for up to thousands of years before being washed into rivers by erosion or released into the atmosphere through soil respiration. Between 1989 and 2008 soil respiration increased by about 0.1% per year. In 2008, the global total of CO_2 released from the soil reached roughly 98 billion tonnes, about 10 times more carbon than humans are now putting into the atmosphere each year by burning fossil fuel. There are a few plausible explanations for this trend, but the most likely explanation is that increasing temperatures have increased rates of decomposition of soil organic matter, which has increased the flow of CO_2. The length of carbon sequestering in soil is dependent on local climatic conditions and thus changes in the course of climate change. From pre-industrial era to 2010, the terrestrial biosphere represented a net source of atmospheric CO_2 prior to 1940, switching subsequently to a net sink.

Oceans

Oceans contain the greatest quantity of actively cycled carbon in this world and are second only to the lithosphere in the amount of carbon they store. The oceans' surface layer holds large amounts of dissolved inorganic carbon that is exchanged rapidly with the atmosphere. The deep layer's concentration of dissolved inorganic carbon (DIC) is about 15% higher than that of the surface layer. DIC is stored in the deep layer for much longer periods of time. Thermohaline circulation exchanges carbon between these two layers.

Carbon enters the ocean mainly through the dissolution of atmospheric carbon dioxide, which is converted into carbonate. It can also enter the oceans through rivers as dissolved organic carbon. It is converted by organisms into organic carbon through photosynthesis and can either be exchanged throughout the food chain or precipitated into the ocean's deeper, more carbon rich layers as dead soft tissue or in shells as calcium carbonate. It circulates in this layer for long periods of time before either being deposited as sediment or, eventually, returned to the surface waters through thermohaline circulation.

Oceanic absorption of CO_2 is one of the most important forms of carbon sequestering limiting the human-caused rise of carbon dioxide in the atmosphere. However, this process is limited by a number of factors. Because the rate of CO_2 dissolution in the ocean is dependent on the weathering of rocks and this process takes place slower than current rates of human greenhouse gas emissions, ocean CO_2 uptake will decrease in the future. CO_2 absorption also makes water more acidic, which affects ocean biosystems. The projected rate of increasing oceanic acidity could slow the biological precipitation of calcium carbonates, thus decreasing the ocean's capacity to absorb carbon dioxide.

Geological Carbon Cycle

The geologic component of the carbon cycle operates slowly in comparison to the other parts of the global carbon cycle. It is one of the most important determinants of the amount of carbon in the atmosphere, and thus of global temperatures.

Most of the earth's carbon is stored inertly in the earth's lithosphere. Much of the carbon stored in the earth's mantle was stored there when the earth formed. Some of it was deposited in the form of organic carbon from the biosphere. Of the carbon stored in the geosphere, about 80% is limestone and its derivatives, which form from the sedimentation of calcium carbonate stored in the shells of marine organisms. The remaining 20% is stored as kerogens formed through the sedimentation and burial of terrestrial organisms under high heat and pressure. Organic carbon stored in the geosphere can remain there for millions of years.

Carbon can leave the geosphere in several ways. Carbon dioxide is released during the metamorphosis of carbonate rocks when they are subducted into the earth's mantle. This carbon dioxide can be released into the atmosphere and ocean through volcanoes and hotspots. It can also be removed by humans through the direct extraction of kerogens in the form of fossil fuels. After extraction, fossil fuels are burned to release energy, thus emitting the carbon they store into the atmosphere.

Human Influence

Since the industrial revolution, human activity has modified the carbon cycle by changing its component's functions and directly adding carbon to the atmosphere.

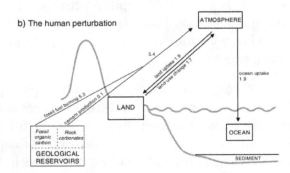

Human activity since the industrial era has changed the balance in the natural carbon cycle. Units are in gigatons.

The largest human impact on the carbon cycle is through direct emissions from burning fossil fuels, which transfers carbon from the geosphere into the atmosphere. The rest of this increase is caused mostly by changes in land-use, particularly deforestation.

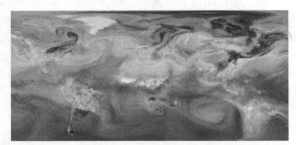

CO_2 in Earth's atmosphere if *half* of global-warming emissions are *not* absorbed.
(NASA computer simulation).

Another direct human impact on the carbon cycle is the chemical process of calcination of limestone for clinker production, which releases CO_2. Clinker is an industrial precursor of cement.

Humans also influence the carbon cycle indirectly by changing the terrestrial and oceanic biosphere. Over the past several centuries, direct and indirect human-caused land use and land cover change (LUCC) has led to the loss of biodiversity, which lowers ecosystems' resilience to environmental stresses and decreases their ability to remove carbon from the atmosphere. More directly, it often leads to the release of carbon from terrestrial ecosystems into the atmosphere. Deforestation for agricultural purposes removes forests, which hold large amounts of carbon, and replaces them, generally with agricultural or urban areas. Both of these replacement land cover types store comparatively small amounts of carbon, so that the net product of the process is that more carbon stays in the atmosphere.

Other human-caused changes to the environment change ecosystems' productivity and their ability to remove carbon from the atmosphere. Air pollution, for example, damages plants and soils, while many agricultural and land use practices lead to higher erosion rates, washing carbon out of soils and decreasing plant productivity.

Humans also affect the oceanic carbon cycle. Current trends in climate change lead to higher ocean temperatures, thus modifying ecosystems. Also, acid rain and polluted runoff from agriculture and industry change the ocean's chemical composition. Such changes can have dramatic effects on highly sensitive ecosystems such as coral reefs, thus limiting the ocean's ability to absorb carbon from the atmosphere on a regional scale and reducing oceanic biodiversity globally.

Arctic methane emissions indirectly caused by anthropogenic global warming also affect the carbon cycle, and contribute to further warming in what is known as climate change feedback.

On 12 November 2015, NASA scientists reported that human-made carbon dioxide (CO_2) continues to increase above levels not seen in hundreds of thousands of years: currently, about half of the carbon dioxide released from the burning of fossil fuels remains in the atmosphere and is not absorbed by vegetation and the oceans.

Alpha and Beta Carbon

The alpha carbon (Cα) in organic molecules refers to the first carbon atom that attaches to a functional group, such as a carbonyl. The second carbon atom is called the beta carbon (Cβ), and the system continues naming in alphabetical order with Greek letters.

Alpha and beta carbons in a skeletal formula. The carbonyl has two β-hydrogens and five α-hydrogens

The nomenclature can also be applied to the hydrogen atoms attached to the carbon atoms. A hydrogen atom attached to an alpha carbon atom is called an alpha-hydrogen atom, a hydrogen atom on the beta-carbon atom is a beta hydrogen atom, and so on.

This naming standard may not be in compliance with IUPAC nomenclature, which encourages that carbons be identified by number, not by Greek letter, but it nonetheless remains very popular, in particular because it is useful in identifying the relative location of carbon atoms to other functional groups.

Organic molecules with more than one functional group can be a source of confusion. Generally the functional group responsible for the name or type of the molecule is the 'reference' group for purposes of carbon-atom naming. For example, the molecules nitrostyrene and phenethylamine are very similar; the former can even be reduced into the latter. However, nitrostyrene's α-carbon atom is adjacent to the styrene group; in phenethylamine this same carbon atom is the β-carbon atom, as phenethylamine (being an amine rather than a styrene) counts its atoms from the opposite "end" of the molecule.

Examples

Skeletal formula of butyric acid with the alpha, beta, and gamma carbons marked

Proteins and Amino Acids

Alpha-carbon (α-carbon) is also a term that applies to proteins and amino acids. It is the backbone carbon before the carbonyl carbon. Therefore, reading along the backbone of a typical protein would give a sequence of $-[N—C\alpha—\text{carbonyl C}]_n-$ etc. (when reading in the N to C direction). The α-carbon is where the different substituents attach to each different amino acid. That is, the groups hanging off the chain at the α-carbon are what give amino acids their diversity. These groups give the α-carbon its stereogenic properties for every amino acid except for glycine. Therefore, the α-carbon is a stereocenter for every amino acid except glycine. Glycine also does not have a β-carbon, while every other amino acid does.

The α-carbon of an amino acid is significant in protein folding. When describing a protein, which is a chain of amino acids, one often approximates the location of each amino acid as the location of its α-carbon. In general, α-carbons of adjacent amino acids in a protein are about 3.8 ångströms (380 picometers) apart.

Enols and Enolates

The α-carbon is important for enol- and enolate-based carbonyl chemistry as well. Chemical transformations affected by the conversion to either an enolate or an enol, in general, lead to the α-carbon acting as a nucleophile, becoming, for example, alkylated in the presence of primary haloalkane. An exception is in reaction with silyl- chlorides, -bromides, and -iodides, where the oxygen acts as the nucleophile to produce silyl enol ether.

References

- Lide, D. R., ed. (2005). CRC Handbook of Chemistry and Physics (86th ed.). Boca Raton (FL): CRC Press. ISBN 0-8493-0486-5.

- Ebbesen, T. W., ed. (1997). Carbon nanotubes—preparation and properties. Boca Raton, Florida: CRC Press. ISBN 0-8493-9602-6.

- Dresselhaus, M. S.; Dresselhaus, G.; Avouris, Ph., eds. (2001). "Carbon nanotubes: synthesis, structures, properties and applications". Topics in Applied Physics. Berlin: Springer. 80. ISBN 3-540-41086-4.

- Heimann, Robert Bertram; Evsyukov, Sergey E. & Kavan, Ladislav (28 February 1999). Carbyne and carbynoid structures. Springer. pp. 1–. ISBN 978-0-7923-5323-2. Retrieved 2011-06-06.

- Jenkins, Edgar (1973). The polymorphism of elements and compounds. Taylor & Francis. p. 30. ISBN 0-423-87500-0. Retrieved 2011-05-01.

- William F McDonough The composition of the Earth in Earthquake Thermodynamics and Phase Transformation in the Earth's Interior. 2000. ISBN 978-0126851854.

- Stefanenko, R. (1983). Coal Mining Technology: Theory and Practice. Society for Mining Metallurgy. ISBN 0-89520-404-5.

- Ostlie, D.A. & Carroll, B.W. (2007). An Introduction to Modern Stellar Astrophysics. Addison Wesley, San Francisco. ISBN 0-8053-0348-0.

- Pauling, L. (1960). The Nature of the Chemical Bond (3rd ed.). Ithaca, NY: Cornell University Press. p. 93. ISBN 0-8014-0333-2.

- Hershey, J. W. (1940). The Book Of Diamonds: Their Curious Lore, Properties, Tests And Synthetic Manufacture. Kessinger Pub Co. p. 28. ISBN 1-4179-7715-9.

8

Allied Fields of Organic and Inorganic Chemistry

Chemical compounds contain within themselves at least one bond. The bond is shared between a carbon atom and a metal. This is known as organometallic chemistry. Some of the allied fields of organic and inorganic chemistry explained in this chapter are bioorganometallic chemistry, medicinal chemistry, biochemistry and bioinorganic chemistry. This text provides a plethora of interdisciplinary topics for better comprehension of organic and inorganic chemistry.

Organometallic Chemistry

Organometallic chemistry is the study of chemical compounds containing at least one bond between a carbon atom of an organic compound and a metal, including alkaline, alkaline earth, transition metal, and other cases. Moreover, some related compounds such as transition metal hydrides and metal phosphine complexes are often included in discussions of organometallic compounds. The field of organometallic chemistry combines aspects of traditional inorganic and organic chemistry.

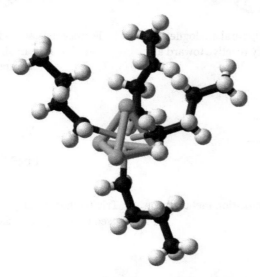

n-Butyllithium, an organometallic compound. Four lithium atoms (in purple) form a tetrahedron, with four butyl groups attached to the faces (carbon is black, hydrogen is white).

Organometallic compounds are widely used both stoichiometrically in research and industrial chemical reactions, as well as in the role of catalysts to increase the rates of such reactions (e.g., as in uses of homogeneous catalysis), where target molecules include polymers, pharmaceuticals, and many other types of practical products.

Organometallic Compounds

Organometallic compounds are distinguished by the prefix "organo-" e.g. organopalladium compounds. Examples of such organometallic compounds include all Gilman reagents, which contain lithium and copper. Tetracarbonyl nickel, and ferrocene are examples of organometallic compounds containing transition metals. Other examples include organomagnesium compounds like iodo(methyl)magnesium MeMgI, dimethylmagnesium (Me_2Mg), and all Grignard reagents; organolithium compounds such as n-butyllithium (n-BuLi), organozinc compounds such as diethylzinc (Et_2Zn) and chloro(ethoxycarbonylmethyl)zinc ($ClZnCH_2C(=O)OEt$); and organocopper compounds such as lithium dimethylcuprate ($Li^+[CuMe_2]^-$).

The term "metalorganics" usually refers to metal-containing compounds lacking direct metal-carbon bonds but which contain organic ligands. Metal beta-diketonates, alkoxides, and dialkylamides are representative members of this class.

In addition to the traditional metals, lanthanides, actinides, and semimetals, elements such as boron, silicon, arsenic, and selenium are considered to form organometallic compounds, e.g. organoborane compounds such as triethylborane (Et_3B).

- Representative Organometallic Compounds

Cobaltocene is a structural analogue of ferrocene, but is highly reactive toward air.

Ferrocene is an archetypal organoiron complex. It is an air-stable, sublimable compound.

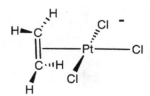

Zeise's salt is an example of a transition metal alkene complex.

Tris(triphenylphosphine)rhodium carbonyl hydride is used in the commercial production of many aldehyde-based fragrances.

Trimethylaluminium is an organometallic compound with a bridging methyl group. It is used in the industrial production of some alcohols

Coordination Compounds with Organic Ligands

Many complexes feature coordination bonds between a metal and organic ligands. The organic ligands often bind the metal through a heteroatom such as oxygen or nitrogen, in which case such compounds are considered coordination compounds. However, if any of the ligands form a direct M-C bond, then complex is usually considered to be organometallic, e.g., $[(C_6H_6)Ru(H_2O)_3]^{2+}$. Furthermore, many lipophilic compounds such as metal acetylacetonates and metal alkoxides are called "metalorganics."

Many organic coordination compounds occur naturally. For example, hemoglobin and myoglobin contain an iron center coordinated to the nitrogen atoms of a porphyrin ring; magnesium is the center of a chlorin ring in chlorophyll. The field of such inorganic compounds is known as bioinorganic chemistry. In contrast to these coordination compounds, methylcobalamin (a form of Vitamin B_{12}), with a cobalt-methyl bond, is a true organometallic complex, one of the few known in biology. This subset of complexes are often discussed within the subfield of bioorganometallic chemistry. Illustrative of the many functions of the B_{12}-dependent enzymes, the MTR enzyme catalyzes the transfer of a methyl group from a nitrogen on N5-methyl-tetrahydrofolate to the sulfur of homocysteine to produce methionine.

The status of compounds in which the canonical anion has a delocalized structure in which the negative charge is shared with an atom more electronegative than carbon, as in enolates, may vary with the nature of the anionic moiety, the metal ion, and possibly the medium; in the absence of direct structural evidence for a carbon–metal bond, such compounds are not considered to be organometallic.

Structure and Properties

The metal-carbon bond in organometallic compounds are generally highly covalent. For highly electropositive elements, such as lithium and sodium, the carbon ligand exhibits carbanionic character, but free carbon-based anions are extremely rare, an example being cyanide.

Concepts and Techniques

As in other areas of chemistry, electron counting is useful for organizing organometallic chemistry. The 18-electron rule is helpful in predicting the stabilities of metal carbonyls and related compounds. Most organometallic compounds do not however follow the 18e rule. Chemical bonding and reactivity in organometallic compounds is often discussed from the perspective of the isolobal principle.

As well as X-ray diffraction, NMR and infrared spectroscopy are common techniques used to determine structure. The dynamic properties of organometallic compounds is often probed with variable-temperature NMR and chemical kinetics.

Organometallic compounds undergo several important reactions:

- oxidative addition and reductive elimination

- transmetalation

- carbometalation

- hydrometalation

- electron transfer

- beta-hydride elimination

- organometallic substitution reaction

- carbon-hydrogen bond activation

- cyclometalation

- migratory insertion

- nucleophilic abstraction

History

Early developments in organometallic chemistry include Louis Claude Cadet's synthesis of methyl arsenic compounds related to cacodyl, William Christopher Zeise's platinum-ethylene complex, Edward Frankland's discovery of dimethyl zinc, Ludwig Mond's discovery of $Ni(CO)_4$, and Victor Grignard's organomagnesium compounds. The abundant and diverse products from coal and petroleum led to Ziegler-Natta, Fischer-Tropsch, hydroformylation catalysis which employ CO, H_2, and alkenes as feedstocks and ligands.

Recognition of organometallic chemistry as a distinct subfield culminated in the Nobel Prizes to Ernst Fischer and Geoffrey Wilkinson for work on metallocenes. In 2005, Yves Chauvin, Robert H. Grubbs and Richard R. Schrock shared the Nobel Prize for metal-catalyzed olefin metathesis.

Organometallic Chemistry Timeline

- 1760 Louis Claude Cadet de Gassicourt investigates inks based on cobalt salts and isolates cacodyl from cobalt mineral containing arsenic

- 1827 William Christopher Zeise produces Zeise's salt; the first platinum/olefin complex

- 1848 Edward Frankland discovers diethylzinc

- 1863 Charles Friedel and James Crafts prepare organochlorosilanes

- 1890 Ludwig Mond discovers nickel carbonyl

- 1899 Introduction of Grignard reaction

- 1899 John Ulric Nef discovers alkynation using sodium acetylides.

- 1900 Paul Sabatier works on hydrogenation organic compounds with metal catalysts. Hydrogenation of fats kicks off advances in food industry.

- 1909 Paul Ehrlich introduces Salvarsan for the treatment of syphilis, an early arsenic based organometallic compound

- 1912 Nobel Prize Victor Grignard and Paul Sabatier

- 1930 Henry Gilman works on lithium cuprates.

- 1951 Walter Hieber was awarded the Alfred Stock prize for his work with metal carbonyl chemistry.

- 1951 Ferrocene is discovered

- 1963 Nobel prize for Karl Ziegler and Giulio Natta on Ziegler-Natta catalyst

- 1965 Discovery of cyclobutadieneiron tricarbonyl

- 1968 Heck reaction

- 1973 Nobel prize Geoffrey Wilkinson and Ernst Otto Fischer on sandwich compounds

- 1981 Nobel prize Roald Hoffmann and Kenichi Fukui for creation of the Woodward-Hoffman Rules

- 2001 Nobel prize W. S. Knowles, R. Noyori and Karl Barry Sharpless for asymmetric hydrogenation

- 2005 Nobel prize Yves Chauvin, Robert Grubbs, and Richard Schrock on metal-catalyzed alkene metathesis

- 2010 Nobel prize Richard F. Heck, Ei-ichi Negishi, Akira Suzuki for palladium catalyzed cross coupling reactions

Scope

Subspecialty areas of organometallic chemistry include:

- Period 2 elements: organolithium chemistry, organoberyllium chemistry, organoborane chemistry,

- Period 3 elements: organomagnesium chemistry, organoaluminum chemistry, organosilicon chemistry

- Period 4 elements: organotitanium chemistry, organochromium chemistry, organomanganese chemistry organoiron chemistry, organocobalt chemistry organonickel chemistry, organocopper chemistry, organozinc chemistry, organogallium chemistry, organogermanium chemistry

- Period 5 elements: organoruthenium chemistry, organopalladium chemistry, organosilver chemistry, organocadmium chemistry, organoindium chemistry, organotin chemistry

- Period 6 elements: organolanthanide chemistry, organoosmium chemistry, organoiridium chemistry, organoplatinum chemistry, organogold chemistry, organomercury chemistry, organothallium chemistry, organolead chemistry

- Period 7 elements: organouranium chemistry

The following is a presentation of elements of the periodic table with known compounds of carbon with other elements.

CH																	He
CLi	CBe											CB	CC	CN	CO	CF	Ne
CNa	CMg											CAl	CSi	CP	CS	CCl	CAr
CK	CCa	CSc	CTi	CV	CCr	CMn	CFe	CCo	CNi	CCu	CZn	CGa	CGe	CAs	CSe	CBr	CKr
CRb	CSr	CY	CZr	CNb	CMo	CTc	CRu	CRh	CPd	CAg	CCd	CIn	CSn	CSb	CTe	CI	CXe
CCs	CBa		CHf	CTa	CW	CRe	COs	CIr	CPt	CAu	CHg	CTl	CPb	CBi	CPo	CAt	Rn
Fr	CRa		Rf	Db	CSg	Bh	Hs	Mt	Ds	Rg	Cn	Nh	Fl	Mc	Lv	Ts	Og

↓

CLa	CCe	CPr	CNd	CPm	CSm	CEu	CGd	CTb	CDy	CHo	CEr	CTm	CYb	CLu
Ac	CTh	CPa	CU	CNp	CPu	CAm	CCm	CBk	CCf	CEs	Fm	Md	No	Lr

Chemical bonds to carbon	
Core organic chemistry	Many uses in chemistry
Academic research, but no widespread use	Bond unknown

Industrial Applications

Organometallic compounds find wide use in commercial reactions, both as homogeneous catalysis and as stoichiometric reagents For instance, organolithium, organomagnesium, and organoaluminium compounds, examples of which are highly basic and highly reducing, are useful stoichiometrically, but also catalyze many polymerization reactions.

Almost all processes involving carbon monoxide rely on catalysts, notable examples being described as carbonylations. The production of acetic acid from methanol and carbon monoxide is catalyzed via metal carbonyl complexes in the Monsanto process and Cativa process. Most synthetic aldehydes are produced via hydroformylation. The bulk of the synthetic alcohols, at least those larger than ethanol, are produced by hydrogenation of hydroformylation-derived aldehydes. Similarly, the Wacker process is used in the oxidation of ethylene to acetaldehyde.

Almost all industrial processes involving alkene-derived polymers rely on organometallic catalysts. The world's polyethylene and polypropylene are produced via both heterogeneously via Ziegler-Natta catalysis and homogeneously, e.g., via constrained geometry catalysts.

Most processes involving hydrogen rely on metal-based catalysts. Whereas bulk hydrogenations, e.g. margarine production, rely on heterogeneous catalysts, For the production of fine chemicals, such hydrogenations rely on soluble organometallic complexes or involve organometallic intermediates. Organometallic complexes allow these hydrogenations to be effected asymmetrically.

A constrained geometry organotitanium complex is a precatalyst for olefin polymerization.

Many semiconductors are produced from trimethylgallium, trimethylindium, trimethylaluminium, and trimethylantimony. These volatile compounds are decomposed along with ammonia, arsine, phosphine and related hydrides on a heated substrate via metalorganic vapor phase epitaxy (MOVPE) process in the production of light-emitting diodes (LEDs).

Environmental Concerns

Natural and contaminant organometallic compounds are found in the environment. Some that are remnants of human use, such as organolead and organomercury compounds, are toxicity hazards. Tetraethyllead was prepared for use as a gasoline additive but has fallen into disuse because of lead's toxicity. Its replacements are other organometallic compounds, such as ferrocene and methylcyclopentadienyl manganese tricarbonyl (MMT). The organoarsenic compound roxarsone is a controversial animal feed additive. In 2006, approximately one million kilograms of t were produced in the U.S alone.

Roxarsone is an organoarsenic compound used as an animal feed.

Bioorganometallic Chemistry

Bioorganometallic chemistry is the study of biologically active molecules that contain carbon directly bonded to metals or metalloids. This area straddles the fields of organometallic chemistry, biochemistry, and medicine. It is subset of bioinorganic chemistry. Naturally occurring bioorganometallics include enzymes and sensor proteins. Also within this realm is the development of new drugs and imaging agents as well as the principles relevant to the toxicology or organometallic compounds.

Naturally Occurring Bioorganometallic Species

Vitamin B_{12} is the preeminent bioorganometallic species. B_{12} is shorthand for a collection of related enzymes which effect numerous reactions involving the making and breaking of C-C and C-H bonds.

Several bioorganometallic enzymes carry out reactions involving carbon monoxide. Carbon monoxide dehydrogenase (CODH) catalyzes the water gas shift reaction which provides CO for the biosynthesis of acetylcoenzyme A. The latter step is effected by the Ni-Fe enzyme acetylCoA synthase. ACS". CODH and ACS often occur together in a tetrameric complex, the CO being transported via a tunnel and the methyl group being provided by methyl cobalamin.

Hydrogenases are bioorganometallic in the sense that their active sites feature Fe-CO functionalities, although the CO ligands are only spectators. The Fe-only hydrogenases have a $Fe_2(\mu\text{-}SR)_2(\mu\text{-}CO)(CO)_2(CN)_2$ active site connected to a 4Fe4S cluster cluster via a bridging thiolate. The active site of the [NiFe]-hydrogenases are described as $(NC)_2(OC)Fe(\mu\text{-}SR)_2Ni(SR)_2$ (where SR is cysteinyl). The "FeS-free" hydrogenases have an undetermined active site containing an $Fe(CO)_2$ center.

Methanogenesis, the biosynthesis of methane, entails as its final step, the scission of a nickel-methyl bond in cofactor F430.

Sensor Proteins

Some [NiFe]-containing proteins are known to sense H_2 and thus regulate transcription.

Copper-containing proteins are known to sense ethylene, which is known to be a hormone relevant to the ripening of fruit. This example illustrates the essential role of organometallic chemistry in nature, as few molecules outside of low-valent transition metal complexes reversibly bind alkenes. Cyclopropenes inhibit ripening by binding to the copper(I) center.

Carbon monoxide occurs naturally and is a transcription factor via its complex with a sensor protein based on ferrous porphyrins.

Organometallics in Medicine

Several organometallic compounds are under study as candidates for diverse therapies. Much work was instigated by the success of cisplatin in chemotherapy. $(C_5H_5)_2TiCl_2$ displays anti-cancer activity; Titanocene Y {bis-[(p-methoxybenzyl)-cyclopentadienyl] titanium(IV) dichloride} is a current anticancer drug candidate. Arene- and cyclopentadienyl complexes are kinetically inert platforms for the design of new radiopharmaceuticals.

Bioorganometallics and Toxicology

Within the realm of bioorganometallic chemistry is the study of the fates of synthetic organometallic compounds. Tetraethyllead has received considerable attention in this regard as has its successors such as methylcyclopentadienyl manganese tricarbonyl. Methylmercury is a particularly infamous case, this cation is produced by the action of vitamin B_{12}-related enzymes on mercury.

Medicinal Chemistry

Medicinal chemistry and pharmaceutical chemistry are disciplines at the intersection of chemistry, especially synthetic organic chemistry, and pharmacology and various other biological specialties, where they are involved with design, chemical synthesis and development for market of pharmaceutical agents, or bio-active molecules (drugs).

Compounds used as medicines are most often organic compounds, which are often divided into the broad classes of small organic molecules (e.g., atorvastatin, fluticasone, clopidogrel) and "biologics" (infliximab, erythropoietin, insulin glargine), the latter of which are most often medicinal preparations of proteins (natural and recombinant antibodies, hormones, etc.). Inorganic and organometallic compounds are also useful as drugs (e.g., lithium and platinum-based agents such as lithium carbonate and cis-platin as well as gallium).

In particular, medicinal chemistry in its most common practice —focusing on small organic molecules—encompasses synthetic organic chemistry and aspects of natural products and computational chemistry in close combination with chemical biology, enzymology and structural biology, together aiming at the discovery and development of new therapeutic agents. Practically speaking, it involves chemical aspects of identification, and then systematic, thorough synthetic alteration of new chemical entities to make them suitable for therapeutic use. It includes synthetic and computational aspects of the study of existing drugs and agents in development in relation to their bioactivities (biological activities and properties), i.e., understanding their structure-activity relationships (SAR). Pharmaceutical chemistry is focused on quality aspects of medicines and aims to assure fitness for purpose of medicinal products.

At the biological interface, medicinal chemistry combines to form a set of highly interdisciplinary sciences, setting its organic, physical, and computational emphases alongside biological areas such as biochemistry, molecular biology, pharmacognosy and pharmacology, toxicology and veterinary and human medicine; these, with project management, statistics, and pharmaceutical business practices, systematically oversee altering identified chemical agents such that after pharmaceutical formulation, they are safe and efficacious, and therefore suitable for use in treatment of disease.

Medicinal Chemistry in The Path of Drug Discovery

Discovery

Discovery is the identification of novel active chemical compounds, often called "hits", which are typically found by assay of compounds for a desired biological activity. Initial hits can come from repurposing existing agents toward a new pathologic processes, and from observations of biologic effects of new or existing natural products from bacteria, fungi, plants, etc. In addition, hits also routinely originate from structural observations of small molecule "fragments" bound to therapeutic targets (enzymes, receptors, etc.), where the fragments serve as starting points to develop more chemically complex forms by synthesis. Finally, hits also regularly originate from *en-masse* testing of chemical compounds against biological targets, where the compounds may be from novel synthetic chemical libraries known to have particular properties (kinase inhibitory activity, diversity or drug-likeness, etc.), or from historic chemical compound collections or libraries created through combinatorial chemistry. While a number of approaches toward the identification and development of hits exist, the most successful techniques are based on chemical and biological intuition developed in team environments through years of rigorous practice aimed solely at discovering new therapeutic agents.

Hit to Lead and Lead Optimization

Further chemistry and analysis is necessary, first to identify the "triage" compounds that do not provide series displaying suitable SAR and chemical characteristics associated with long-term potential for development, then to improve remaining hit series with regard to the desired primary activity, as well as secondary activities and physiochemical properties such that the agent will be useful when administered in real patients. In this regard, chemical modifications can improve the recognition and binding geometries (pharmacophores) of the candidate compounds, and so their affinities for their targets, as well as improving the physicochemical properties of the molecule that underlie necessary pharmacokinetic/pharmacodynamic (PK/PD), and toxicologic profiles (stability toward metabolic degradation, lack of geno-, hepatic, and cardiac toxicities, etc.) such that the chemical compound or biologic is suitable for introduction into animal and human studies.

Process Chemistry and Development

The final synthetic chemistry stages involve the production of a lead compound in suitable quantity and quality to allow large scale animal testing, and then human clinical trials. This involves the optimization of the synthetic route for bulk industrial production, and discovery of the most suitable drug formulation. The former of these is still the bailiwick of medicinal chemistry, the latter brings in the specialization of formulation science (with its components of physical and polymer chemistry and materials science). The synthetic chemistry specialization in medicinal chemistry aimed at adaptation and optimization of the synthetic route for industrial scale syntheses of hundreds of kilograms or more is termed process synthesis, and involves thorough knowledge of acceptable synthetic practice in the context of large scale reactions (reaction thermodynamics, economics, safety, etc.). Critical at this stage is the transition to more stringent GMP requirements for material sourcing, handling, and chemistry.

Training in Medicinal Chemistry

Medicinal chemistry is by nature an interdisciplinary science, and practitioners have a strong background in organic chemistry, which must eventually be coupled with a broad understanding of biological concepts related to cellular drug targets. Scientists in medicinal chemistry work are principally industrial scientists, working as part of an interdisciplinary team that uses their chemistry abilities, especially, their synthetic abilities, to use chemical principles to design effective therapeutic agents. The length of training is intense with practitioners often required to attain a 4-year bachelor's followed by a 4-6 year Ph.D. in organic chemistry. Most training regimens include a postdoctoral fellowship period of 2 or more years after receiving a Ph.D. in chemistry making the length of training ranging from *10-12* years of college education. However, employment opportunities at the Master's level also exist in the pharmaceutical industry, and at that and the Ph.D. level there are further opportunities for employment in academia and government. Many medicinal chemists, particularly in academia and research, also earn a Pharm.D (doctor of pharmacy). Some of these PharmD/PhD researchers are RPh's (Registered Pharmacists).

Graduate level programs in medicinal chemistry can be found in traditional medicinal chemistry or pharmaceutical sciences departments, both of which are traditionally associated with schools of pharmacy, and in some chemistry departments. However, the majority of working medicinal chemists have graduate degrees (MS, but especially Ph.D.) in organic chemistry, rather than medicinal chemistry, and the preponderance of positions are in discovery, where the net is necessarily cast widest, and most broad synthetic activity occurs.

In discovery of small molecule therapeutics, an emphasis on training that provides for breadth of synthetic experience and "pace" of bench operations is clearly present (e.g., for individuals with pure synthetic organic and natural products synthesis in Ph.D. and post-doctoral positions, ibid.). In the medicinal chemistry specialty areas associated with the design and synthesis of chemical libraries or the execution of process chemistry aimed at viable commercial syntheses (areas generally with fewer opportunities), training paths are often much more varied (e.g., including focused training in physical organic chemistry, library-related syntheses, etc.).

As such, most entry-level workers in medicinal chemistry, especially in the U.S., do not have formal training in medicinal chemistry but receive the necessary medicinal chemistry and pharmacologic background after employment—at entry into their work in a pharmaceutical company, where the company provides its particular understanding or model of "medichem" training through active involvement in practical synthesis on therapeutic projects. (The same is somewhat true of computational medicinal chemistry specialties, but not to the same degree as in synthetic areas.)

Biochemistry

Biochemistry, sometimes called biological chemistry, is the study of chemical processes within and relating to living organisms. By controlling information flow through biochemical signaling and the flow of chemical energy through metabolism, biochemical processes give rise to the complexity of life. Over the last decades of the 20th century, biochemistry has become so successful at explaining living processes that now almost all areas of the life sciences from botany to medicine to genetics are engaged in biochemical research. Today, the main focus of pure biochemistry is on understanding how biological molecules give rise to the processes that occur within living cells, which in turn relates greatly to the study and understanding of tissues, organs, and whole organisms—that is, all of biology.

Biochemistry is closely related to molecular biology, the study of the molecular mechanisms by which genetic information encoded in DNA is able to result in the processes of life. Depending on the exact definition of the terms used, molecular biology can be thought of as a branch of biochemistry, or biochemistry as a tool with which to investigate and study molecular biology.

Much of biochemistry deals with the structures, functions and interactions of biological macro-molecules, such as proteins, nucleic acids, carbohydrates and lipids, which provide the structure of cells and perform many of the functions associated with life. The chemistry of the cell also depends on the reactions of smaller molecules and ions. These can be inorganic, for example water and metal ions, or organic, for example the amino acids, which are used to synthesize proteins. The mechanisms by which cells harness energy from their environment via chemical reactions are known as metabolism. The findings of biochemistry are applied primarily in medicine, nutrition, and agriculture. In medicine, biochemists investigate the causes and cures of diseases. In nutrition, they study how to maintain health and study the effects of nutritional deficiencies. In agriculture, biochemists investigate soil and fertilizers, and try to discover ways to improve crop cultivation, crop storage and pest control.

History

At its broadest definition, biochemistry can be seen as a study of the components and composition of living things and how they come together to become life, and the history of biochemistry may therefore go back as far as the ancient Greeks. However, biochemistry as a specific scientific discipline has its beginning sometime in the 19th century, or a little earlier, depending on which aspect of biochemistry is being focused on. Some argued that the beginning of biochemistry may have been the discovery of the first enzyme, diastase (today called amylase), in 1833 by Anselme Payen, while others considered Eduard Buchner's first demonstration of a complex biochemical process alcoholic fermentation in cell-free extracts in 1897 to be the birth of biochemistry. Some might also point as its beginning to the influential 1842 work by Justus von Liebig, *Animal chemistry, or, Organic chemistry in its applications to physiology and pathology*, which presented a chemical theory of metabolism, or even earlier to the 18th century studies on fermentation and respiration by Antoine Lavoisier. Many other pioneers in the field who helped to uncover the layers of complexity of biochemistry have been proclaimed founders of modern biochemistry, for example Emil Fischer for his work on the chemistry of proteins, and F. Gowland Hopkins on enzymes and the dynamic nature of biochemistry.

Gerty Cori and Carl Cori jointly won the Nobel Prize in 1947 for their discovery of the Cori cycle at RPMI.

The term "biochemistry" itself is derived from a combination of biology and chemistry. In 1877, Felix Hoppe-Seyler used the term (*biochemie* in German) as a synonym for physiological chemistry in the foreword to the first issue of *Zeitschrift für Physiologische Chemie* (Journal of Physiological

Chemistry) where he argued for the setting up of institutes dedicated to this field of study. The German chemist Carl Neuberg however is often cited to have coined the word in 1903, while some credited it to Franz Hofmeister.

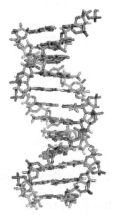

DNA structure (1D65)

It was once generally believed that life and its materials had some essential property or substance (often referred to as the "vital principle") distinct from any found in non-living matter, and it was thought that only living beings could produce the molecules of life. Then, in 1828, Friedrich Wöhler published a paper on the synthesis of urea, proving that organic compounds can be created artificially. Since then, biochemistry has advanced, especially since the mid-20th century, with the development of new techniques such as chromatography, X-ray diffraction, dual polarisation interferometry, NMR spectroscopy, radioisotopic labeling, electron microscopy, and molecular dynamics simulations. These techniques allowed for the discovery and detailed analysis of many molecules and metabolic pathways of the cell, such as glycolysis and the Krebs cycle (citric acid cycle).

Another significant historic event in biochemistry is the discovery of the gene and its role in the transfer of information in the cell. This part of biochemistry is often called molecular biology. In the 1950s, James D. Watson, Francis Crick, Rosalind Franklin, and Maurice Wilkins were instrumental in solving DNA structure and suggesting its relationship with genetic transfer of information. In 1958, George Beadle and Edward Tatum received the Nobel Prize for work in fungi showing that one gene produces one enzyme. In 1988, Colin Pitchfork was the first person convicted of murder with DNA evidence, which led to the growth of forensic science. More recently, Andrew Z. Fire and Craig C. Mello received the 2006 Nobel Prize for discovering the role of RNA interference (RNAi), in the silencing of gene expression.

Starting Materials: The Chemical Elements of Life

Around two dozen of the 92 naturally occurring chemical elements are essential to various kinds of biological life. Most rare elements on Earth are not needed by life (exceptions being selenium and iodine), while a few common ones (aluminum and titanium) are not used. Most organisms share element needs, but there are a few differences between plants and animals. For example, ocean algae use bromine, but land plants and animals seem to need none. All animals require sodium, but some plants do not. Plants need boron and silicon, but animals may not (or may need ultra-small amounts).

The main elements that compose the human body are shown from most abundant (by mass) to least abundant.

Just six elements—carbon, hydrogen, nitrogen, oxygen, calcium, and phosphorus—make up almost 99% of the mass of living cells, including those in the human body. In addition to the six major elements that compose most of the human body, humans require smaller amounts of possibly 18 more.

Biomolecules

The four main classes of molecules in biochemistry (often called biomolecules) are carbohydrates, lipids, proteins, and nucleic acids. Many biological molecules are polymers: in this terminology, *monomers* are relatively small micromolecules that are linked together to create large macromolecules known as *polymers*. When monomers are linked together to synthesize a biological polymer, they undergo a process called dehydration synthesis. Different macromolecules can assemble in larger complexes, often needed for biological activity.

Carbohydrates

Glucose, a monosaccharide

A molecule of sucrose (glucose + fructose), a disaccharide

Amylose, a polysaccharide made up of several thousand glucose units

The function of carbohydrates includes energy storage and providing structure. Sugars are carbohydrates, but not all carbohydrates are sugars. There are more carbohydrates on Earth than any other known type of biomolecule; they are used to store energy and genetic information, as well as play important roles in cell to cell interactions and communications.

The simplest type of carbohydrate is a monosaccharide, which among other properties contains carbon, hydrogen, and oxygen, mostly in a ratio of 1:2:1 (generalized formula $C_nH_{2n}O_n$, where n is at least 3). Glucose ($C_6H_{12}O_6$) is one of the most important carbohydrates, others include fructose ($C_6H_{12}O_6$), the sugar commonly associated with the sweet taste of fruits, and deoxyribose ($C_5H_{10}O_4$).

A monosaccharide can switch from the acyclic (open-chain) form to a cyclic form, through a nucleophilic addition reaction between the carbonyl group and one of the hydroxyls of the same molecule. The reaction creates a ring of carbon atoms closed by one bridging oxygen atom. The resulting molecule has an hemiacetal or hemiketal group, depending on whether the linear form was an aldose or a ketose. The reaction is easily reversed, yielding the original open-chain form.

Conversion between the furanose, acyclic, and pyranose forms of D-glucose.

In these cyclic forms, the ring usually has 5 or 6 atoms. These forms are called furanoses and pyranoses, respectively — by analogy with furan and pyran, the simplest compounds with the same carbon-oxygen ring (although they lack the double bonds of these two molecules). For example, the aldohexose glucose may form a hemiacetal linkage between the hydroxyl on carbon 1 and the oxygen on carbon 4, yielding a molecule with a 5-membered ring, called glucofuranose. The same reaction can take place between carbons 1 and 5 to form a molecule with a 6-membered ring, called glucopyranose. Cyclic forms with a 7-atom ring (the same of oxepane), rarely encountered, are called heptoses.

When two monosaccharides undergo dehydration synthesis whereby a molecule of water is released, as two hydrogen atoms and one oxygen atom are lost from the two monosaccharides. The new molecule, consisting of two monosaccharides, is called a *disaccharide* and is conjoined together by a glycosidic or ether bond. The reverse reaction can also occur, using a molecule of water to split up a disaccharide and break the glycosidic bond; this is termed *hydrolysis*. The most well-known disaccharide is sucrose, ordinary sugar (in scientific contexts, called *table sugar* or *cane sugar* to differentiate it from other sugars). Sucrose consists of a glucose molecule and a fructose molecule joined together. Another important disaccharide is lactose, consisting of a glucose molecule and a galactose molecule. As most humans age, the production of lactase, the enzyme that hydrolyzes lactose back into glucose and galactose, typically decreases. This results in lactase deficiency, also called *lactose intolerance.*

When a few (around three to six) monosaccharides are joined, it is called an *oligosaccharide* (*oligo-* meaning "few"). These molecules tend to be used as markers and signals, as well as having some other uses. Many monosaccharides joined together make a polysaccharide. They can be joined together in one long linear chain, or they may be branched. Two of the most common poly-saccharides are cellulose and glycogen, both consisting of repeating glucose monomers. Examples are *Cellulose* which is an important structural component of plant's cell walls, and *glycogen*, used as a form of energy storage in animals.

Sugar can be characterized by having reducing or non-reducing ends. A reducing end of a carbohydrate is a carbon atom that can be in equilibrium with the open-chain aldehyde (aldose) or keto form (ketose). If the joining of monomers takes place at such a carbon atom, the free hydroxy group of the pyranose or furanose form is exchanged with an OH-side-chain of another sugar, yielding a full acetal. This prevents opening of the chain to the aldehyde or keto form and renders the modified residue non-reducing. Lactose contains a reducing end at its glucose moiety, whereas the galactose moiety form a full acetal with the C4-OH group of glucose. Saccharose does not have a reducing end because of full acetal formation between the aldehyde carbon of glucose (C1) and the keto carbon of fructose (C2).

Lipids

Structures of some common lipids. At the top are cholesterol and oleic acid. The middle structure is a triglyceride composed of oleoyl, stearoyl, and palmitoyl chains attached to a glycerol backbone. At the bottom is the common phospholipid, phosphatidylcholine.

Lipids comprises a diverse range of molecules and to some extent is a catchall for relatively water-insoluble or nonpolar compounds of biological origin, including waxes, fatty acids, fatty-acid derived phospholipids, sphingolipids, glycolipids, and terpenoids (e.g., retinoids and steroids). Some lipids are linear aliphatic molecules, while others have ring structures. Some are aromatic, while others are not. Some are flexible, while others are rigid.

Lipids are usually made from one molecule of glycerol combined with other molecules. In triglycerides, the main group of bulk lipids, there is one molecule of glycerol and three fatty acids. Fatty acids are considered the monomer in that case, and may be saturated (no double bonds in the carbon chain) or unsaturated (one or more double bonds in the carbon chain).

Most lipids have some polar character in addition to being largely nonpolar. In general, the bulk of their structure is nonpolar or hydrophobic ("water-fearing"), meaning that it does not interact well with polar solvents like water. Another part of their structure is polar or hydrophilic ("water-loving") and will tend to associate with polar solvents like water. This makes them amphiphilic molecules (having both hydrophobic and hydrophilic portions). In the case of cholesterol, the polar group is a mere -OH (hydroxyl or alcohol). In the case of phospholipids, the polar groups are considerably larger and more polar, as described below.

Lipids are an integral part of our daily diet. Most oils and milk products that we use for cooking and eating like butter, cheese, ghee etc., are composed of fats. Vegetable oils are rich in various polyunsaturated fatty acids (PUFA). Lipid-containing foods undergo digestion within the body and are broken into fatty acids and glycerol, which are the final degradation products of fats and lipids. Lipids, especially phospholipids, are also used in various pharmaceutical products, either as co-solubilisers (e.g., in parenteral infusions) or else as drug carrier components (e.g., in a liposome or transfersome).

Proteins

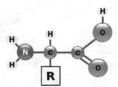

The general structure of an α-amino acid, with the amino group on the left and the carboxyl group on the right.

Proteins are very large molecules – macro-biopolymers – made from monomers called amino acids. An amino acid consists of a carbon atom bound to four groups. One is an amino group, $-NH_2$, and one is a carboxylic acid group, $-COOH$ (although these exist as $-NH_3^+$ and $-COO^-$ under physiologic conditions). The third is a simple hydrogen atom. The fourth is commonly denoted "$-R$" and is different for each amino acid. There are 20 standard amino acids, each containing a carboxyl group, an amino group, and a side-chain (known as an "R" group). The "R" group is what makes each amino acid different, and the properties of the side-chains greatly influence the overall three-dimensional conformation of a protein. Some amino acids have functions by themselves or in a modified form; for instance, glutamate functions as an important neurotransmitter. Amino acids can be joined via a peptide bond. In this dehydration synthesis, a water molecule is removed and the peptide bond connects the nitrogen of one amino acid's amino group to the carbon of the other's carboxylic acid group. The resulting molecule is called a *dipeptide*, and short stretches of amino acids (usually, fewer than thirty) are called *peptides* or polypeptides. Longer stretches merit the title *proteins*. As an example, the important blood serum protein albumin contains 585 amino acid residues.

Generic amino acids (1) in neutral form, (2) as they exist physiologically, and (3) joined together as a dipeptide.

Some proteins perform largely structural roles. For instance, movements of the proteins actin and myosin ultimately are responsible for the contraction of skeletal muscle. One property many proteins have is that they specifically bind to a certain molecule or class of molecules—they may be *extremely* selective in what they bind. Antibodies are an example of proteins that attach to one specific type of molecule. In fact, the enzyme-linked immunosorbent assay (ELISA), which uses antibodies, is one of the most sensitive tests modern medicine uses to detect various biomolecules. Probably the most important proteins, however, are the enzymes. Virtually every reaction in a living cell requires an enzyme to lower the activation energy of the reaction. These molecules recognize specific reactant molecules called *substrates*; they then catalyze the reaction between them. By lowering the activation energy, the enzyme speeds up that reaction by a rate of 10^{11} or more; a reaction that would normally take over 3,000 years to complete spontaneously might take less than a second with an enzyme. The enzyme itself is not used up in the process, and is free to catalyze the same reaction with a new set of substrates. Using various modifiers, the activity of the enzyme can be regulated, enabling control of the biochemistry of the cell as a whole.

A schematic of hemoglobin. The red and blue ribbons represent the protein globin; the green structures are the heme groups.

The structure of proteins is traditionally described in a hierarchy of four levels. The primary structure of a protein simply consists of its linear sequence of amino acids; for instance, "alanine-glycine-tryptophan-serine-glutamate-asparagine-glycine-lysine-...". Secondary structure is concerned with local morphology (morphology being the study of structure). Some combinations of amino acids will tend to curl up in a coil called an α-helix or into a sheet called a β-sheet; some α-helixes can be seen in the hemoglobin schematic above. Tertiary structure is the entire three-dimensional shape of the protein. This shape is determined by the sequence of amino acids. In fact, a single change can change the entire structure. The alpha chain of hemoglobin contains 146 amino acid residues; substitution of the glutamate residue at position 6 with a valine residue changes the behavior of hemoglobin so much that it results in sickle-cell disease. Finally, quaternary structure is concerned with the structure of a protein with multiple peptide subunits, like hemoglobin with its four subunits. Not all proteins have more than one subunit.

Ingested proteins are usually broken up into single amino acids or dipeptides in the small intestine, and then absorbed. They can then be joined to make new proteins. Intermediate products of glycolysis, the citric acid cycle, and the pentose phosphate pathway can be used to make all twenty amino acids, and most bacteria and plants possess all the necessary enzymes to synthesize them. Humans and other mammals, however, can synthesize only half of them. They cannot synthesize isoleucine, leucine, lysine, methionine, phenylalanine, threonine, tryptophan, and valine. These are the essential amino acids, since it is essential to ingest them. Mammals do possess the enzymes to synthesize alanine, asparagine, aspartate, cysteine, glutamate, glutamine, glycine, proline, serine, and tyrosine, the nonessential amino acids. While they can synthesize arginine and histidine, they cannot produce it in sufficient amounts for young, growing animals, and so these are often considered essential amino acids.

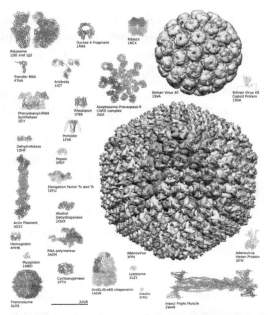

Examples of protein structures from the Protein Data Bank

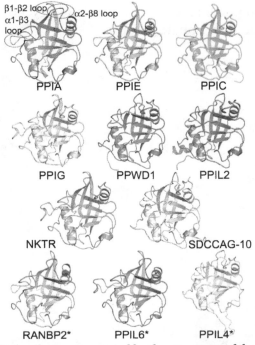

Members of a protein family, as represented by the structures of the isomerase domains.

If the amino group is removed from an amino acid, it leaves behind a carbon skeleton called an α-keto acid. Enzymes called transaminases can easily transfer the amino group from one amino acid (making it an α-keto acid) to another α-keto acid (making it an amino acid). This is important in the biosynthesis of amino acids, as for many of the pathways, intermediates from other biochemical pathways are converted to the α-keto acid skeleton, and then an amino group is added, often via transamination. The amino acids may then be linked together to make a protein.

A similar process is used to break down proteins. It is first hydrolyzed into its component amino acids. Free ammonia (NH_3), existing as the ammonium ion (NH_4^+) in blood, is toxic to life forms. A suitable method for excreting it must therefore exist. Different tactics have evolved in different animals, depending on the animals' needs. Unicellular organisms, of course, simply release the ammonia into the environment. Likewise, bony fish can release the ammonia into the water where it is quickly diluted. In general, mammals convert the ammonia into urea, via the urea cycle.

In order to determine whether two proteins are related, or in other words to decide whether they are homologous or not, scientists use sequence-comparison methods. Methods like sequence alignments and structural alignments are powerful tools that help scientists identify homologies between related molecules. The relevance of finding homologies among proteins goes beyond forming an evolutionary pattern of protein families. By finding how similar two protein sequences are, we acquire knowledge about their structure and therefore their function.

Nucleic Acids

Nucleic acids, so called because of its prevalence in cellular nuclei, is the generic name of the family of biopolymers. They are complex, high-molecular-weight biochemical macromolecules that can convey genetic information in all living cells and viruses. The monomers are called nucleotides, and each consists of three components: a nitrogenous heterocyclic base (either a purine or a pyrimidine), a pentose sugar, and a phosphate group.

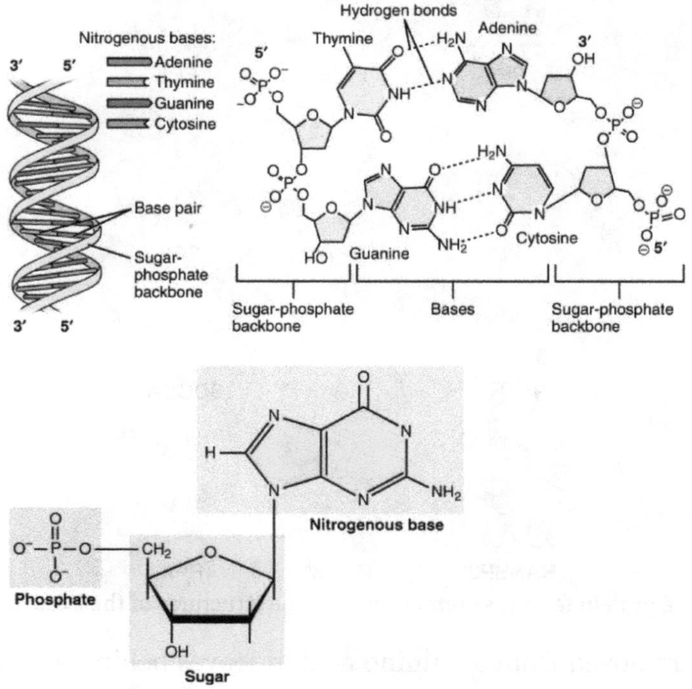

The structure of deoxyribonucleic acid (DNA), the picture shows the monomers being put together.

Structural elements of common nucleic acid constituents. Because they contain at least one phosphate group, the compounds marked *nucleoside monophosphate*, *nucleoside diphosphate* and *nucleoside triphosphate* are all nucleotides (not simply phosphate-lacking nucleosides).

The most common nucleic acids are deoxyribonucleic acid (DNA) and ribonucleic acid (RNA). The phosphate group and the sugar of each nucleotide bond with each other to form the backbone of the nucleic acid, while the sequence of nitrogenous bases stores the information. The most common nitrogenous bases are adenine, cytosine, guanine, thymine, and uracil. The nitrogenous bases of each strand of a nucleic acid will form hydrogen bonds with certain other nitrogenous bases in a complementary strand of nucleic acid (similar to a zipper). Adenine binds with thymine and uracil; Thymine binds only with adenine; and cytosine and guanine can bind only with one another.

Aside from the genetic material of the cell, nucleic acids often play a role as second messengers, as well as forming the base molecule for adenosine triphosphate (ATP), the primary energy-carrier molecule found in all living organisms. Also, the nitrogenous bases possible in the two nucleic acids are different: adenine, cytosine, and guanine occur in both RNA and DNA, while thymine occurs only in DNA and uracil occurs in RNA.

Metabolism

Carbohydrates as Energy Source

Glucose is the major energy source in most life forms. For instance, polysaccharides are broken down into their monomers (glycogen phosphorylase removes glucose residues from glycogen). Disaccharides like lactose or sucrose are cleaved into their two component monosaccharides.

Glycolysis (Anaerobic)

Glucose is mainly metabolized by a very important ten-step pathway called glycolysis, the net result of which is to break down one molecule of glucose into two molecules of pyruvate. This also produces a net two molecules of ATP, the energy currency of cells, along with two reducing equivalents of converting NAD^+ (nicotinamide adenine dinucleotide:oxidised form) to NADH (nicotinamide adenine dinucleotide:reduced form). This does not require oxygen; if no oxygen is available (or the cell cannot use oxygen), the NAD is restored by converting the pyruvate to lactate (lactic acid) (e.g., in humans) or to ethanol plus carbon dioxide (e.g., in yeast). Other monosaccharides like galactose and fructose can be converted into intermediates of the glycolytic pathway.

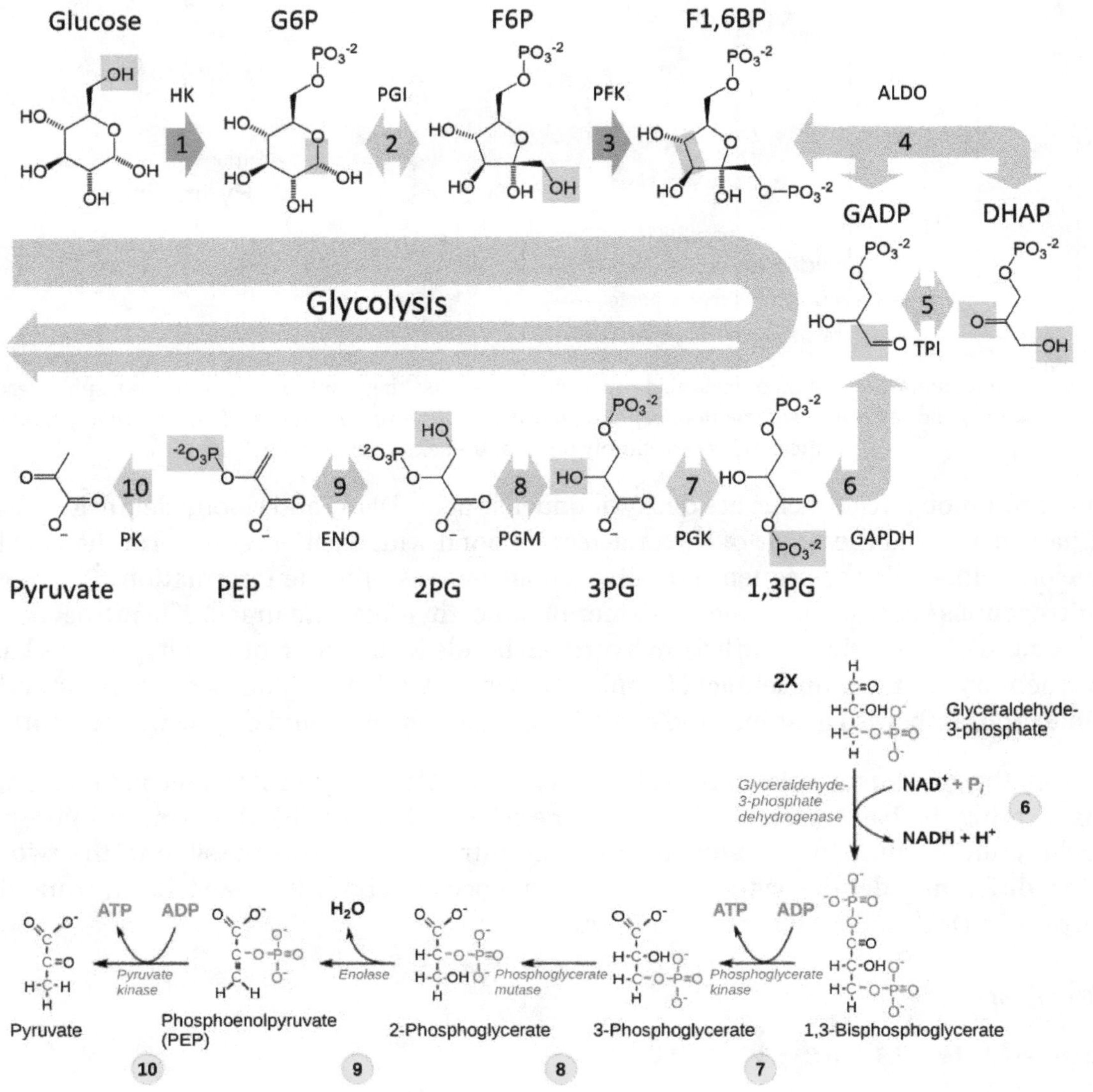

Glucose　　G6P　　F6P　　F1,6BP

HK　1　PGI　2　PFK　3　ALDO　4

Glycolysis

GADP　DHAP

TPI　5

Pyruvate　PEP　2PG　3PG　1,3PG

PK　10　ENO　9　PGM　8　PGK　7　GAPDH　6

2X　Glyceraldehyde-3-phosphate

Glyceraldehyde-3-phosphate dehydrogenase　　$NAD^+ + P_i$　6　　$NADH + H^+$

ATP　ADP　H_2O　ATP　ADP

Pyruvate kinase　Enolase　Phosphoglycerate mutase　Phosphoglycerate kinase

Pyruvate　Phosphoenolpyruvate (PEP)　2-Phosphoglycerate　3-Phosphoglycerate　1,3-Bisphosphoglycerate

10　9　8　7

The metabolic pathway of glycolysis converts glucose to pyruvate by via a series of intermediate metabolites. Each chemical modification (red box) is performed by a different enzyme. Steps 1 and 3 consume ATP (blue) and steps 7 and 10 produce ATP (yellow). Since steps 6-10 occur twice per glucose molecule, this leads to a net production of ATP.

Aerobic

In aerobic cells with sufficient oxygen, as in most human cells, the pyruvate is further metabolized. It is irreversibly converted to acetyl-CoA, giving off one carbon atom as the waste product carbon dioxide, generating another reducing equivalent as NADH. The two molecules acetyl-CoA (from one molecule of glucose) then enter the citric acid cycle, producing two more molecules of ATP, six more NADH molecules and two reduced (ubi)quinones (via $FADH_2$ as enzyme-bound cofactor), and releasing the remaining carbon atoms as carbon dioxide. The produced NADH and quinol molecules then feed into the enzyme complexes of the respiratory chain, an electron transport system transferring the electrons ultimately to oxygen and conserving the released energy in the form of a proton gradient over a membrane (inner mitochondrial membrane in eukaryotes). Thus, oxygen is reduced to water and the original electron acceptors NAD^+ and quinone are regenerated. This is why humans breathe in oxygen and breathe out carbon dioxide. The energy released from transferring the electrons from high-energy states in NADH and quinol is conserved first as proton gradient and converted to ATP via ATP synthase. This generates an additional *28* molecules of ATP (24 from the 8 NADH + 4 from the 2 quinols), totaling to 32 molecules of ATP conserved per degraded glucose (two from glycolysis + two from the citrate cycle). It is clear that using oxygen to completely oxidize glucose provides an organism with far more energy than any oxygen-independent metabolic feature, and this is thought to be the reason why complex life appeared only after Earth's atmosphere accumulated large amounts of oxygen.

Gluconeogenesis

In vertebrates, vigorously contracting skeletal muscles (during weightlifting or sprinting, for example) do not receive enough oxygen to meet the energy demand, and so they shift to anaerobic metabolism, converting glucose to lactate. The liver regenerates the glucose, using a process called gluconeogenesis. This process is not quite the opposite of glycolysis, and actually requires three times the amount of energy gained from glycolysis (six molecules of ATP are used, compared to the two gained in glycolysis). Analogous to the above reactions, the glucose produced can then undergo glycolysis in tissues that need energy, be stored as glycogen (or starch in plants), or be converted to other monosaccharides or joined into di- or oligosaccharides. The combined pathways of glycolysis during exercise, lactate's crossing via the bloodstream to the liver, subsequent gluconeogenesis and release of glucose into the bloodstream is called the Cori cycle.

Relationship to Other "Molecular-scale" Biological Sciences

Researchers in biochemistry use specific techniques native to biochemistry, but increasingly combine these with techniques and ideas developed in the fields of genetics, molecular biology and biophysics. There has never been a hard-line among these disciplines in terms of content and technique. Today, the terms *molecular biology* and *biochemistry* are nearly interchangeable. The following figure is a schematic that depicts one possible view of the relationship between the fields:

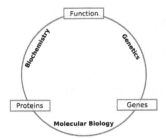

Schematic relationship between biochemistry, genetics, and molecular biology.

- *Biochemistry* is the study of the chemical substances and vital processes occurring in living organisms. Biochemists focus heavily on the role, function, and structure of biomolecules. The study of the chemistry behind biological processes and the synthesis of biologically active molecules are examples of biochemistry.

- *Genetics* is the study of the effect of genetic differences on organisms. Often this can be inferred by the absence of a normal component (e.g., one gene). The study of "mutants" – organisms with a changed gene that leads to the organism being different with respect to the so-called "wild type" or normal phenotype. Genetic interactions (epistasis) can often confound simple interpretations of such "knock-out" or "knock-in" studies.

- *Molecular biology* is the study of molecular underpinnings of the process of replication, transcription and translation of the genetic material. The central dogma of molecular biology where genetic material is transcribed into RNA and then translated into protein, despite being an oversimplified picture of molecular biology, still provides a good starting point for understanding the field. This picture, however, is undergoing revision in light of emerging novel roles for RNA.

- *Chemical biology* seeks to develop new tools based on small molecules that allow minimal perturbation of biological systems while providing detailed information about their function. Further, chemical biology employs biological systems to create non-natural hybrids between biomolecules and synthetic devices (for example emptied viral capsids that can deliver gene therapy or drug molecules).

Bioinorganic Chemistry

Bioinorganic chemistry is a field that examines the role of metals in biology. Bioinorganic chemistry includes the study of both natural phenomena such as the behavior of metalloproteins as well as artificially introduced metals, including those that are non-essential, in medicine and toxicology. Many biological processes such as respiration depend upon molecules that fall within the realm of inorganic chemistry. The discipline also includes the study of inorganic models or mimics that imitate the behaviour of metalloproteins.

As a mix of biochemistry and inorganic chemistry, bioinorganic chemistry is important in elucidating the implications of electron-transfer proteins, substrate bindings and activation, atom and group transfer chemistry as well as metal properties in biological chemistry.

Composition of Living Organisms

About 99% of mammals' mass are the elements carbon, nitrogen, calcium, sodium, chlorine, potassium, hydrogen, phosphorus, oxygen and sulfur. The organic compounds (proteins, lipids and carbohydrates) contain the majority of the carbon and nitrogen and most of the oxygen and hydrogen is present as water. The entire collection of metal-containing biomolecules in a cell is called the metallome.

History

Paul Ehrlich used organoarsenic ("arsenicals") for the treatment of syphilis, demonstrating the relevance of metals, or at least metalloids, to medicine, that blossomed with Rosenberg's discovery of the anti-cancer activity of cisplatin (cis-$PtCl_2(NH_3)_2$). The first protein ever crystallized was urease, later shown to contain nickel at its active site. Vitamin B_{12}, the cure for pernicious anemia was shown crystallographically by Dorothy Crowfoot Hodgkin to consist of a cobalt in a corrin macrocycle. The Watson-Crick structure for DNA demonstrated the key structural role played by phosphate-containing polymers.

Themes in Bioinorganic Chemistry

Several distinct systems are of identifiable in bioinorganic chemistry. Major areas include:

Metal Ion Transport and Storage

This topic covers a diverse collection of ion channels, ion pumps (e.g. NaKATPase), vacuoles, siderophores, and other proteins and small molecules which control the concentration of metal ions in the cells. One issue is that many metals that are metabolically required are not readily available owing to solubility or scarcity. Organisms have developed a number of strategies for collecting such elements and transporting them.

Enzymology

Many reactions in life sciences involve water and metal ions are often at the catalytic centers (active sites) for these enzymes, i.e. these are metalloproteins. Often the reacting water is a ligand. Examples of hydrolase enzymes are carbonic anhydrase, metallophosphatases, and metalloproteinases. Bioinorganic chemists seek to understand and replicate the functi on of these metalloproteins.

Metal-containing electron transfer proteins are also common. They can be organized into three major classes: iron-sulfur proteins (such as rubredoxins, ferredoxins, and Rieske proteins), blue copper proteins, and cytochromes. These electron transport proteins are complementary to the non-metal electron transporters nicotinamide adenine dinucleotide (NAD) and flavin adenine dinucleotide (FAD). The nitrogen cycle make extensive use of metals for the redox interconversions.

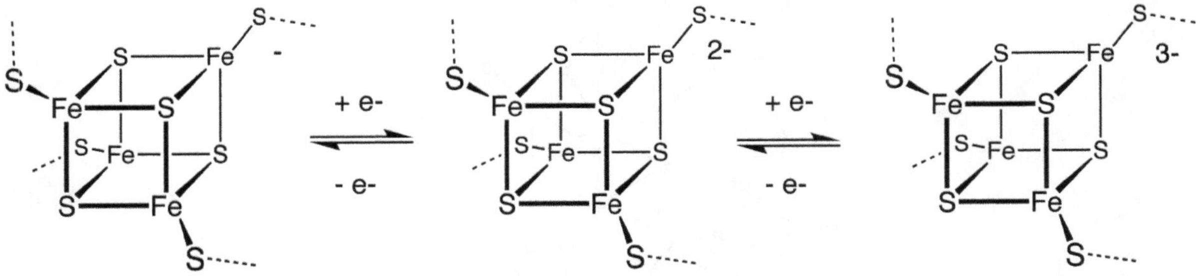

4Fe-4S clusters serve as electron-relays in proteins.

Oxygen Transport and Activation Proteins

Aerobic life make extensive use of metals such as iron, copper, and manganese. Heme is utilized by red blood cells in the form of hemoglobin for oxygen transport and is perhaps the most recognized metal system in biology. Other oxygen transport systems include myoglobin, hemocyanin, and hemerythrin. Oxidases and oxygenases are metal systems found throughout nature that take advantage of oxygen to carry out important reactions such as energy generation in cytochrome c oxidase or small molecule oxidation in cytochrome P450 oxidases or methane monooxygenase. Some metalloproteins are designed to protect a biological system from the potentially harmful effects of oxygen and other reactive oxygen-containing molecules such as hydrogen peroxide. These systems include peroxidases, catalases, and superoxide dismutases. A complementary metalloprotein to those that react with oxygen is the oxygen evolving complex present in plants. This system is part of the complex protein machinery that produces oxygen as plants perform photosynthesis.

Myoglobin is a prominent subject in bioinorganic chemistry, with particular attention to the iron-heme complex that is anchored to the protein.

Bioorganometallic Chemistry

Bioorganometallic systems feature metal-carbon bonds as structural elements or as intermediates. Bioorganometallic enzymes and proteins include the hydrogenases, FeMoco in nitrogenase, and methylcobalamin. These naturally occurring organometallic compounds. This area is more focused on the utilization of metals by unicellular organisms. Bioorganometallic compounds are significant in environmental chemistry.

Structure of FeMoco, the catalytic center of nitrogenase.

Metals in Medicine

A number of drugs contain metals. This theme relies on the study of the design and mechanism of action of metal-containing pharmaceuticals, and compounds that interact with endogenous metal ions in enzyme active sites. The most widely used anti-cancer drug is cisplatin. MRI contrast agent commonly contain gadolinium. Lithium carbonate has been used to treat the manic phase of bipolar disorder. Gold antiarthritic drugs, e.g. auranofin have been commerciallized. Carbon monoxide-releasing molecules are metal complexes have been developed to suppress inflammation by releasing small amounts of carbon monoxide. The cardiovascular and neuronal importance of nitric oxide has been examined, including the enzyme nitric oxide synthase.

Environmental Chemistry

Environmental chemistry traditionally emphasizes the interaction of heavy metals with organisms. Methylmercury has caused major disaster called Minamata disease. Arsenic poisoning is a widespread problem owing largely to arsenic contamination of groundwater, which affects many millions of people in developing countries. The metabolism of mercury- and arsenic-containing compounds involves cobalamin-based enzymes.

Biomineralization

Biomineralization is the process by which living organisms produce minerals, often to harden or stiffen existing tissues. Such tissues are called mineralized tissues. Examples include silicates in algae and diatoms, carbonates in invertebrates, and calcium phosphates and carbonates in vertebrates.Other examples include copper, iron and gold deposits involving bacteria. Biologically-formed minerals often have special uses such as magnetic sensors in magnetotactic bacteria (Fe_3O_4), gravity sensing devices ($CaCO_3$, $CaSO_4$, $BaSO_4$) and iron storage and mobilization ($Fe_2O_3 \cdot H_2O$ in the protein ferritin). Because extracellular iron is strongly involved in inducing calcification, its control is essential in developing shells; the protein ferritin plays an important role in controlling the distribution of iron.

Types of Inorganic Elements in Biology

Alkali and Alkaline Earth Metals

Like many antibiotics, monensin-A is an ionophore that tighlty bind Na^+ (shown in yellow).

The abundant inorganic elements act as ionic electrolytes. The most important ions are sodium, potassium, calcium, magnesium, chloride, phosphate, and the organic ion bicarbonate. The maintenance of precise gradients across cell membranes maintains osmotic pressure and pH. Ions are also critical for nerves and muscles, as action potentials in these tissues are produced by the exchange of electrolytes between the extracellular fluid and the cytosol. Electrolytes enter and leave cells through proteins in the cell membrane called ion channels. For example, muscle contraction depends upon the movement of calcium, sodium and potassium through ion channels in the cell membrane and T-tubules.

Transition Metals

The transition metals are usually present as trace elements in organisms, with zinc and iron being most abundant. These metals are used in some proteins as cofactors and are essential for the activity of enzymes such as catalase and oxygen-carrier proteins such as hemoglobin. These cofactors are bound tightly to a specific protein; although enzyme cofactors can be modified during catalysis, cofactors always return to their original state after catalysis has taken place. The metal micronutrients are taken up into organisms by specific transporters and bound to storage proteins such as ferritin or metallothionein when not being used. Cobalt is essential for the functioning of vitamin B12.

Main Group Compounds

Many other elements aside from metals are bio-active. Sulfur and phosphorus are required for all life. Phosphorus almost exclusively exists as phosphate and its various esters. Sulfur exists in a variety of oxidation states, ranging from sulfate (SO_4^{2-}) down to sulfide (S^{2-}). Selenium is a trace element involved in proteins that are antioxidants. Cadmium is important because of its toxicity.

References

- Crabtree, Robert H. (2009). The Organometallic Chemistry of the Transition Metals (5th ed.). New York, NY: John Wiley and Sons. pp. 2, 560, and passim. ISBN 0470257628. Retrieved 23 May 2016.

- Oliveira, José; Elschenbroich, Christoph (2006). Organometallics (3., completely rev. and extended ed.). Weinheim: Wiley-VCH-Verl. ISBN 978-3-527-29390-2.

- Berg, Jeremy M.; Lippard, Stephen J. (1994). Principles of bioinorganic chemistry ([Pbk. ed.]. ed.). Mill Valley: University Science Books. ISBN 0-935702-73-3.

- Leeuwen, Piet W.N.M. van (2004). Homogeneous catalysis : understanding the art. Dordrecht: Springer. ISBN 978-1-4020-3176-2.

- Astrid Sigel, Helmut Sigel and Roland K.O. Sigel, ed. (2009). Metal-carbon bonds in enzymes and cofactors. Metal Ions in Life Sciences. 6. Royal Society of Chemistry. ISBN 978-1-84755-915-9.

- Amsler, Mark (1986). The Languages of Creativity: Models, Problem-solving, Discourse. University of Delaware Press. ISBN 978-0874132809.

- Ben-Menahem, Ari (2009). Historical Encyclopedia of Natural and Mathematical Sciences. Springer. p. 2982. ISBN 978-3-540-68831-0.

- Burton, Feldman (2001). The Nobel Prize: A History of Genius, Controversy, and Prestige. Arcade Publishing. ISBN 978-1559705929.

- Eldra P. Solomon; Linda R. Berg; Diana W. Martin (2007). Biology, 8th Edition, International Student Edition. Thomson Brooks/Cole. ISBN 978-0495317142.

- Finkel, Richard; Cubeddu, Luigi; Clark, Michelle (2009). Lippencott's Illustrated Reviews: Pharmacology (4th ed.). Lippencott Williams & Wilkins. ISBN 978-0-7817-7155-9.

- Krebs, Jocelyn E.; Goldstein, Elliott S.; Lewin, Benjamin; Kilpatrick, Stephen T. (2012). Essential Genes. Jones & Bartlett Publishers. ISBN 978-1-4496-1265-8.

- Helvoort, Ton van (2000). Arne Hessenbruch, ed. Reader's Guide to the History of Science. Fitzroy Dearborn Publishing. ISBN 188496429X.

9

Organic Synthesis: An Integrated Study

Chemical synthesis has a branch of study known as organic synthesis. It deals with organic compounds which are more complex than inorganic compounds. The techniques discussed within this text are retrosynthetic analysis, enantioselective synthesis, electro synthesis and drug design. This chapter will provide an integrated understanding of organic synthesis.

Organic Synthesis

Organic synthesis is a special branch of chemical synthesis and is concerned with the construction of organic compounds via organic reactions. Organic molecules often contain a higher level of complexity than purely inorganic compounds, so that the synthesis of organic compounds has developed into one of the most important branches of organic chemistry. There are several main areas of research within the general area of organic synthesis: *total synthesis*, *semisynthesis*, and *methodology*.

Total Synthesis

A total synthesis is the complete chemical synthesis of complex organic molecules from simple, commercially available (petrochemical) or natural precursors. Total synthesis may be accomplished either via a linear or convergent approach. In a *linear* synthesis—often adequate for simple structures—several steps are performed one after another until the molecule is complete; the chemical compounds made in each step are called *synthetic intermediates*. For more complex molecules, a convergent synthetic approach may be preferable, one that involves individual preparation of several "pieces" (key intermediates), which are then combined to form the desired product.

Robert Burns Woodward, who received the 1965 Nobel Prize for Chemistry for several total syntheses (e.g., his 1954 synthesis of strychnine), is regarded as the father of modern organic synthesis. Some latter-day examples include Wender's, Holton's, Nicolaou's, and Danishefsky's total syntheses of the anti-cancer therapeutic, paclitaxel (trade name, Taxol).

Methodology and Applications

Each step of a synthesis involves a chemical reaction, and reagents and conditions for each of these reactions must be designed to give an adequate yield of pure product, with as little work as possible. A method may already exist in the literature for making one of the early synthetic intermediates, and this method will usually be used rather than an effort to "reinvent the wheel". However, most intermediates are compounds that have never been made before, and these will normally be made using general methods developed by methodology researchers.

To be useful, these methods need to give high yields, and to be reliable for a broad range of substrates. For practical applications, additional hurdles include industrial standards of safety and purity.

Methodology research usually involves three main stages: *discovery*, *optimisation*, and studies of *scope and limitations*. The *discovery* requires extensive knowledge of and experience with chemical reactivities of appropriate reagents. *Optimisation* is a process in which one or two starting compounds are tested in the reaction under a wide variety of conditions of temperature, solvent, reaction time, etc., until the optimum conditions for product yield and purity are found. Finally, the researcher tries to extend the method to a broad range of different starting materials, to find the scope and limitations. Total syntheses are sometimes used to showcase the new methodology and demonstrate its value in a real-world application. Such applications involve major industries focused especially on polymers (and plastics) and pharmaceuticals.

Stereoselective Synthesis

Most complex natural products are chiral, and the bioactivity of chiral molecules varies with the enantiomer. Historically, total syntheses targeted racemic mixtures, mixtures of both possible enantiomers, after which the racemic mixture might then be separated via chiral resolution.

In the later half of the twentieth century, chemists began to develop methods of stereoselective catalysis and kinetic resolution whereby reactions could be directed to produce only one enantiomer rather than a racemic mixture. Early examples include stereoselective hydrogenations (e.g., as reported by William Knowles and Ryōji Noyori), and functional group modifications such as the asymmetric epoxidation of Barry Sharpless; for these specific achievements, these workers were awarded the Nobel Prize in Chemistry in 2001. Such reactions gave chemists a much wider choice of enantiomerically pure molecules to start from, where previously only natural starting materials could be used. Using techniques pioneered by Robert B. Woodward and new developments in synthetic methodology, chemists became more able to take simple molecules through to more complex molecules without unwanted racemisation, by understanding stereocontrol, allowing final target molecules to be synthesised pure enantiomers (i.e., without need for resolution). Such techniques are referred to as *stereoselective synthesis*.

Synthesis Design

Elias James Corey brought a more formal approach to synthesis design, based on retrosynthetic analysis, for which he won the Nobel Prize for Chemistry in 1990. In this approach, the synthesis is planned backwards from the product, using standard rules. The steps "breaking down" the parent structure into achievable component parts are shown in a graphical scheme that uses *retrosynthetic arrows*.

More recently, and less widely accepted, computer programs have been written for designing a synthesis based on sequences of generic "half-reactions".

Retrosynthetic Analysis

Retrosynthetic analysis is a technique for solving problems in the planning of organic synthe-ses. This is achieved by transforming a target molecule into simpler precursor structures without assumptions regarding starting materials. Each precursor material is examined using the same method. This procedure is repeated until simple or commercially available structures are reached. E.J. Corey formalized this concept in his book *The Logic of Chemical Synthesis*.

The power of retrosynthetic analysis becomes evident in the design of a synthesis. The goal of retrosynthetic analysis is structural simplification. Often, a synthesis will have more than one pos-sible synthetic route. Retrosynthesis is well suited for discovering different synthetic routes and comparing them in a logical and straightfoward fashion. A database may be consulted at each stage of the analysis, to determine whether a component already exists in the literature. In that case, no further exploration of that compound would be required.

Definitions

Disconnection

A retrosynthetic step involving the breaking of a bond to form two (or more) synthons.

Retron

A minimal molecular substructure that enables certain transformations.

Retrosynthetic tree

A directed acyclic graph of several (or all) possible retrosyntheses of a single target.

Synthon

An idealized molecular fragment. A synthon and the corresponding commercially available synthetic equivalent are shown below:

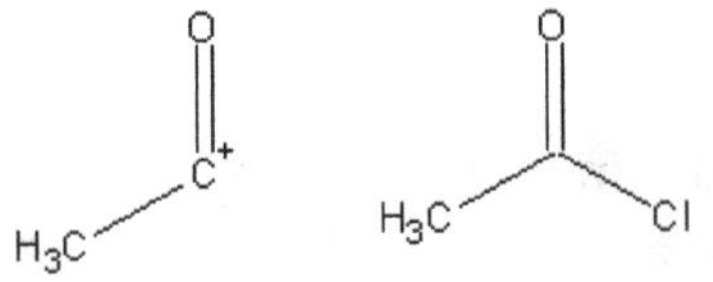

Target

The desired final compound.

Transform

The reverse of a synthetic reaction; the formation of starting materials from a single prod-uct.

Example

An example will allow the concept of retrosynthetic analysis to be easily understood.

In planning the synthesis of phenylacetic acid, two synthons are identified. A nucleophilic "-COOH" group, and an electrophilic "$PhCH_2^+$" group. Of course, both synthons do not exist per se; synthetic equivalents corresponding to the synthons are reacted to produce the desired product. In this case, the cyanide anion is the synthetic equivalent for the $^-$COOH synthon, while benzyl bromide is the synthetic equivalent for the benzyl synthon.

The synthesis of phenylacetic acid determined by retrosynthetic analysis is thus:

$$PhCH_2Br + NaCN \rightarrow PhCH_2CN + NaBr$$

$$PhCH_2CN + 2\,H_2O \rightarrow PhCH_2COOH + NH_3$$

In fact, phenylacetic acid has been synthesized from benzyl cyanide, itself prepared by the analogous reaction of benzyl chloride with sodium cyanide.

Strategies

Functional Group Strategies

Manipulation of functional groups can lead to significant reductions in molecular complexity.

Stereochemical Strategies

Numerous chemical targets have distinct stereochemical demands. Stereochemical transformations (such as the Claisen rearrangement and Mitsunobu reaction) can remove or transfer the desired chirality thus simplifying the target.

Structure-goal Strategies

Directing a synthesis toward a desirable intermediate can greatly narrow the focus of an analysis. This allows bidirectional search techniques.

Transform-based Strategies

The application of transformations to retrosynthetic analysis can lead to powerful reductions in molecular complexity. Unfortunately, powerful transform-based retrons are rarely present in complex molecules, and additional synthetic steps are often needed to establish their presence.

Topological Strategies

The identification one or more key bond disconnections may lead to the identification of key substructures or difficult to identify rearrangement transformations in order to identify the key structures .

- Disconnections that preserve ring structures are encouraged.

- Disconnections that create rings larger than 7 members are discouraged.

Enantioselective Synthesis

In the Sharpless dihydroxylation reaction the chirality of the product can be controlled by the "AD-mix" used. This is an example of enantioselective synthesis using asymmetric induction
Key: R_L = Largest substituent; R_M = Medium-sized substituent; R_S = Smallest substituent

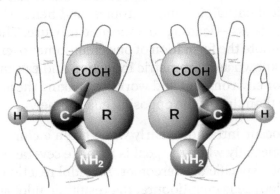

Enantioselective synthesis, also called chiral synthesis or asymmetric synthesis, is a form of chemical synthesis. It is defined by IUPAC as: a chemical reaction (or reaction sequence) in which one or more new elements of chirality are formed in a substrate molecule and which produces the stereoisomeric (enantiomeric or diastereoisomeric) products in unequal amounts.

Two enantiomers of a generic alpha amino acid

Put more simply: it is the synthesis of a compound by a method that favors the formation of a specific enantiomer or diastereomer.

Enantioselective synthesis is a key process in modern chemistry and is particularly important in the field of pharmaceuticals, as the different enantiomers or diastereomers of a molecule often have different biological activity.

Overview

Many of the building blocks of biological systems, such as sugars and amino acids, are produced exclusively as one enantiomer. As a result of this living systems possess a high degree of chemical chirality and will often react differently with the various enantiomers of a given compound. Examples of this selectivity include:

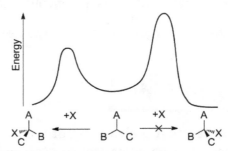

A Gibbs free energy plot of an enantioselective addition reaction.

- Flavour: the artificial sweetener aspartame has two enantiomers. L-aspartame tastes sweet, yet D-aspartame is tasteless

- Odor: R-($-$)-carvone smells like spearmint yet S-(+)-carvone, smells like caraway.

- Drug effectiveness: the antidepressant drug Citalopram is sold as a racemic mixture. However, studies have shown that only the (S)-(+) enantiomer is responsible for the drug's beneficial effects.

- Drug safety: Dpenicillamine is used in chelation therapy and for the treatment of rheumatoid arthritis. However Lpenicillamine is toxic as it inhibits the action of pyridoxine, an essential B vitamin.

As such enantioselective synthesis is of great importance; but it can also be difficult to achieve. Enantiomers possess identical enthalpies and entropies, and hence should be produced in equal amounts by an undirected process – leading to a racemic mixture. The solution is to introduce a chiral feature which will promote the formation of one enantiomer over another via interactions at the transition state. This is known as asymmetric induction and can involve chiral features in the substrate, reagent, catalyst or environment and works by making the activation energy required to form one enantiomer lower than that of the opposing enantiomer.

Asymmetric induction can occur intramolecularly when given a chiral starting material. This behaviour can be exploited, especially when the goal is to make several consecutive chiral centres to give a specific enantiomer of a specific diastereomer. An aldol reaction, for example, is inherently diastereoselective; if the aldehyde is enantiopure, the resulting aldol adduct is diastereomerically and enantiomerically pure.

Approaches

Enantioselective Catalysis

In general, enantioselective catalysis (known traditionally as asymmetric catalysis) refers to the use of chiral coordination complexes as catalysts. This approach is very commonly encountered, as it is effective for a broader range of transformations than any other method of enantioselective synthesis. The catalysts are typically rendered chiral by using chiral ligands, however it is also possible to generate chiral-at-metal complexes using simpler achiral ligands. Most enantioselective catalysts are effective at low concentrations making them well suited to industrial scale synthesis; as even exotic and expensive catalysts can be used affordably. Perhaps the most versatile example of enantioselective synthesis is asymmetric hydrogenation, which is able to reduce a wide variety of functional groups.

With only 75 natural metals in existence (and not all of these showing extensive catalytic activities) the design of new catalysts is very much dominated by the development of new classes of ligands. Certain ligands, often referred to as 'privileged ligands', have been found to be effective in a wide range of reactions; examples include BINOL, Salen and BOX. However, most catalysts are rarely general, requiring certain functional groups in the substrate to form the transition state complex correctly: arbitrary structures cannot be used. For example, Noyori asymmetric hydrogenation with BINAP/Ru requires a β-ketone, although another catalyst, BINAP/diamine-Ru, widens the scope to α,β-olefins and aromatics.

Chiral Auxiliaries

A chiral auxiliary is an organic compound which couples to the starting material to form new compound which can then undergo enantioselective reactions via intramolecular asymmetric induction. At the end of the reaction the auxiliary is removed, under conditions that will not cause racemization of the product. It is typically then recovered for future use.

Chiral auxiliaries must be used in stoichiometric amounts to be effective and require additional synthetic steps to append and remove the auxiliary. However, in some cases the only available stereoselective methodology relies on chiral auxiliaries and these reactions tend to be versatile and very well-studied, allowing the most time-efficient access to enantiomerically pure products. Additionally, the products of auxiliary-directed reactions are diastereomers, which enables their facile separation by methods such as column chromatography or crystallization.

Biocatalysis

Biocatalysis makes use of biological compounds, ranging from isolated enzymes to living cells, to perform chemical transformations. The advantages of these reagents include very high ee's and reagent specificity, as well as mild operating conditions and low environmental impact. Biocatalysts are more commonly used in industry than in academic research; for example in the production of statins. The high reagent specificity can be a problem however; as it often requires that a wide range of biocatalysts be screened before an effective reagent is found.

Enantioselective Organocatalysis

Organocatalysis refers to a form of catalysis, where the rate of a chemical reaction is increased by an organic compound consisting of carbon, hydrogen, sulfur and other non-metal elements. When the organocatalyst is chiral enantioselective synthesis can be achieved; for example a number of carbon–carbon bond forming reactions become enantioselective in the presence of proline with the aldol reaction being a prime example. Organocatalysis often employs natural compounds and secondary amines as chiral catalysts; these are inexpensive and environmentally friendly, as no metals are involved.

Chiral Pool Synthesis

Chiral pool synthesis is one of the simplest and oldest approaches for enantioselective synthesis. A readily available chiral starting material is manipulated through successive reactions, often using achiral reagents, to obtain the desired target molecule. This can meet the criteria for enantioselective synthesis when a new chiral species is created, such as in an S_N2 reaction.

Chiral pool synthesis is especially attractive for target molecules having similar chirality to a relatively inexpensive naturally occurring building-block such as a sugar or amino acid. However, the number of possible reactions the molecule can undergo is restricted and tortuous synthetic routes may be required (e.g. Oseltamivir total synthesis). This approach also requires a stoichiometric amount of the enantiopure starting material, which can be expensive if it is not naturally occurring.

Alternative Approaches

Alternatives to enantioselective synthesis usually involve the isolation of one enantiomer from a racemic mixture by any of a number of methods. If the cost in time and money of making such racemic mixtures is low (or if both enantiomers may find use) then this approach may remain cost-effective. Common methods of separation are based around chiral resolution or kinetic resolution.

Separation and Analysis of Enantiomers

The two enantiomers of a molecule possess the same physical properties (e.g. melting point, boiling point, polarity etc.) and so behave identically to each other. As a result, they will migrate with an identical R_f in thin layer chromatography and have identical retention times in HPLC and GC. Their NMR and IR spectra are identical.

This can make it very difficult to determine whether a process has produced a single enantiomer (and crucially which enantiomer it is) as well as making it hard to separate enantiomers from a reaction which has not been 100% enantioselective. Fortunately, enantiomers behave differently in the presence of other chiral materials and this can be exploited to allow their separation and analysis.

Enantiomers do not migrate identically on chiral chromatographic media, such as quartz or standard media that has been chirally modified. This forms the basis of chiral column chromatography, which can be used on a small scale to allow analysis via GC and HPLC, or on a large scale to separate chirally impure materials. However this process can require large amount of chiral packing material which can be expensive. A common alternative is to use a chiral derivatizing agent to convert the enantiomers into a diastereomers, in much the same way as chiral auxiliaries. These have different physical properties and hence can be separated and analysed using conventional methods. Special chiral derivitizing agents known as 'chiral resolution agents' are used in the NMR spectroscopy of stereoisomers, these typically involve coordination to chiral europium complexes such as $Eu(fod)_3$ and $Eu(hfc)_3$.

The enantiomeric excess of a substance can also be determined using certain optical methods. The oldest method for doing this is to use a polarimeter to compare the level of optical rotation in the product against a 'standard' of known composition. It is also possible to perform ultraviolet-visible spectroscopy of stereoisomers by exploiting the Cotton effect.

One of the most accurate ways of determining the chirality of compound is to determine its absolute configuration by Xray Crystallography. However this is a labour-intensive process which requires that a suitable single crystal be grown.

History

Inception (1815–1905)

In 1815 the French physicist Jean-Baptiste Biot showed that certain chemicals could rotate the plane of a beam of polarised light, a property called optical activity. The nature of this property remained a mystery until 1848, when Louis Pasteur proposed that it had a molecular basis originating from some form of *dissymmetry*, with the term *chirality* being coined by Lord Kelvin a year later. The origin of chirality itself was finally described in 1874, when Jacobus Henricus van 't Hoff and Joseph Le Bel independently proposed the tetrahedral geometry of carbon; structural models prior to this work had been two-dimensional, and van 't Hoff Le Bel theorized that the arrangement of groups around this tetrahedron could dictate the optical activity of the resulting compound.

Marckwald's brucine-catalyzed enantioselective decarboxylation of 2-ethyl-2-methylmalonic acid, resulting in a slight excess of the levorotary form of the 2-methylbutyric acid product.

In 1894 Hermann Emil Fischer outlined the concept of asymmetric induction; in which he correctly ascribed selective the formation of D-glucose by plants to be due to the influence of optically active substances within chlorophyll. Fischer also successfully performed what would now be regarded as the first example of enantioselective synthesis, by enantioselectively elongating sugars via a process which would eventually become the Kiliani–Fischer synthesis.

Brucine, an alkaloid natural product related to strychnine, used successfully as an organocatalyst by Marckwald in 1904.

The first enantioselective chemical synthesis is most often attributed to Willy Marckwald, Universität zu Berlin, for a brucine-catalyzed enantioselective decarboxylation of *2-ethyl-2-methylmalonic acid* reported in 1904. A slight excess of the levorotary form of the product of the reaction, 2-methylbutyric acid, was produced; as this product is also a natural product—e.g., as a side chain of lovastatin formed by its diketide synthase (LovF) during its biosynthesis—this result constitutes the first recorded total synthesis with enantioselectivity, as well other firsts (as Koskinen notes, first "example of asymmetric catalysis, enantiotopic selection, and organocatalysis"). This observation is also of historical significance, as at the time enantioselective synthesis could only be understood in terms of vitalism. Natural and artificial compounds were fundamentally different, it was argued, and chirality could only exist in natural compounds. Unlike Fischer, Marckwald had performed an enantioselective reaction upon an achiral, *un-natural* starting material, albeit with a chiral organocatalyst (as we now understand this chemistry).

Early Work (1905–1965)

The development of enantioselective synthesis was initially slow, largely due to the limited range of techniques available for their separation and analysis. Diastereomers possess different physical properties, allowing separation by conventional means, however at the time enantiomers could only be separated by spontaneous resolution (where enantiomers separate upon crystallisation) or kinetic resolution (where one enantiomer is selectively destroyed). The only tool for analysing enantiomers was optical activity using a polarimeter, a method which provides no structural data.

It was not until the 1950s that major progress really began. Driven in part by chemists such as R. B. Woodward and Vladimir Prelog but also by the development of new techniques. The first of these was Xray Crystallography, which was used to determine the absolute configuration of an organic compound by Johannes Bijvoet in 1951. Chiral chromatography was introduced a year later by Dalgliesh, who used paper chromatography to separate chiral amino acids. Although Dalgliesh was not the first to observe such separations, he correctly attributed the separation of enantiomers to differential retention by the chiral cellulose. This was expanded upon in 1960, when Klem and Reed first reported the use of chirally-modified silica gel for chiral HPLC chromatographic separation.

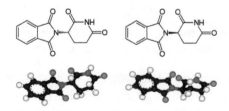

The two enantiomers of thalidomide:
Left: (*S*)-thalidomide
Right: (*R*)-thalidomide

Thalidomide

While it had long been known that the different enantiomers of a drug could have different activities, this was not accounted for in early drug design and testing. However following the thalidomide disaster the development and licensing of drugs changed dramatically.

First synthesized in 1953, thalidomide was widely prescribed for morning sickness from 1957 to 1962, but was soon found to be seriously teratogenic, eventually causing birth defects in more than 10,000 babies. The disaster prompted many counties to introduce tougher rules for the testing and licensing of drugs, such as the Kefauver-Harris Amendment (U.S.) and Directive 65/65/EEC1 (E.U.).

Early research into the teratogenic mechanism, using mice, suggested that one enantiomer of thalidomide was teratogenic while the other possessed all the therapeutic activity. This theory was later shown to be incorrect and has now been superseded by a body of research. However it raised the importance of chirality in drug design, leading to increased research into enantioselective synthesis.

Modern Age (Since 1965)

The Cahn–Ingold–Prelog priority rules (often abbreviated as the CIP system) were first published in 1966; allowing enantiomers to be more easily and accurately described. The same year saw first successful enantiomeric separation by gas chromatography an important development as the technology was in common use at the time.

Metal catalysed enantioselective synthesis was pioneered by William S. Knowles, Ryōji Noyori and K. Barry Sharpless; for which they would receive the 2001 Nobel Prize in Chemistry. Knowles and Noyori began with the development of asymmetric hydrogenation, which they developed independently in 1968. Knowles replaced the achiral triphenylphosphine ligands in Wilkinson's catalyst with chiral phosphine ligands. This experimental catalyst was employed in an asymmetric hydrogenation with a modest 15% enantiomeric excess. Knowles was also the first to apply enantioselective metal catalysis to industrial-scale synthesis; while working for the Monsanto Company he developed an enantioselective hydrogenation step for the production of L-DOPA, utilising the DIPAMP ligand.

Knowles: Asymmetric hydrogenation (1968)   Noyori: Enantioselective cyclopropanation (1968)

Noyori devised a copper complex using a chiral Schiff base ligand, which he used for the metal-carbenoid cyclopropanation of styrene. In common with Knowles' findings, Noyori's results for the enantiomeric excess for this first-generation ligand were disappointingly low: 6%. However continued research eventually led to the development of the Noyori asymmetric hydrogenation reaction.

The Sharpless oxyamination

Sharpless complemented these reduction reactions by developing a range of asymmetric oxidations (Sharpless epoxidation, Sharpless asymmetric dihydroxylation, Sharpless oxyamination) during the 1970s to 1980's. With the asymmetric oxyamination reaction, using osmium tetroxide, being the earliest.

During the same period, methods were developed to allow the analysis of chiral compounds by NMR; either using chiral derivatizing agents, such as Mosher's acid, or europium based shift reagents, of which $Eu(DPM)_3$ was the earliest.

Chiral auxiliaries were introduced by E.J. Corey in 1978 and featured prominently in the work of Dieter Enders. Around the same time enantioselective organocatalysis was developed, with pioneering work including the Hajos–Parrish–Eder–Sauer–Wiechert reaction. Enzyme-catalyzed enantioselective reactions became more and more common during the 1980s, particularly in industry, with their applications including asymmetric ester hydrolysis with pig-liver esterase. The emerging technology of genetic engineering has allowed the tailoring of enzymes to specific processes, permitting an increased range of selective transformations. For example, in the asymmetric hydrogenation of statin precursors.

Electrosynthesis

Electrosynthesis in chemistry is the synthesis of chemical compounds in an electrochemical cell. The main advantage of electrosynthesis over an ordinary redox reaction is avoidance of the potential wasteful other half-reaction and the ability to precisely tune the required potential. Electrosynthesis is actively studied as a science and also has many industrial applications. Electrooxidation is studied not only for synthesis but also for efficient removal of certain harmful organic compounds in wastewater.

Experimental Setup

The basic setup in electrosynthesis is a galvanic cell, a potentiostat and two electrodes. Good electrosynthetic conditions use a solvent and electrolyte combination that minimizes electrical resistance. Protic conditions often use alcohol-water or dioxane-water solvent mixtures with an electrolyte such as a soluble salt, acid or base. Aprotic conditions often use an organic solvent such as acetonitrile or dichloromethane with electrolytes such as lithium perchlorate or tetrabutylammonium acetate. Electrodes are selected which provide favorable electron transfer properties towards the substrate while maximizing the activation energy for side reactions. This activation energy is often related to an overpotential of a competing reaction. For example, in aqueous conditions the competing reactions in the cell are the formation of oxygen at the anode and hydrogen at the cathode. In this case a graphite anode and lead cathode could be used effectively because of their high overpotentials for oxygen and hydrogen formation respectively. Many other materials can be used as electrodes. Other examples include platinum, magnesium, mercury (as a liquid pool in the reactor), stainless steel or reticulated vitreous carbon. Some reactions use a sacrificial electrode is used which is consumed during the reaction like zinc or lead. The two basic cell types are undivided cell or divided cell type. In divided cells the cathode and anode chambers are separated with a semiporous membrane. Common membrane materials include sintered glass, porous porcelain, polytetrafluoroethene or polypropylene. The purpose of the divided cell is to permit the diffusion of ions while restricting the flow of the products and reactants. This is important when unwanted side reactions are possible. An example of a reaction requiring a divided cell is the reduction of nitrobenzene to phenylhydroxylamine, where the latter chemical is susceptible to oxidation at the anode.

Reactions

Organic oxidations take place at the anode with initial formation of radical cations as reactive intermediates. Compounds are reduced at the cathode to radical anions. The initial reaction takes place at the surface of the electrode and then the intermediates diffuse into the solution where they participate in secondary reactions.

The yield of an electrosynthesis is expressed both in terms the chemical yield and current efficiency. Current efficiency is the ratio of Coulombs consumed in forming the products to the total number of Coulombs passed through the cell. Side reactions decrease the current efficiency.

The potential drop between the electrodes determines the rate constant of the reaction. Electrosynthesis is carried out with either constant potential or constant current. The reason one chooses one over the other is due to a trade off of ease of experimental conditions versus current efficiency. Constant potential uses current more efficiently because the current in the cell decreases with time due to the depletion of the substrate around the working electrode (stirring is usually necessary to decrease the diffusion layer around the electrode). This is not the case under constant current conditions however. Instead as the substrate's concentration decreases the potential across the cell increases in order to maintain the fixed reaction rate. This consumes current in side reactions produced outside the target voltage.

Anodic Oxidations

- The most well-known electrosynthesis is the Kolbe electrolysis, in which two carboxylic acids decarboxylate, and the remaining structures bond together:

$$2 \; R\text{-}C(=O)\text{-}O^- \xrightarrow[-2\,CO_2]{-2\,e^-} R\text{-}R$$

- A variation is called the non-Kolbe reaction when a heteroatom (nitrogen or oxygen) is present at the α-position. The intermediate oxonium ion is trapped by a nucleophile usually solvent.

- Amides can be oxidized to *N*-acyliminium ions, which can be captured by various nucleophiles, for example:

  This reaction type is called a Shono oxidation. An example is the α-methoxylation of *N*-carbomethoxypyrrolidine

- Oxidation of a carbanion can lead to a coupling reaction for instance in the electrosynthesis of the tetramethyl ester of ethanetetracarboxylic acid from the corresponding malonate ester

- α-amino acids form nitriles and carbon dioxide via oxidative decarboxylation at AgO anodes (the latter is formed *in-situ* by oxidation of Ag_2O):

- Cyanoacetic acid from cathodic reduction of carbon dioxide and anodic oxidation of acetonitrile.

Cathodic Reductions

- In the Markó–Lam deoxygenation, an alcohol could be almost instantaneously deoxygenated by electroreducing their toluate ester.

- The cathodic hydroisomerization of activated olefins is applied industrially in the synthesis of adiponitrile from 2 equivalents of acrylonitrile:

- The cathodic reduction of arene compounds to the 1,4-dihydro derivatives is similar to a

Birch reduction. Examples from industry are the reduction of phthalic acid:

and the reduction of 2-methoxynaphthalene:

- The Tafel rearrangement, named for Julius Tafel, was at one time an important method for the synthesis of certain hydrocarbons from alkylated ethyl acetoacetate, a reaction accompanied by the rearrangement reaction of the alkyl group:

- The cathodic reduction of a nitrile to a primary amine in a divided cell:

- Cathodic reduction of a nitroalkene can give the oxime in good yield. At higher negative reduction potentials, the nitroalkene can be reduced further, giving the primary amine but with lower yield.

- An electrochemical carboxylation of a para-isobutylbenzyl chloride to Ibuprofen is promoted under supercritical carbon dioxide.

- Cathodic reduction of a carboxylic acid (oxalic acid) to an aldehyde (glyoxylic acid, shows as the rare aldehyde form) in a divided cell:

- An electrocatalysis by a copper complex helps reduce carbon dioxide to oxalic acid; this conversion uses carbon dioxide as a feedstock to generate oxalic acid.

Electrofluorination

In organofluorine chemistry, many perfluorinated compounds are prepared by electrochemical synthesis, which is conducted in liquid HF at voltages near 5–6 V using Ni anodes. The method was invented in the 1930s. Amines, alcohols, carboxylic acids, and sulfonic acids are converted to the perfluorinated derivatives using this technology. A solution or suspension of the hydrocarbon in hydrogen fluoride is electrolyzed at 5–6 V to produce high yields of the perfluorinated product.

Drug Design

Drug design, often referred to as rational drug design or simply rational design, is the inventive process of finding new medications based on the knowledge of a biological target. The drug is most commonly an organic small molecule that activates or inhibits the function of a biomolecule such as a protein, which in turn results in a therapeutic benefit to the patient. In the most basic sense, drug design involves the design of molecules that are complementary in shape and charge to the biomolecular target with which they interact and therefore will bind to it. Drug design frequently but not necessarily relies on computer modeling techniques. This type of modeling is sometimes referred to as computer-aided drug design. Finally, drug design that relies on the knowledge of the three-dimensional structure of the biomolecular target is known as structure-based drug design. In addition to small molecules, biopharmaceuticals and especially therapeutic antibodies are an increasingly important class of drugs and computational methods for improving the affinity, selectivity, and stability of these protein-based therapeutics have also been developed.

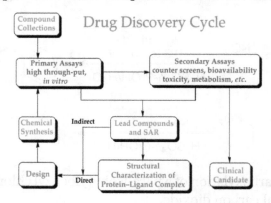

The phrase "drug design" is to some extent a misnomer. A more accurate term is ligand design (i.e., design of a molecule that will bind tightly to its target). Although design techniques for prediction of binding affinity are reasonably successful, there are many other properties, such as bioavailability, metabolic half-life, side effects, etc., that first must be optimized before a ligand can become a safe and efficacious drug. These other characteristics are often difficult to predict with rational design techniques. Nevertheless, due to high attrition rates, especially during clinical phases of drug development, more attention is being focused early in the drug design process on selecting candidate drugs whose physicochemical properties are predicted to result in fewer complications during development and hence more likely to lead to an approved, marketed drug. Furthermore, in vitro experiments complemented with computation methods are increasingly used in early drug discovery to select compounds with more favorable ADME (absorption, distribution, metabolism, and excretion) and toxicological profiles.

Drug Targets

A biomolecular target (most commonly a protein or nucleic acid) is a key molecule involved in a particular metabolic or signaling pathway that is associated with a specific disease condition or pathology or to the infectivity or survival of a microbial pathogen. Potential drug targets are not necessarily disease causing but must by definition be disease modifying. In some cases, small molecules will be designed to enhance or inhibit the target function in the specific disease modifying pathway. Small molecules (for example receptor agonists, antagonists, inverse agonists, or modulators; enzyme activators or inhibitors; or ion channel openers or blockers) will be designed that are complementary to the binding site of target. Small molecules (drugs) can be designed so as not to affect any other important "off-target" molecules (often referred to as antitargets) since drug interactions with off-target molecules may lead to undesirable side effects. Due to similarities in binding sites, closely related targets identified through sequence homology have the highest chance of cross reactivity and hence highest side effect potential.

Most commonly, drugs are organic small molecules produced through chemical synthesis, but biopolymer-based drugs (also known as biopharmaceuticals) produced through biological processes are becoming increasingly more common. In addition, mRNA-based gene silencing technologies may have therapeutic applications.

Rational Drug Discovery

In contrast to traditional methods of drug discovery (known as forward pharmacology), which rely on trial-and-error testing of chemical substances on cultured cells or animals, and matching the apparent effects to treatments, rational drug design (also called reverse pharmacology) begins with a hypothesis that modulation of a specific biological target may have therapeutic value. In order for a biomolecule to be selected as a drug target, two essential pieces of information are required. The first is evidence that modulation of the target will be disease modifying. This knowledge may come from, for example, disease linkage studies that show an association between mutations in the biological target and certain disease states. The second is that the target is "druggable". This means that it is capable of binding to a small molecule and that its activity can be modulated by the small molecule.

Once a suitable target has been identified, the target is normally cloned and produced and purified. The purified protein is then used to establish a screening assay. In addition, the three-dimensional structure of the target may be determined.

The search for small molecules that bind to the target is begun by screening libraries of potential drug compounds. This may be done by using the screening assay (a "wet screen"). In addition, if the structure of the target is available, a virtual screen may be performed of candidate drugs. Ideally the candidate drug compounds should be "drug-like", that is they should possess properties that are predicted to lead to oral bioavailability, adequate chemical and metabolic stability, and minimal toxic effects. Several methods are available to estimate druglikeness such as Lipinski's Rule of Five and a range of scoring methods such as lipophilic efficiency. Several methods for predicting drug metabolism have also been proposed in the scientific literature.

Due to the large number of drug properties that must be simultaneously optimized during the design process, multi-objective optimization techniques are sometimes employed. Finally because of the limitations in the current methods for prediction of activity, drug design is still very much reliant on serendipity and bounded rationality.

Computer-aided Drug Design

The most fundamental goal in drug design is to predict whether a given molecule will bind to a target and if so how strongly. Molecular mechanics or molecular dynamics is most often used to estimate the strength of the intermolecular interaction between the small molecule and its biological target. These methods are also used to predict the conformation of the small molecule and to model conformational changes in the target that may occur when the small molecule binds to it. Semi-empirical, ab initio quantum chemistry methods, or density functional theory are often used to provide optimized parameters for the molecular mechanics calculations and also provide an estimate of the electronic properties (electrostatic potential, polarizability, etc.) of the drug candidate that will influence binding affinity.

Molecular mechanics methods may also be used to provide semi-quantitative prediction of the binding affinity. Also, knowledge-based scoring function may be used to provide binding affinity estimates. These methods use linear regression, machine learning, neural nets or other statistical techniques to derive predictive binding affinity equations by fitting experimental affinities to computationally derived interaction energies between the small molecule and the target.

Ideally, the computational method will be able to predict affinity before a compound is synthesized and hence in theory only one compound needs to be synthesized, saving enormous time and cost. The reality is that present computational methods are imperfect and provide, at best, only qualitatively accurate estimates of affinity. In practice it still takes several iterations of design, synthesis, and testing before an optimal drug is discovered. Computational methods have accelerated discovery by reducing the number of iterations required and have often provided novel structures.

Drug design with the help of computers may be used at any of the following stages of drug discovery:

1. hit identification using virtual screening (structure- or ligand-based design)

2. hit-to-lead optimization of affinity and selectivity (structure-based design, QSAR, etc.)

3. lead optimization of other pharmaceutical properties while maintaining affinity

In order to overcome the insufficient prediction of binding affinity calculated by recent scoring functions, the protein-ligand interaction and compound 3D structure information are used for analysis. For structure-based drug design, several post-screening analyses focusing on protein-ligand interaction have been developed for improving enrichment and effectively mining potential candidates:

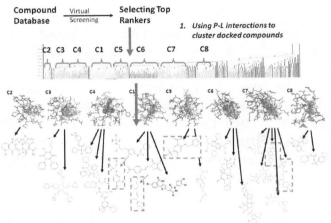

Flowchart of a Usual Clustering Analysis for Structure-Based Drug Design

- Consensus scoring

    o Selecting candidates by voting of multiple scoring functions

    o May lose the relationship between protein-ligand structural information and scoring criterion

- Cluster analysis

    o Represent and cluster candidates according to protein-ligand 3D information

    o Needs meaningful representation of protein-ligand interactions.

Types

There are two major types of drug design. The first is referred to as ligand-based drug design and the second, structure-based drug design.

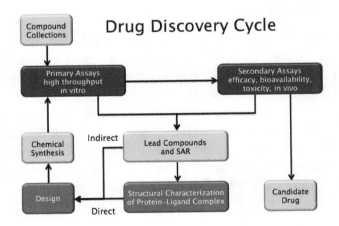

Drug discovery cycle highlighting both ligand-based (indirect) and structure-based (direct) drug design strategies.

Ligand-based

Ligand-based drug design (or indirect drug design) relies on knowledge of other molecules that bind to the biological target of interest. These other molecules may be used to derive a pharmacophore model that defines the minimum necessary structural characteristics a molecule must possess in order to bind to the target. In other words, a model of the biological target may be built based on the knowledge of what binds to it, and this model in turn may be used to design new molecular entities that interact with the target. Alternatively, a quantitative structure-activity relationship (QSAR), in which a correlation between calculated properties of molecules and their experimentally determined biological activity, may be derived. These QSAR relationships in turn may be used to predict the activity of new analogs.

Structure-based

Structure-based drug design (or direct drug design) relies on knowledge of the three dimensional structure of the biological target obtained through methods such as x-ray crystallography or NMR spectroscopy. If an experimental structure of a target is not available, it may be possible to create a homology model of the target based on the experimental structure of a related protein. Using the structure of the biological target, candidate drugs that are predicted to bind with high affinity and selectivity to the target may be designed using interactive graphics and the intuition of a medicinal chemist. Alternatively various automated computational procedures may be used to suggest new drug candidates.

Current methods for structure-based drug design can be divided roughly into three main categories. The first method is identification of new ligands for a given receptor by searching large databases of 3D structures of small molecules to find those fitting the binding pocket of the receptor using fast approximate docking programs. This method is known as virtual screening. A second category is de novo design of new ligands. In this method, ligand molecules are built up within the constraints of the binding pocket by assembling small pieces in a stepwise manner. These pieces can be either individual atoms or molecular fragments. The key advantage of such a method is that novel structures, not contained in any database, can be suggested. A third method is the optimization of known ligands by evaluating proposed analogs within the binding cavity.

Binding Site Identification

Binding site identification is the first step in structure based design. If the structure of the target or a sufficiently similar homolog is determined in the presence of a bound ligand, then the ligand should be observable in the structure in which case location of the binding site is trivial. However, there may be unoccupied allosteric binding sites that may be of interest. Furthermore, it may be that only apoprotein (protein without ligand) structures are available and the reliable identification of unoccupied sites that have the potential to bind ligands with high affinity is non-trivial. In brief, binding site identification usually relies on identification of concave surfaces on the protein that can accommodate drug sized molecules that also possess appropriate "hot spots" (hydrophobic surfaces, hydrogen bonding sites, etc.) that drive ligand binding.

Scoring Functions

Structure-based drug design attempts to use the structure of proteins as a basis for designing new ligands by applying the principles of molecular recognition. Selective high affinity binding to the target is generally desirable since it leads to more efficacious drugs with fewer side effects. Thus, one of the most important principles for designing or obtaining potential new ligands is to predict the binding affinity of a certain ligand to its target (and known antitargets) and use the predicted affinity as a criterion for selection.

One early general-purposed empirical scoring function to describe the binding energy of ligands to receptors was developed by Böhm. This empirical scoring function took the form:

$$\Delta G_{bind} = \Delta G_0 + \Delta G_{hb} \Sigma_{h-bonds} + \Delta G_{ionic} \Sigma_{ionic-int} + \Delta G_{lipophilic} \left| A \right| + \Delta G_{rot} NROT$$

where:

- ΔG_0 – empirically derived offset that in part corresponds to the overall loss of translational and rotational entropy of the ligand upon binding.

- ΔG_{hb} – contribution from hydrogen bonding

- ΔG_{ionic} – contribution from ionic interactions

- ΔG_{lip} – contribution from lipophilic interactions where $\left| A_{lipo} \right|$ is surface area of lipophilic contact between the ligand and receptor

- ΔG_{rot} – entropy penalty due to freezing a rotatable in the ligand bond upon binding

A more general thermodynamic "master" equation is as follows:

$$\Delta G_{bind} = -RT \ln K_d$$
$$[1.3ex] K_d = \frac{[\text{Ligand}][\text{Receptor}]}{[\text{Complex}]}$$
$$[1.3ex] \Delta G_{bind} = \Delta G_{desolvation} + \Delta G_{motion} + \Delta G_{configuration} + \Delta G_{interaction}$$

where:

- desolvation – enthalpic penalty for removing the ligand from solvent

- motion – entropic penalty for reducing the degrees of freedom when a ligand binds to its receptor

- configuration – conformational strain energy required to put the ligand in its "active" conformation

- interaction – enthalpic gain for "resolvating" the ligand with its receptor

The basic idea is that the overall binding free energy can be decomposed into independent components that are known to be important for the binding process. Each component reflects a certain kind of free energy alteration during the binding process between a ligand and its target receptor. The Master Equation is the linear combination of these components. According to Gibbs free energy equation, the relation between dissociation equilibrium constant, K_d, and the components of free energy was built.

Various computational methods are used to estimate each of the components of the master equation. For example, the change in polar surface area upon ligand binding can be used to estimate the desolvation energy. The number of rotatable bonds frozen upon ligand binding is proportional to the motion term. The configurational or strain energy can be estimated using molecular mechanics calculations. Finally the interaction energy can be estimated using methods such as the change in non polar surface, statistically derived potentials of mean force, the number of hydrogen bonds formed, etc. In practice, the components of the master equation are fit to experimental data using multiple linear regression. This can be done with a diverse training set including many types of ligands and receptors to produce a less accurate but more general "global" model or a more restricted set of ligands and receptors to produce a more accurate but less general "local" model.

Examples

A particular example of rational drug design involves the use of three-dimensional information about biomolecules obtained from such techniques as X-ray crystallography and NMR spectroscopy. Computer-aided drug design in particular becomes much more tractable when there is a high-resolution structure of a target protein bound to a potent ligand. This approach to drug discovery is sometimes referred to as structure-based drug design. The first unequivocal example of the application of structure-based drug design leading to an approved drug is the carbonic anhydrase inhibitor dorzolamide, which was approved in 1995.

Another important case study in rational drug design is imatinib, a tyrosine kinase inhibitor designed specifically for the *bcr-abl* fusion protein that is characteristic for Philadelphia chromosome-positive leukemias (chronic myelogenous leukemia and occasionally acute lymphocytic leukemia). Imatinib is substantially different from previous drugs for cancer, as most agents of chemotherapy simply target rapidly dividing cells, not differentiating between cancer cells and other tissues.

Additional examples include:

- Many of the atypical antipsychotics

- Cimetidine, the prototypical H_2-receptor antagonist from which the later members of the class were developed

- Selective COX-2 inhibitor NSAIDs

- Enfuvirtide, a peptide HIV entry inhibitor

- Nonbenzodiazepines like zolpidem and zopiclone

- Raltegravir, an HIV integrase inhibitor

- SSRIs (selective serotonin reuptake inhibitors), a class of antidepressants

- Zanamivir, an antiviral drug

Case Studies

- 5-HT3 antagonists

- Acetylcholine receptor agonists

- Angiotensin receptor antagonists

- Bcr-Abl tyrosine-kinase inhibitors

- Cannabinoid receptor antagonists

- CCR5 receptor antagonists

- Cyclooxygenase 2 inhibitors

- Dipeptidyl peptidase-4 inhibitors

- HIV protease inhibitors

- NK1 receptor antagonists

- Non-nucleoside reverse transcriptase inhibitors

- Nucleoside and nucleotide reverse transcriptase inhibitors

- PDE5 inhibitors

- Proton pump inibitors

- Renin inhibitors

- Triptans

- TRPV1 antagonists

- c-Met inhibitors

Criticism

It has been argued that the highly rigid and focused nature of rational drug design suppresses serendipity in drug discovery. Because many of the most significant medical discoveries have been inadvertent, the recent focus on rational drug design may limit the progress of drug discovery. Furthermore, the rational design of a drug may be limited by a crude or incomplete understanding of the underlying molecular processes of the disease it is intended to treat.

References

- Clayden, Jonathan; Greeves, Nick; Warren, Stuart; Wothers, Peter (2001). Organic Chemistry (1st ed.). Oxford University Press. ISBN 978-0-19-850346-0.Page 1226

- N. Jacobsen, Eric; Pfaltz, Andreas; Yamamoto, Hisashi (1999). Comprehensive asymmetric catalysis 1-3. Berlin: Springer. ISBN 9783540643371.

- Evans, D. A.; Helmchen, G.; Rüping, M. (2007). "Chiral Auxiliaries in Asymmetric Synthesis". In Christmann, M. Asymmetric Synthesis – The Essentials. Wiley-VCH Verlag GmbH & Co. pp. 3–9. ISBN 978-3-527-31399-0.

- Faber, Kurt (2011). Biotransformations in organic chemistry a textbook (6th rev. and corr. ed.). Berlin: Springer-Verlag. ISBN 9783642173936.

- Gröger, Albrecht Berkessel; Harald (2005). Asymmetric organocatalysis – from biomimetic concepts to applications in asymmetric synthesis (1. ed., 2. reprint. ed.). Weinheim: Wiley-VCH. ISBN 3-527-30517-3.

- Koskinen, Ari M.P. (2013). Asymmetric synthesis of natural products (Second ed.). Hoboken, N.J.: Wiley. pp. 17, 28–29. ISBN 1118347331.

- Grimshaw, James (2000). Electrochemical Reactions and Mechanisms in Organic Chemistry. Amsterdam: Elsevier Science. pp. 1–7, 282, & 310. ISBN 9780444720078.

- Madsen U, Krogsgaard-Larsen P, Liljefors T (2002). Textbook of Drug Design and Discovery. Washington, DC: Taylor & Francis. ISBN 0-415-28288-8.

- Reynolds CH, Merz KM, Ringe D, eds. (2010). Drug Design: Structure- and Ligand-Based Approaches (1 ed.). Cambridge, UK: Cambridge University Press. ISBN 978-0521887236.

- Wu-Pong S, Rojanasakul Y (2008). Biopharmaceutical drug design and development (2nd ed.). Totowa, NJ Humana Press: Humana Press. ISBN 978-1-59745-532-9.

- Hopkins AL (2011). "Chapter 25: Pharmacological space". In Wermuth CG. The Practice of Medicinal Chemistry (3 ed.). Academic Press. pp. 521–527. ISBN 978-0-12-374194-3.

- Kirchmair J (2014). Drug Metabolism Prediction. Wiley's Methods and Principles in Medicinal Chemistry. 63. Wiley-VCH. ISBN 978-3-527-67301-8.

Permissions

Index